*The modern revolution in geology and
scientific change*

Drifting continents and shifting theories

H. E. Le GRAND

Department of History
University of M

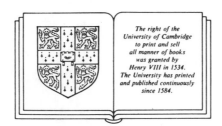

The right of the
University of Cambridge
to print and sell
all manner of books
was granted by
Henry VIII in 1534.
The University has printed
and published continuously
since 1584.

CAMBRIDGE UNIVERSITY PRESS
Cambridge
New York   Port Chester
Melbourne   Sydney

Published by the Press Syndicate of the University of Cambridge
The Pitt Building, Trumpington Street, Cambridge CB2 1RP
40 West 20th Street, New York, NY 10011, USA
10 Stamford Road, Oakleigh, Melbourne 3166, Australia

First published 1988
Reprinted 1990

Printed in Great Britain at the University Press, Cambridge

British Library cataloguing in publication data

Le Grand, H. E.
Drifting continents and shifting theories.
1. Continental drift
I. Title
551.1'36

Library of Congress cataloguing in publication data

Le Grand, H. E. (Homer Eugene), 1944–
Drifting continents and shifting theories/H. E. Le Grand.
p.   cm.
Bibliography: p.
Includes index.
ISBN 0 521 32210 3.   ISBN 0 521 31105 5 (paperback)
1. Continental drift.   2. Science–Philosophy.   3. Science–Methodology.   I. Title.
QE511.5.L44   1988
551.1'36'09–dc19   88-30254   CIP

SE

Contents

	Acknowledgements	v
1	Introduction	1
2	Solid as a rock? Geology before Wegener	17
3	Wegener and his theory: antecedents and arguments	37
4	Debate around the earth	55
5	Specialties, localism and problem-solving	80
6	Interregnum: competion and stalemate	100
7	Palacomagnetism, Drift and polar wandering	138
8	Patterns and puzzles from the sea	170
9	Shifting theories	188
10	The 'revolution' proclaimed	229
11	Theories of scientific change and the modern revolution in geology	267
	Bibliography	279
	Index	305

To
the women in my life

Acknowledgements

This book is a resting-place on a journey of exploration and reflection which began with my narrowly historical investigations of eighteenth-century chemistry, traversed thickets of philosophy and sociology, and has led to the concerns and approaches contained in this study. That journey is not yet finished. There were many along the way from whom I have received assistance; this is deeply appreciated. Many have also provided advice; this is also appreciated. Sometimes I have accepted it, although it has not always guided me in directions they might have wished.

Henry Frankel first aroused my interest in the history of continental drift and in using that to critique several theories of scientific change. This interest remained dormant, however, until a year spent in the Department of Géologie Fondamentale at the University of Lille I. Michel Waterlot's bound-less enthusiasm reawakened my interest. My subsequent research was greatly facilitated by Special Studies programs funded by the University of Melbourne. Material support was also provided by the Faculty of Arts of the University of Melbourne and the Center for the Study of Science in Society in Blacksburg, Virginia. Much of my history and many of my arguments were first inflicted upon undergraduate and honours students in my classes at Melbourne, where teaching and research have not yet been divorced.

Richard Ziemacki encouraged me to prepare a book based on my research and teaching interests and he and Susan Sternberg at Cambridge University Press have never flagged in their support of the project. Whatever the merit of this study, I am sure that it is greater than it would otherwise have been thanks to the encouragement and the insightful criticisms of Roy Porter, who discharged meticulously and gracefully the onerous task of overseeing for the Press the transition from (very) rough draft to final product. Substan-tial parts of an early version of this book were written while visiting the Center for the Study of Science in Society and the Department of History in Blacksburg. I particularly thank Larry and Rachel Laudan, Art Donovan, Harold Livesay and Phyllis Albritton for making my stay fruitful and pleasurable. Conversations with Mott Greene and other colleagues in the US were a valuable source of information and ideas. Detailed criticism on various drafts or portions thereof were given by Jan Sapp, Howard Sankey, Robert Stafford and Peter Riggs, colleagues in the Department of History and Philosophy of Science at Melbourne, and by several colleagues in the School of Earth Sciences, notably Neil Archbold and Kerry Hegarty. I count myself fortunate to be immersed in an environment of intellectual stimulation, co-operation and informed criticism.

I express my appreciation to the many libraries and archives which facilitated my research: the Baillieu Library (especially the Inter-Library Loans section) and the Earth Science Library at the University of Melbourne; the archives and libraries of the University of Chicago, the University of London, Cambridge University, VPI and SU, the University of California (Berkeley) and the Bibliothèque Nationale; the Library of Congress, the particularly extensive library of the US Geological Survey in Reston; and the archives of the Scripps Institution of Oceanography. I trust that the reader will share my appreciation of Anne Pottage's skill in preparing the illustrations.

Finally, I owe more than I can express to my family, especially to Brenda, Chris, Chip and Alex, for their patience and love. I also promise to fix the boards around the bottom of the house – someday!

1

Introduction

Alfred L. Wegener proposed in 1912 that all the continents had once been united, had broken apart, and had drifted through the ocean floor to their current locations. This 'Drift' theory[1] conflicted with the two prevailing views. 'Permanentists' believed that continents and ocean basins had remained mostly unchanged in location and configuration since their formation. 'Contractionists' believed that due to the gradual contraction of the earth, dry land had become ocean floor and vice versa. Some lateral motion within a landmass could also take place, as evidenced by such folded mountain ranges as the Alps, but not the displacement of continents. Controversy and competition ensued among the proponents of the rival approaches.

The stakes were clear. If Drift were accepted, a revolution in geology would result. Opponents and proponents agreed on this. R.T. Chamberlin (1928: 87), an opponent, conceded 'If we are to believe in Wegener's hypothesis we must forget everything which has been learned in the last 70 years and start all over again'. Alexander du Toit (1937: 3), a champion of Drift, asserted that 'the principles advocated by the supporters of continental drift form generally the antithesis of those currently held. The differences between the two doctrines are indeed fundamental and the acceptance of the one must largely exclude the other'. Partisans of different theories appealed to different facts and classes of facts, interpreted the same data in different ways, argued for different geological methods, and put forward distinctively different views of the history and current state of the earth.

Wegener and his allies lost the early battles but Drift ultimately triumphed. From the mid-1920s to the mid-1960s most geologists worked within Permanentist or Contractionist frameworks. Few adhered to Drift. From the mid-1950s, two developments took place. First, some groups of geologists concentrated on new phenomena and geophysical data which had come to light since Wegener. Second, new versions of Drift were put forward. There were similarities between Wegener's theory and these; there were also important differences. The new versions quickly won acceptance. By the early 1970s the 'modern revolution' in geology was complete: the plate tectonics version of Drift, in which the surface of the earth was composed of slowly-moving slabs of crust, was firmly entrenched as the new orthodoxy.

This revolution in theories of the earth accompanied and contributed to a change in the image of geology as a science. Geology in the first half of the nineteenth century was a prestigious and popular science. By the turn of the century, the spotlight had shifted to physics, chemistry and biology; geology became an honest, hard-working but poor relation in the glamorous family of the sciences. It was perceived as a historical science which borrowed its theories from and deferred to other disciplines. The caricatured academic geologist was an elderly professor mumbling an interminable catalogue of the common fossils of the Carboniferous, occasionally stirring a cloud of dust as he produced a specimen. The field geologist was a robust outdoorsman dashing about the Gobi Desert collecting dinosaur eggs[2] or hacking his way through Amazonian jungle in search of oil. By the early 1970s geophysics had become the focal point of an 'Earth Science' which had its own global theory and which laid claim to a position of equality with the other sciences, to an expanded place in curricula and to increased research funding.[3]

The folk-tale of Drift is the stuff of myth and legend in which Cinderella, after years of abuse from her vain step-sisters, is visited by her Fairy Geophysicist, is touched by the Magnetic Wand, goes to the Ball and marries the Prince. Even a less mythical history of this transformation of perceptions of the earth and of perceptions of the discipline is full of drama, of success and failure, of struggle and perseverance, of imagination and plodding hard work, of conflicts of theories and conflicts of personalities. This history forms the narrative and connecting thread of our study. Each chapter begins with a narration of part of the Drift story; each concludes with a 'Voice-Over' containing some reflections on that story. In these Voice-Overs we shall discuss many issues arising from the 'modern revolution' in geology. Some of these we may resolve; others we may be able only to define or clarify or resolve in part. The transformation of geology is excellent territory for exploring ideas about scientific practices, about how these may change, about how scientists may evaluate and choose among competing theories and research programs, and about how scientific change occurs. Many geologists caught up in this 'revolution' interpreted the changes they witnessed or brought about in terms of Thomas S. Kuhn's (1962; 1970) vision of a Scientific Revolution as a convulsive replacement of one world-view by another. Among other topics, we shall critically examine not only specific aspects of this interpretation but also other recent accounts of science and scientific changes as they illuminate or are illuminated by the Drift story.

Voice-Over

Geologists were concerned with many areas of research and debate other than such global theories as Drift. Those who participated in the disputes over Drift comprised only a small minority. By concentrating on the Drift story, we skew the history of geology, especially for the first half of this century. The plate tectonics version now dominates the discipline and is generally accepted by geologists. Nonetheless, Drift is not selected for study because it is 'successful'. Had we written a history in 1962, when Drift had only a small but enthusiastic following, that history could still have raised many issues about the nature of scientific change. We can imagine accounts in which Drift is interpreted as a research program which was put forward as a challenge to the ruling views, which was defeated but which nonetheless secured a precarious toe-hold among a few geologists and geophysicists. Or, perhaps Drift would be presented as a curiosity, a piece of 'deviant' science, a study of which might give us insight into 'normal' science. But, Drifters did win out, a 'revolution' did occur and Drift has become orthodoxy. This gives Drift added value in exploring some recent views on scientific change.[4]

The views of T.S. Kuhn (1970), Imre Lakatos (1978), Paul Feyerabend (1975), Larry Laudan (1977) and much of the assortment of 'social' theories were formulated in the context of an existing body of historical studies of science. Although these views encompass the general development of science, their focus was often on the work of Copernicus, Galileo, Newton, Lavoisier, Darwin, Einstein and the major changes in science – the 'revolutions' – with which they are eponymously associated. It would be surprising if the general schemes of Kuhn and others, which incorporated models of theory change, failed to provide at least a gross account of some of these revolutions. This aroma of devising explanations to fit cases might be dispelled by re-examining these models in light of more fine-grained histories of those archetypal revolutions. Another path, and the one we shall follow, is to evaluate these models and other ideas about scientific practice with respect to an instance of change which was not part of the 'problem situation' for the formulation of those models and ideas. None of the architects of theories of science addressed the development of Drift.[5]

In the nineteenth and early twentieth centuries comprehensive theories were proposed to account for the special character of scientific knowledge, to lay down canons of scientific practice and, sometimes, to explain the historical development of a science. Some of these were proposed by scientists themselves; e.g., William Whewell, Pierrre Duhem and Ernst Mach. Within geology, more limited theories and methodologies of science were rife, particularly 'inductivism', 'hypothetico-deductivism' and the method of 'multiple working hypotheses'.[6] In the early 1960s – about the same time that geologists were entering a critical phase in the Drift story – most of the older theories about science were being undermined, rebutted or even refuted by criticism based on philosophy, history and contemporary scientific practice. Inductivism, 'empiricism', 'logical positivism' were, among other views, claimed to be inadequate and inaccurate as descriptions of or prescriptions for science.

Kuhn in 1962 proposed a view of science which had some correspondences with the revolution in which geologists were soon to be engaged. He developed it in the context of criticisms of older views and intended it to describe both 'normal science' (in which scientists solve empirical puzzles within the framework of an over-arching 'paradigm') and 'scientific revolutions' (in which scientists abandon a paradigm and simultaneously accept another, incompatible one). Several geologists, aware of the shortcomings of older views of science as they applied to their work, and of an apparent fit between Kuhn's model of theory change (as they understood it) and the transformation of their discipline (as they understood it and its history),[7] adopted his model to explicate that transformation.[8]

Kuhn raised a number of important issues but our examination of Drift raises serious problems for a Kuhnian account. For example, rather than a scientific community being governed by one monolithic paradigm, the study of geology from the nineteenth century down to the present seems to have been characterized by competition between a number of rival 'big' theories. A Kuhnian might respond that geology acquired its first 'paradigm' only in the 1970s. Not only would this response seem counter-intuitive, it would also rule off as exceptions any episodes which do not seem to conform to one's preferred theory of science. After all, Kuhn himself (1970: 10) identified Lyell's *Principles of Geology* of 1830 as paradigmatic. Moreover, a case for 'theoretical pluralism' as the normal state of affairs in other sciences could be made. Another response might be to suggest that the community of geologists needs to be broken up into such smaller units as specialities. This is constructive, for as we shall see, specialization in several senses played an important part in the reaction of many geologists to Drift. However, as we shall also see, even within specialities there was pluralism and often agnosticism with respect to global theories.

Most of the general views put forward since Kuhn agree that theories are part of larger, more complex conceptual structures and that competition among these 'big' theories is the rule rather than the exception. Wegener's 'theory' embraced evidence, methods, aims, standards and many theoretical notions and in effect constituted a general approach to geology which competed with other 'big' theories; e.g., 'Contractionism', 'Permanentism' and 'Expansionism'. There is, however, a marked divergence between those theorists who read Kuhn as espousing a partially arational, relativistic view of science and react by adumbrating 'rational' theories of science and those who read Kuhn as opening the way for a social analysis of science.

Imre Lakatos represented the rationalist reaction. He proposed (1978) that scientists worked in rival Research Programmes.[9] Each Programme has a 'hard core' of unchangeable and unchallengeable general assumptions and consists of a succession of theories. Each theory embodies the hard core and has a greater verified empirical content than its predecessor. A Programme is 'progressing' if its adherents predict novel facts which are subsequently confirmed. It is 'stagnating' if its adherents be always reacting to rather than anticipating empirical discoveries. Many criticisms have been made of Lakatos's account. In practice, it seems impossible to define precisely the hard core, other than retrospectively in terms of that which has not been discarded or modified over time. He gave precise criteria to identify whether Programmes were progressing or stagnating but their application to actual scientific practice and to the geology case seems impracticable. Finally, he did not enamour himself to historians by suggesting (1978: 120–121) that if the 'real' history did not agree with a 'rational' account, the 'real' history should be put in footnotes to a 'rational' history.

Larry Laudan (1977) proposed a view which resembles those of both Kuhn and Lakatos. He sides with Kuhn in claiming that science is problem- or puzzle-solving. Indeed, he claims that the fundamental aim of science is to maximize the scope and number of solved empirical and conceptual problems while minimizing the number of anomalies and conceptual problems generated. He agrees with Lakatos in arguing that normally a scientific field or discipline is guided by a number of competing 'Research Traditions' rather than one paradigm. A Research Tradition is akin to some of Kuhn's definitions of paradigm: it includes general assumptions about what the world is like, about what sorts of things make up the world and about what methods we might use to find out about the world. Like Lakatos's Programme, it serves as an umbrella for a set of successive theories (each an expression and refinement of that Tradition) all of which bear a resemblance to one another by virtue of sharing a basic common view as to what constitutes valid knowledge in a field of science and how we can acquire that knowledge. Understood in this way, a Research Tradition evolves over time: the theories which express it change over time. There is competition between Research Traditions and, within a Research Tradition, competition between theories. Scientists may have a strong preference for one Tradition or one theory within a Tradition but still explore, try out or, as Laudan puts it, *pursue* another without giving up their original allegiance (*acceptance*). They are not trapped within a single paradigm.

Laudan extends Kuhn's discussion of science as a problem-solving activity. Kuhn (1970: 25–34) seemed to be concerned chiefly with empirical problems: problems in the area of theory–data agreement. These did figure in the response to Drift but Laudan argues scientists also deal with what he terms 'conceptual problems'.[10] Conceptual problems may be 'internal' or 'external'. Internal ones include inconsistencies or circularities within a theory or Tradition. External ones include conflicts between a theory or a Tradition and a belief in some other branch of science which is thought to be well-founded; e.g., a conflict between a theory in geology and a theory in physics; or what is taken to be the proper method of science; e.g., a theory cast in hypothetico-deductive form in a field in which most scientists champion inductivism; or a 'world-view'; e.g., uniformitarian geology and a literal interpretation of the biblical account of Noah's flood. Laudan also presents an alternative to Kuhn's notion of 'anomaly' as a puzzle within a paradigm which resists repeated attempts at solution. An accumulation of anomalies is the requisite for a Kuhnian 'crisis state' in which scientists become insecure about the validity and efficacy of their paradigm and which prepares the way for a subsequent revolution. Kuhn's anomalies are usually empirical in character. Laudan categorizes all problems – empirical and conceptual – as 'solved' or 'unsolved' or 'anomalous'. If scientists using competing theories or Traditions fail to solve a particular problem, it is simply an 'unsolved' problem and counts neither for nor against any of the contenders. If, however, a problem is solved in one of the theories or Traditions, then for that one it becomes a 'solved' problem and for its competitors, until they succeed in solving it, an 'anomalous' problem. In Laudan's view, since the aim is to maximize the number of solved problems while minimizing the number of conceptual problems and anomalies, anomalies are important in evaluating rivals and in bringing about scientific change.

What is a 'scientific revolution' for Laudan? It is not a sudden 'paradigm switch'. A 'revolution' occurs when a new research program (or a new version of an established program) is formed which has a high initial rate of progress (and perhaps which creates anomalies for its competitors) such that scientists who are working in rival Traditions can no longer ignore it. They may decide to switch their allegiance and accept the new one; they may continue to accept the older ones but decide to pursue the new one (just to see how useful it may be in solving problems), or they may alter their own Tradition to try to meet the challenge. Scientists are constantly making comparative evaluations of rival Traditions and theories in respect of their problem-solving ability; scientific revolutions are dramatic and decisive instances of this process. A revolution may not involve the whole community but only a small portion of it; e.g., the leaders in the field or maybe the more prolific and ingenious problem-solvers. This picture of competition among programs and of an ebb and flow of support describes the modern revolution in geology better than Kuhn's convulsive upheavals.

Laudan argues that he provides a measuring-stick for assessing rival Traditions and theories. The theory which is the most effective problem-solver while generating the fewest anomalies and conceptual problems in a given historical situation is the most progressive; similarly, the Tradition which best fits those criteria is the most progressive. All of these matters are, however, to be judged in light of the knowledge, beliefs, techniques, methods and standards of the day – not by what we accept now. In this sense external concerns such as theology, metaphysics, or ethics may be part of a rational appraisal of theories or Traditions. Here Laudan differs from Lakatos: Laudan is not looking for some absolute standard or objectivity but comparative good performance relative to a specific historical context. Scientific rationality, crudely put, consists in choosing theories or Traditions which are the most progressive in the above sense. Although this may be more constructive than Lakatos's approach, it too suffers from difficulties in detailed application. It would seem to entail counting up the solved and anomalous problems for each theory or Tradition and assigning some weighting which would reflect the relative importance of those problems. It is not at all clear that proponents of the different – or even the same – theories or traditions would agree that a particular problem has been solved or that it is anomalous or, if it be anomalous, what degree of epistemic threat it might pose to a given view. As we shall see, solutions put forward by Drifters were frequently rejected by their opponents and vice versa. Proponents of a program on occasion simply ignored anomalies thrown up by rivals. Sometimes, the very existence of a problem was disputed.[11]

Despite differences among Kuhn and his 'rationalist' critics, there are substantial areas of agreement (cf. L. Laudan *et al.*, 1986: 154–6). There is agreement, for example, that 'theories' are part of larger structures; that these theories or structures are rarely given up simply because they seem to run afoul of 'facts'; that there is not a sharp distinction between theories and facts; that empirical data alone do not determine scientists' attitudes towards theories or research programs; that these attitudes often reflect judgments about promise rather than past performance; and that these judgments are not absolute but relative to a historical context. We shall return to these areas of agreement in the concluding chapter; for most of the 'Voice-Over' sections we shall be stressing differences.

At the opposite pole to Lakatos's and Laudan's reactions to Kuhn are those who interpret Kuhn as underwriting or sympathizing with a 'social analysis' of science. Until recently sociological studies did not extend to scientific knowledge. Scientists, like bus drivers or legislators, might be subject to study as sociological groups but not the content of their research. Broader socio-political concerns might affect the rate and direction of scientific research; e.g., national defense policies might funnel money into high energy physics at the expense of other areas of research. Otherwise, science was an autonomous activity in which, for example, judgments on theories were made on the basis of objective scientific criteria. Within the past fifteen years, however, some social analysts have argued that scientific knowledge, like other forms of knowledge, is a social construction with no independent objective validity. The label 'social analysis' covers a plethora of changing positions. These views taken together – or, for the most part, even separately – do not yet constitute a detailed, coherent interpretation of scientific change. However, this has not been the aim of much of their writings. Their concern, and one which is shared to some extent by Kuhn and Laudan, is to combat the view that the development of science comes through the triumph of truth over error or that scientific knowledge is essentially divorced from the historical context in which it is produced. To this end, many case studies have been amassed in which it is shown that the social context of scientific practice can be of crucial importance in the shaping, reception and application of scientific knowledge (see Shapin, 1982).

Most social analysts would endorse many of the 'rationalist' views mentioned above, though perhaps with different emphases. These would include Kuhn's ideas of 'paradigms' as self-consistent, internally coherent but incompatible world-views or webs of belief;[12] the blurring or erasure of distinctions between 'facts' and 'theories'; and the argument that for any set of data there are many possible theories consistent with it. The inferences drawn range from the claim that much of the science is subject to dispute and negotiation in which social interests may sometimes be of paramount importance to the claim that science is fundamentally a social construct, the form and content of which is socially determined.[13] Of the diverse approaches under the rubric of the 'social analysis of science' we will focus on two: the 'interest' and the 'internal struggle' interpretations.

The most prominent proponents of the interest approach are David Bloor, Barry Barnes, Steven Shapin and Donald A. McKenzie.[14] Barnes (1977; 1982; Barnes & McKenzie, 1979; Barnes & Shapin, 1979), Shapin (1981; 1982) and McKenzie (1981) base their analyses on two propositions. First, scientific and social concepts form a web of belief through a complex network of associations (cf. Hesse, 1974). There is no demarcation between 'social' and 'scientific' knowledge. Social and scientific beliefs are interlinked, though the connections may be complex, and changes in one part of the web affect other parts. This notion is similar to the global sense of 'paradigm' in Kuhn's account: the entire constellation of beliefs held by a group of scientists (Kuhn, 1970: 175), though Kuhn appears to refer only to scientific beliefs. The second proposition is that scientists are motivated by 'interest'; that is, scientists carry out their research, choose among rival research programs or theories, and engage in other cognitive and social activities on the basis of their 'social' and 'scientific interests'. Scientific interests might include one's expertise in particular techniques or one's greater skill in theorization as opposed to experimentation. Social interests range from narrower ones which might include one's reputation in the field, and institutional affiliations, to broader ones which might include religious or political or other social beliefs or one's economic or social position in the larger society (cf. Barnes, 1982; Shapin, 1982). In the earlier, polemical writings of this group, the importance of wider social interests (interests which extended beyond the notional boundaries of science) were particularly stressed. There was a tendency to short-circuit the connection between these social interests and scientific activities: to assume a one-to-one relationship, for example, between social class and specific theories (e.g., McKenzie, 1981). More attention has been given recently to the operation of interests within the social system of science, especially in those fields of science which have achieved some degree of apparent autonomy from the wider social setting and in which the research has no obvious social import. Broader social interests in these fields do not directly determine the work of scientists, instead they are mediated by the complex of narrower social and scientific interests operating within those fields.[15] There are some suggestive case studies (e.g., A. Pickering, 1984) but this approach has yet to yield a comprehensive theory of scientific development, much less one to which most of its proponents would assent.

Social interests in their broader form did not seem directly to favor proponents or opponents of Drift. The boundary between a scientific field of study and the society in which it is embedded may have lesser or greater permeability and this degree of permeability may change over time. In the first half of the nineteenth century, the boundary between geology and broader social concerns was relatively permeable, at least in the Anglo-American world. This was evident in the use of geology to buttress religious or social beliefs and, inferentially, the existing social order.[16] In the twentieth century broader social interests have played a less obvious role within geology.[17] Geologists themselves sometimes invoked various forms of social accounting (Mulkay & Gilbert, 1982a) and social interests to explain opposition to their preferred views. John Leuba, an early French defender of Drift, wrote (1925: 208) that the reason all geologists had not immediately accepted Drift was that 'one is inclined to accept only the solitary ideas put forward by specialists. But Wegener is not a geologist, but a geophysicist: the resistance of certain geologists has no other reasons.' Much later, Edward Bullard (1975a: 5) gave a different social interpretation[18] of opposition to Drift or indeed to any novel theory:

> There is always a strong inclination for a body of professionals to oppose an unorthodox view. Such a group has a considerable investment in orthodoxy; they have learned to interpret a large body of data in terms of the old view, and they have prepared lectures and perhaps written books with the old background. To think the whole subject through again when one is no longer young is not easy and involves admitting a partially misspent youth. Further, if one endeavours to change one's views in midcareer, one may be wrong. . . . Clearly it is more prudent to keep quiet, to be a moderate defender of orthodoxy, or to maintain that all is doubtful, sit on the fence, and wait in statesmanlike ambiguity for more data. . . .

The remarks of Leuba and Bullard could be regarded as a means of explaining the erroneous (compared with their own) judgments reached by other geologists in terms of interests that lie within the discipline. As we shall see, 'internal' interests were ubiquitous in the development of and controversies over Drift theories.

The 'internal struggle' account of science focusses squarely on the play of interests within science. If the growth of science is not to be directly dependent on broader social interests, then what may be crucial is not the external social order but the operation of interests within science. Bourdieu (1975), Rudwick (1985), Latour & Woolgar (1979) Latour (1987) and Sapp (1987) are among those who advocate this interpretation. Science is an arena of competition in which the competitors lose, gain or maintain authority or credibility within a scientific field.[19] The competence, authority or credibility of scientists is constituted by their technical capacity and social power as recognized by their fellow scientists, especially by their competitors (Bourdieu, 1975: 19). Technical capacity and social power are inseparable: gaining a prestigious chair or membership in an academy is not purely a matter of politics; gaining acceptance for a new theory is not purely a matter of the objective merits of the theory. Such categories would be at best meaningless; at worst, systematically misleading (cf. Latour, 1987). Social power within science derives from recognized achievements; judgments of a knowledge claim are always 'contaminated' by an awareness of the authority the claimant possesses in the field (Bourdieu, 1975: 20–21). Latour and Woolgar (1979) have adopted, criticized and augmented Borudieu's analysis. Scientists are engaged in a struggle for credibility within the field. Scientists accrue credibility and authority by promoting their own knowledge claims and demoting their competitors' on a scale which ranges from 'illfounded speculation' (in the judgment of others) to 'a fact of nature' or 'the established theory' (in the judgment of others). The construction of 'facts' and 'theories' occurs through a complex process of negotiation with colleagues and competitors which is simultaneously social and technical (Latour & Woolgar, 1979: 77–87, 189–208; cf. Rudwick, 1985 and Latour, 1987). A key point in the 'internal struggle' model is that recognition, authority or credibility must be extracted from one's competitors. It is not given readily or willingly. It is given only if the claimant establish a case which meets their standards of validity. It follows that the more a knowledge claim threatens previously-established patterns of authority, the more stoutly it will be resisted and the more strictly the prevailing standards will be applied.[20]

The play of competing interests in the field ensures that 'progress' will occur. The interest of the established practitioners is the maintenance of orthodoxy: resistance to major changes in theory, techniques, methodology, aims and standards. For newcomers the threshold for entry is raised by established practitioners who seek to maintain their authority in the field. Newcomers may take the safe path of solving routine problems using routine approaches, in which case there is a cumulation of scientific knowledge consisting of extensions of orthodoxy. The sum of technical capacity and social power represented within the field increases: this has parallels with Kuhnian 'normal science'. However, Bourdieu claims that change also occurs through small, permanent revolutions rather than the major, discontinuous ones portrayed by Kuhn (Bordieu, 1975: 29, 31–4). Newcomers may adopt the high risk strategy of trying to become 'instant' authorities through proposing unorthodox theories or otherwise challenging accepted beliefs and practices: if accepted, the result is a 'revolution'. The strategy is a high risk one because of the technical capacity and social power vested in the current orthodoxy and its sustainers. Or, perhaps, newcomers may attempt to create a new field (discipline, speciality, sub-speciality) in which they would be the authorities: this would presumably bring about 'progress' in depth or breadth.[21] These same strategies are open to established practitioners as they seek to gain further authority, competence or credibility. The old saw 'Nothing ventured; nothing gained' applies with equal force to them, although their venture capital of existing credibility is likely to be greater than that of newcomers. Proponents of this analysis argue that it does not rely on the explicit motivations of scientists: whether they describe what they are doing as searching for truth or solving problems or striving for fame and glory or building bank accounts, their careers and the development of the field in which they are engaged both can be understood in terms of 'internal struggle'. Indeed, there should ideally be no distinction between a 'cognitive' and a 'social' history of the same episode in science. A theory or a piece of data is not accepted on the basis of some inherent truth. Latour (1987: 258) expresses this as a rule: 'Since the settlement of a controversy is the *cause* of Nature's representation . . . we can never use Nature to explain how and why a controversy has been settled.' The 'struggle' between ideas and the 'struggle' between scientists are indistinguishable and the outcome is determined by the social and technical taken together (Bourdieu, 1975: 21).

We shall see in the Drift story facets which are amenable to 'interests' or 'internal struggle' analyses – and facets which are amenable to analysis by 'rationalistic' theories of scientific change. There seems to be an unbridgeable chasm between proponents of 'social' and 'rationalist' views of science. Salvoes may be exchanged (e.g., Bloor, 1981; Laudan, 1981a) but otherwise there seems to be little interest in communication, much less in compromise. We observe, somewhat tongue-in-cheek, that it may be in their interests to maintain their struggle. We could interpret it as a struggle for a monopoly of authority in the field of interpreting science. Indeed, we could carry this a step further and suggest that broader social interests may operate. Those who favor a discontinuous development of science and lay stress on sudden revolutions – taking their cue from Kuhn, whose model was put forward in the politically turbulent 1960s and who drew an explicit analogy between political and scientific revolutions – may tend to fall on the left side of the political spectrum. Those who favor a gradualist, evolutionary model and lay stress on objectivity and rationality may tend to fall on the right. However, this does not mean that we should not explore the extent to which these different approaches are assimilable or at least complementary to one another.

Astute readers long before now will ask themselves what is my view of science and more particularly what general theory of science – if any – do I think best comprehends the modern transformation of geology? Evaluating various models of science is by no means the exclusive focus of the 'Voice-Overs': many topics are discussed which have little immediate relevance to that task. Nonetheless, much of the discussion does bear on such models and their usefulness with respect to the revolution in geology as it is presented in the historical sections. This raises an obvious problem. Any account of this episode or of any other instance of scientific change is not, of course, a neutral recital of the facts. Historians' selections of facts, the ways they arrange and present them and their interpretations within which 'facts' find their meanings and significances are laden with implicit or explicit assumptions about the nature of science and scientific change. There is, therefore, a potential circularity in my constructing a historical account and using that to explore and evaluate general theories of science. I follow this *mea culpa* with an *ex culpa*. First, my views have not remained fixed even while researching this book but have changed over time precisely through an extended process of comparing general views of science with the historical development of science. My training as a historian of science occurred in an intellectual environment which was pre-Kuhnian and pre-social analysis. However, I subsequently employed Kuhn's model for several years. My growing familiarity with several historical episodes led me to question Kuhn's model and I briefly flirted with a 'Lakatosian' view. Detailed discussions with a Lakatosian over a particular episode eventually resulted in our both abandoning that interpretation. My study of Drift has led me to favor some aspects of both Laudan's rationalist model and social analyses. My own experience suggests that there is a dialectic between one's views about science and the facts one unearths in a detailed historical investigation. Neither do neutral, objective historical facts determine a general theory about science nor does a preference for a general theory determine the facts. Our preferences and preconceptions give us filters, not blinders.

My views are eclectic. I find value in Laudan's approach, in part because at a general level it seems to me to answer many of the criticisms I had of Kuhn and in part because at a more specific level some features of his model mesh relatively well with my research in and present understanding of the modern transformation of geology. I would single out especially his emphasis on theoretical pluralism, his discussion and classification of 'problems' and 'anomalies', his defence of comparative 'good performance' as a reason for choosing one program or theory rather than another and his distinction between 'acceptance' and 'pursuit'. Even though I agree with Laudan's views on many issues by no means do I accept his general theory as unproblematic: I have already indicated some general reservations and I am sure that the reader will find much in the narrative that does not easily square with Laudan's account. As one instance, a range of technical, specialist and local interests were of importance in the Drift story yet these are not addressed in his scheme. These interests lend themselves to an 'internal struggle' or possibly an 'interest' interpretation. I am not yet persuaded that an interpretation in this vein cannot be reconciled with what I believe to be some of the useful elements from Laudan's model. In this sense, my own views could be characterized as not only eclectic but also provisional and exploratory rather than dogmatic. I hope that this attitude is accurately reflected in the remainder of the 'Voice-Over' sections. These are intended to encourage reflection and critical discussion, not to close them off; to raise questions, not to answer them. The aim is to develop a better understanding of the process and the form of knowledge we call science which is so influential in our culture, not to propose or prove or disprove some preselected general theory of science.

Notes

1 I use 'Drift' generically to refer to global theories which invoke large-scale lateral displacements of the crust as a central geological process.

2 Roy Chapman Andrews's many books, especially *All About Dinosaurs* and the narrative of his adventures in the Gobi, almost persuaded one impressionable ten-year-old to become a palaeontologist.

3 See Wood (1985) for a racy account which weaves together changes in theory with this transformation of the discipline. *Cf.* on geology as description *versus* geology as explanation, R. Laudan (1982b).

4 I assume that the reader is familiar with the more traditional general theories of science and also that of Kuhn; I give only a brief résumé for some of the 'post-Kuhnian' views. For short introductions for the non-specialist, see Losee (1980) or Chalmers (1982). L. Laudan *et al.* (1986) breaks down several recent philosophical theories into sets of succinct theses and discusses areas of agreement and disagreement. Mulkay (1979) introduces some 'social' and sociological perspectives.

5 A further advantage of using the Drift story to suggest and critique ideas about the nature of science is that it is a modern episode in an established science. Unlike, say, the eighteenth-century 'Chemical Revolution', Drift is immune to the charge that is does not correspond to modern scientific practice. Such charges are sometimes raised by first-year students and 'discourse analysts'.

6 For a brief history of general theories and methodologies, see Losee (1980); for a more critical discussion, see L. Laudan (1981b); for some case studies on the multifarious roles of scientific method, see Schuster & Yeo (1986); for the method of multiple hypotheses, see Pyne (1978) and R. Laudan (1980b).

7 This often involved paying little heed to the history of Drift prior to the 1960s, thereby presenting the victory of Drift as sudden and unexpected rather than the

outcome of a prolonged struggle. This was particularly the case for British and North American geologists, for Drift had been far less popular there than elsewhere.

8 Some geologists who have taken this view are J. Tuzo Wilson (1964, 1968c, 1976), Menard (1971), Cox (1973), Hallam (1973), Marvin (1973), McArthur & Pestana (1974), Wyllie (1976), Pestana (1979) and Wood (1980). The Kuhnian interpretation has been criticized by, among others, R. Laudan (1980a) and Hallam (1983). Frankel (1979a, b) has sketched the revolution in terms of Lakatos' and Laudan's theories of science.

9 'Research Programme' refers to Lakatos's technical use of the term as opposed to the more customary usage.

10 We shall see that a distinction between 'theories' and 'fact' or 'empirical' and 'conceptual' is highly problematic. Feyerabend (1981: 60–1, n.3) argues that conceptual problems had not been ignored by many other rationalist writers prior to Laudan. There is merit in this claim but Laudan is systematic in his taxonomy and discussion of conceptual problems and does assign them great importance.

11 Readers wishing a far less sympathetic treatment of Laudan's model should see Feyerabend's (1981) stinging review of L. Laudan (1977).

12 But see Barnes & McKenzie (1979: 49–52) who suggest that even if incommensurability exists at a formal philosophical level, at a practical level scientists can understand and develop competence within more than one paradigm at a time.

13 *Cf.* Mulkay (1979) for an introduction to some of these matters.

14 Their writings do not reflect a single, dogmatic party line. They have themselves different 'interests' and would differ among themselves over many aspects of the approach here attributed to them.

15 See Barnes & Shapin (1979) for a brief discussion of how a scientific or technical interest analysis might be adapted to Kuhn's paradigms.

16 *Cf.* Gillispie, (1951). Secord (1986: 314) claims that Victorian geologists believed that 'technical commitments could be (indeed should be) supported and sustained by parallel social, religious, and philosophical ones'. Rudwick (1985: esp. 438–55) emphasizes, however, the importance of social interests within the community of 'gentlemanly specialists' involved in the Devonian controversy.

17 One exception was the boom in marine geology and geophysics in the United States which began in the late 1940s (see Chapter 9). It could also be argued that the economic interests represented by oil and mining companies impinged upon the internal workings of the discipline. This affected recruitment into the discipline and possibly university training and choices of research projects but the effects on such matters as theory development and theory appraisal seem to have been minimal.

18 Stewart (1986), writing in the 'interest' framework, takes Bullard's comment to illustrate the major reason why Drift was resisted by prominent geologists until the 1950s since, he claims, Drift is the 'correct' theory and some of the evidence advanced by Wegener is still accepted. His claim is very problematic (cf. R. Laudan, 1987b) in light of the nature of the disputes over Drift and the changes in 'theory' and 'evidence' in the half-century after Wegener.

19 What constitutes a scientific field is itself the subject of competition and negotiation among scientists. It may correspond to a discipline; e.g., geology, in which scientists compete with one another to establish the legitimate aims, methods and

preferred theories of the discipline. These same processes may occur within a speciality. The formation of new disciplines or specialties is also a strategy which may be adopted by scientists to further their interests (cf. Rudwick, 1972; 1985; for examples in biology, see Sapp, 1983; 1987).

20 Of course, such standards are open to negotiation and change. One implication is that a 'scientific revolution' likely involves not only a conceptual transformation but also a transformation of the social order within the field.

21 Proponents of the 'internal struggle' model may not share Bourdieu's views on progress. Indeed, the idea of 'progress' in science is a protean one: its meaning shifts according to the model in which it is used and sometimes even within the same model (See, e.g., Kuhn, 1970: 160–73). The reader may wish to construct a definition: I shall not provide one, other than to endorse L. Laudan's (1984: 64–6) stricture that we should always discuss progress in relation to specific aims or goals.

2

Solid as a rock? Geology before Wegener

What was the state of geology when Wegener in 1912 put forward Drift? Until recently, the history of geology was often told by British and American authors as a series of victories by British geologists over their French and German counterparts. The most notable was the victory in the 1830s of Charles Lyell and 'uniformitarianism' over European and British proponents of 'catastrophism'. Lyell's *Principles of Geology* became the examplar for geological theory and practice and uniformitarianism became the basis for the discipline. Lyell laid the foundations of this interpretation in his own polemical history. This systematically misrepresented his opponents as catastrophists and served to legitimate his theories. Archibald Geikie (1901), one of the most famous British geologists at the turn of the century, reaffirmed this interpretation and it has since been followed by many later geologists and historians (Ospovat, 1976).

This history may be crudely summarized. Near the end of the eighteenth century, the dominant program was that of Abraham Werner, a German geologist and world-famous teacher at the Freiberg School of Mines. Werner, like most of his contemporaries, argued for great global catastrophes as the prime agents of geological change. Some 'Mosaic geologists' harmonized this idea with Scripture and fused theology with geological system-building. In 1795 the Scottish natural philosopher James Hutton proposed the first scientific theory of geology. His earth was a self-renewing machine which did not need God's intervention: the forces of erosion gradually wore down the land but new land was gradually built up by volcanic action and the heat of the earth's interior. The only forces invoked to explain geological features and the history of the earth were forces and agents which he could observe in operation. There were no global cataclysms. This doctrine, uniformitarianism, is captured in the slogan 'The present is the key to the past'. Hutton's own presentation was ineffectual, allegedly because of his unreadable style. However, in the hands of supporters and popularizers, especially the Scottish natural philosopher John

Playfair, it was sufficient to rout Wernerians. This was the first British victory: Hutton over Werner, inductivism over deduction from speculative hypotheses, science over scripture! The victory was symbolized by the founding of the Geological Society of London in 1807. The founders made their cornerstone the careful study of rocks and strata, the multiplication of piecemeal observations and careful inductive reasoning. Theories were untrustworthy so long as detailed knowledge of the surface of the earth – and any knowledge of the interior of the earth – was lacking.[1] This attitude persisted, particularly in Britain and North America, well into the twentieth century. Description, not theory, was needed.

The uniformitarian and inductive approach was challanged in the 1820s and 1830s by a resurgent catastrophism. European scientists and some British theologians enlisted under a banner which portrayed the earth's history as a process of development and gradual perfection, punctuated by massive upheavals, ending with the creation of man. Uniformitarians were put on the defensive. To the rescue in 1830 rode Charles Lyell. He cast himself in the role of the Newton of geology as the title of his major work, *Principles of Geology*, implied. He upheld and developed the main tenets of Hutton. He pushed a strict uniformitarianism according to which the geological agencies which shaped the earth are identical not only in kind but also in magnitude with those which we can now observe. His earth underwent ceaseless cycles of uplift and erosion always produced by the same forces. As opposed to the developing earth of catastrophists, his was fundamentally a static one in which cyclical change reduced to no change at all. He closely tied his system to the available geological data and used it to explain in detail the past history and current geological features of the earth. He claimed to have arrived at his theory inductively, a stance which appealed to the program earlier set by the Geological Society and to the empiricist flavor of British science generally. His devotion to the facts, the scope and detail of his synthesis, and the mode of his presentation overcame the combination of revealed religion and speculative geology which had threatened to derail the progress of geology. Lyell's geology triumphed first in Britain and then around the rest of the world. From the 1840s the Lyellian program endured virtually unchallenged. This was the second great victory. Lyell's geology formed the basis for subsequent researchers; catastrophism was banished to the lunatic fringe.

This history – even the burlesqued version I have given – is not wholly wrong and had currency among geologists themselves. However, it misrep-resents the history of geology before Wegener in several crucial respects. First, there is an assumption that grand theoretical debates were the centrepiece of nineteenth-century geology. Rudwick (1985) and Secord (1986) argue persuasively that most Victorian geologists took as their central task the identification, classification, ordering and mapping of strata. Emphasis on painstaking field work, careful description and cautious induction was appropriate to this aim; global theorizing was of minor importance. The stratigraphical column was perhaps a more influential export than Lyell's geology. Second, Lyell's tendentious claims about his predecessors and contemporary opponents are highly misleading. His geological opponents are more fairly described as 'progressionists' rather than 'catastrophists'. Most accepted the great antiquity of the earth and their conclusions of an evolving earth, of massive extinctions and abrupt emergence of species were based upon a staunch empiricism. They took as their evidence the record of the rocks as it appeared; Lyell argued (less empirically) that the record of the rocks was imperfect.[2] Third, Lyell's strict uniformitarianism did not triumph in either Britain or Europe: non-uniformitarian views were espoused by many geologists from the 1830s to the present. Finally, this history reflects a natural focus by English-speaking geologists and historians on British developments to the exclusion of what was going on elsewhere. Over the past 15 years a different history has been constructed.[3]

This 'new' history suggests that few geologists actively engaged in disputes over global theories and that many regarded them as largely irrelevant to their daily work. Geology prior to Wegener was characterized by theoretical pluralism: the existence of a number of geophysical theories and theories of the earth's history which were mutually incompatible. Most of them were also irreconcilable with strict uniformitarianism. At the beginning of the twentieth century geology was a science riven by controversies over methods, data, theories and even the aims of the discipline. There were significant divergences between European and Anglo-American approaches to geology. There was not one approach but several. The most important were 'Permanentism' and 'Contractionism' but within those two there were competing versions.

Permanentism

Anglo-American geologists after Lyell often regarded themselves as uniformitarians, though perhaps not so strict as Lyell. Many were 'actualists'; i.e., they would admit of variations in the degree but not the kind of geological agents.[4] Earthquakes or volcanoes might, for example, have been much more powerful in the past than the present. The President of the Geological Section of the British Association for the Advancement of Science, Grenville Cole, argued (1915) that strict uniformitarianism did not take into account the evolution of the earth from an incandescent mass to its present state. Geikie, despite his encomium to Lyell, warned in 1903 (i, 3) that geologists should not too hastily accept either the assumption that there had been no forces in the past other than those we can observe today or that those past agents acted in the same way and with the same power as those of today. There was a general belief, at least until the early years of this century, that the earth had gradually contracted as it cooled. However inconsistent this may have been with Lyell's own uniformitarianism, and however much we might see this as a concession to his opponents, British and American geologists perceived themselves to be closer to uniformitarianism than were their European counterparts. The rhetoric of uniformitarianism continued to be powerful in the Permanentist tradition.[5]

The proper aim of geology was a detailed description of the earth rather than a search for hidden causes. The work of different national geological surveys and the effort expended in preparing large-scale geological maps are evidence of the importance attached to this aim. Gradual change through erosion and uplift did occur but this was very slow and probably cyclical. The method to be followed was inductivism: 'the interpretation of the direct contemporary evidence of the rocks' (Gregory, 1930: 959). Great emphasis was laid upon the importance of data-gathering, of careful mapping and detailed regional studies.[6] The dangers of speculative theorizing were to be avoided: 'the British mind preferred facts that could be established by observation to the uncertain products of speculation' (Gregory, 1915: ix). Given our ignorance of the interior and much of the surface of the earth (e.g., the ocean floors were mostly *terra incognita*) grand synthetic theories should be distrusted. There were influential dissenters from this view. Thomas C. Chamberlin at the University of Chicago inveighed in the early years of this century against mere data collection as 'the method of colorless observation' (see Pyne, 1978; R. Laudan, 1980b). He proposed the 'method of multiple hypotheses' according to which all field workers should be familiar with the major theories concerning the problems which they were investigating and should constantly evaluate the relative merits of those theories with respect to their usefulness in making sense of field observations. His opinion was widely endorsed, especially in North America. Even so, for most geologists in this tradition the earth was essentially static and stable.

The Permanentist program had as its central tenet continental permanence. It was developed to its fullest extent in North America but had a following throughout the English-speaking world and some support in Europe as well. The continents were formed in remote geological times as the earth had gradually cooled and contracted. Since then, they had been permanent features of the earth's surface. There had been minor elevations and subsidences producing from time to time shallow inland seas and the submergence of continental margins, but the catchcry of this approach was 'Once a continent, always a continent; once an ocean, always an ocean' (Dana, 1873; 1881). Permanentists thus broke with the Lyellian view of a free interchange between ocean floors and continents.

James D. Dana, Professor of Natural History and Geology at Yale University from 1850 to 1892, was the most famous exponent of this program. Dana had travelled widely in the Pacific and elsewhere but concentrated on such North American problems as the structure and formation of the Appalachians. The continents were stable but were slowly growing by accretion at their margins. The Appalachians were an annexed strip of the Atlantic ocean floor. As the earth gradually cooled, the crust along the Atlantic coast had sagged. This geosyncline slowly filled with sediment, then was compressed, perhaps by the continued subsidence of the ocean floor, and uplifted to form the Appalachians (see Dott, 1974, 1978). In outline Dana's version of Permanentism was very popular, especially among American geologists, for it seemed readily applicable to North American landforms. There was, however, no unanimity on details. Chamberlin (1928), for example, opposed the view that the earth had cooled from a nebular state. His field studies of the effects of past glacial action in Wisconsin and his awareness that there had been glacial epochs in sunny South America convinced him that the earth could not be steadily cooling. He postulated a 'cold earth' formed by the accretion of asteroids and contracting through gravity rather than heat loss. Though he had a different cosmogony, he reached similar conclusions: the permanence of the old continental cores, gradual growth on the continental margins, the relative permanence of ocean basins, and so on.

Abundant evidence was marshalled in favor of Permanentism. There were fossil deposits of marine flora and fauna found on most continents. These had earlier been used to argue for the submergence of old and the formation of new continents. However, it was now claimed that these deposits were composed of shallow-water species – not deep-ocean species. This conflicted with the idea of new continents emerging from the deep-ocean basins and supported the notion of continents being more permanent structures which might occasionally be partially flooded. On the geophysical side, Permanentism accorded with 'isostasy' which had been conceptualized in the 1850s and further developed in the 1870s and 1880s. In the 1840s surveyors near the Himalayas had problems obtaining consistency among their measurements. It was first thought that the difficulty was due to a deflection of surveyors' plumb-bobs by the gravitational mass of the mountains. However, when this was taken into account, the actual deflection was far less than expected. This implied that the mountains and the crust beneath them were lighter than expected. George Biddle Airy, the Astronomer Royal of England, proposed that mountains and the continents themselves floated in a denser underlying substrate. Analogous to icebergs, mountains had lighter 'roots' extending much deeper into the substrate than plains. Clarence Dutton, an American geologist, in 1889 coined the term 'isostasy' for this explanation and applied it to several phenomena. For example, erosion would result in a slow local elevation of the crust as tons of soil were removed from its surface. Other versions of isostasy rejected isostatic equilibrium for such local features as mountains and plains and proposed that isostasy held for the general bulk of a continent. By the end of the century, extensive gravity surveys in many countries underpinned an acceptance of some form of isostasy. The concept of isostasy in turn supported a belief in the permanence of continents and ocean basins. If the ocean floors differed in composition from the continents, interchange was physically impossible: lighter continents could not permanently sink into the denser substrate. The lighter continental crustal material was termed 'sal' (later, 'sial', an acronym for its composition which was dominated by silica and alumina); the denser substrate which also formed the ocean floors, 'sima' (silica and magnesia).

Permanentism was not without problems in the early years of this century. What was the motive force in the 'geosynclinal model' of orogeny (mountain-building); i.e., granted that geosynclines might be formed through sags in a collapsing earth, what force or forces uplifted mountains and mountain chains? With respect to the Appalachians, what was the source of the sediment which filled the Atlantic geosyncline? Did this imply the existence of a former continent to the east of North America? Moreover, the theory of orogeny developed for the Appalachians was not easily transposable to the more complex structure of the Alps with their folding and overfolding. Orogeny seems, however, to have been less important to British and American geologists than to Europeans. The more serious problem was connected with the distribution of fossil and living species: 'palaeobiogeography' and 'biogeography'. If the ocean basins and the continents were permanent, how were similarities in species separated by ocean barriers to be explained? If one accepted some form of Darwinian evolutionary theory, solutions in terms of multiple centers of origin and parallel evolution would be ruled out.

Toward the end of the nineteenth century, the idea of a slowly contracting earth came under attack on geological grounds. Early in the twentieth century it seemed to be refuted, as it had once been supported, by physics. This did not pose a serious threat to Permanentism: a shrinking earth had at best a small role in the program. Indeed, if the earth had not contracted at all, this would be more ammunition for Permanentists to fire at their opponents. At a methodological level this would have reinforced the attitude of many geologists in the English-speaking world that grand general theories were not necessary and perhaps even deleterious. If what was thought to be the well-established theory of a contracting earth were in error, was this not a plain lesson that the edifice of geology should be built upon the solid rock of narrow 'inductivism' rather than on the shifting sands of 'theories' and speculative 'hypotheses'? This strengthened atheoretic or even anti-theoretic stance was expressed in North America and Britain by a retreat from global theories and global earth history to low-level theories employed within specialities and by the gathering of more data and preparation of more accurate descriptions.[7]

New versions of Permanentism in the early decades of this century often dispensed with a contracting earth. The size and state of the earth had remained unchanged for most of its history; similarly, the continents and oceans had undergone no more than superficial changes. If no significant changes have occurred, then there is no need for a global theory of geological change. These theories of Permanentism were antidotes to the disease of excessive theorizing. To accept the permanence of the earth in its present general conformation was effectively to limit theory to the explanation of minor alterations. Geology was not a study of processes, except on a lilliputian scale; it was a historical science with causes left out. From this perspective, Permanentism was congenial to those who put primacy on the gathering of facts and on accurate description as the proper aim of the discipline. To make an exaggerated contrast, many Permanentist geologists, if confronted with the Alps, would think it sufficient to describe and classify the strata observed; European geologists, working in a different program, would take the primary task to be the explication of how such a complex structure could have been formed.

Contractionism

A rival approach, much stronger in Europe but with advocates in the English-speaking world looked to Werner as its progenitor and tended to catastrophism. This program, developed by Cuvier, Elie de Beaumont, Henry De la Beche (the first director of the British Geological Survey), several

German geologists and culminating in the work of Eduard Suess, gave far greater emphasis to the evolutionary nature of the earth and of the processes which had shaped it. The conditions now encountered in and on the earth are not the same as those of the remote past: the earth has cooled and this has profoundly changed geological agents in kind, degree, and mode of operation. Since circumstances have changed, a naive inductivism will not lead to geological certainties: we must be concerned with large-scale processes and we must engage in some speculation and theorizing in order to determine what facts and what sorts of facts are relevant to geological reconstructions and to the solution of both historical and contemporary geological problems. Geologists cannot be overly narrow fieldworkers; they must draw on a wealth of information and theories from other sciences in order to make sense of these processes.

European geology was dominated from mid-century by the problem of orogeny, especially the formation and structure of the Alps. Any general theory of the earth was expected to give a plausible account of the Alps. Lyell's theory was of little interest and his uniformitarianism held little appeal: how could the Alps possibly be explained by very slow elevation through earthquakes and volcanic action? Surely the complex patterns of folding and crustal shortening demonstrated the importance of horizontal motions of the crust and the earth's contraction. In the 1850s and 1860s global theories were advanced which had as a major aim the solution of the problem of the Alps but none managed to gain general support. In the closing years of the century, however, there was an emergent consensus on the comprehensive, global theory of Suess. For a few heady years the 'Suessian synthesis' came close to becoming the master theory for a major part of the geological community.

Suess's global theory, which exemplified the European tradition of Contractionism and crustal plasticity and mobility, was fundamentally catastrophic. The history of the earth was divided into epochs of 'rapid' change and 'stability'. The earth was very different from what it had been in the past or what it would be in the future. The present continents were fragments of larger palaeocontinents, sections of which had broken up and subsided, forming parts of ocean basins and isolating once-joined landmasses. Rather than the continents growing at the expense of the oceans, Suess's oceans were growing at the expense of the continents. The crust of the earth was highly mobile: portions of it could shift laterally for dozens or even hundreds of kilometers. In *The Origin of the Alps* (1875) he depicted mountain chains as successive waves breaking on unyielding obstacles. First, the mountains formed during the Caledonian orogeny (about 400 million years ago (MYA)) swept up to crash on the basement rocks; then, the Variscan mountains (about 300 MYA) moved *en echelon* until dammed by the Caledonians; then, the majestic Alps (about 25 MYA) moved as a group hundreds of kilometers from the south until dammed by the Variscans. Evidence of great catastrophes, of dramatic change, he found in the existence and form of the mountains themselves. For Suess the riddle of mountain structure was the central riddle of geology (See Greene, 1982: 167–91).

The motor of these processes was the contraction of the earth. Earth history was not a history of cycles but of development, even of decay. Suess rejected the Lyellian doctrine of erosion and uplift; he denied any significant upward movement of any kind. As the earth lost its heat, a rigid crust formed. As the earth continued to cool and shrink, this crust wrinkled, folded and subsided. Crustal movements were either the direct effects of subsidence or the horizontal expression of stresses in the crust as the interior of the earth shrank away from its rigid and thickening crust. These processes were not continuous: the crust collapsed sporadically, unequally and suddenly. The Alps were the lateral expression of a collapse of the crust to the south. Continents and other landmasses sank beneath the waves. Crustal material under the oceans subsided, creating depressions into which seas drained, exposing new landmasses. Continents and ocean basins were evanescent and did not differ fundamentally in structure or composition.

Suess's theory was far more than a theory of the Alps: in *The Face of the Earth* (1883–1904) he drew upon the existing literature to extend his system in detail to regional geology. This method differed from that embraced by his Permanentist counterparts. His synthesis was precisely that, a synthesis of disparate theoretical views and of published materials. It was 'arm-chair' geology as some of his critics were to complain. Suess supported and documented his theoretical views not with extensive fieldwork but with extensive library work. His tools were not a sketchpad and hammer but a library card and a creative imagination. He did not devalue field studies; rather, he regarded them as a means to an end instead of an end in themselves. A chaotic heap of facts did not constitute a science.

The synthesis was firmly grounded – for the moment – on the theory and observation. It incorporated a contracting earth. Many Permanentists accepted this but Suess made it central to his theory. A cooling earth had been strongly argued for by physicists when Suess constructed his synthesis and therefore the theory seemed almost to be entailed by physics. The theory seemed well supported by palaeobiogeography and biogeography: it could explain why many plant and animal species, both extinct and living, were

spread across intervening ocean basins. For example, the similarities between Australian and some African and Indian species were explained as
follows: these three continents were once parts of the palaeocontinent
'Gondwana'; when the rest of Gondwana sank into the Indian Ocean, these
three sections remained. The similarities between some African and South
American species could be accounted for by the postulation of either a
landbridge or a whole continent located in the Atlantic which had since
subsided. Evidence of intercontinental connections was evidence for Suess's
theory and against Permanentism. Marcel Bertrand, a French partisan of
Suess, in the 1880s argued that many of the mountain chains in the New
World were continuations of those in the Old World: the links between the
two had sunk beneath the Atlantic (See Greene, 1985: 202–9). Structural
and stratigraphic similarities between the opposing coastlines of the Atlantic were further evidence for the Suessian global view.

Suess's theory was further refined in Switzerland and France. The Swiss
geologist Maurice Lugeon, for example, won considerable renown for his
theory that the rearward folds of the Alps had overriden earlier folds so that
the younger folds were now farther from the original subsidence in the south
than the older folds. Studies of the Alps and surrounding districts steadily
raised the estimate of displacement and crustal shortening from tens of
kilometers to hundreds of kilometers to perhaps over 1000 kilometers.[8] The
Suessian synthesis soon gained support in much of Europe and among some
American and British geologists.

By the early years of this century the grand synthesis had begun to
collapse and fragment. Some geologists pointed out that the thousands of
kilometers of crustal shortening and overfolding required to explain all of the
earth's mountains far exceeded any reasonable estimate of cooling and
consequent contraction. For the Alps, some estimated that 1200 kilometers
of crust had been wrinkled, folded and compressed into 150 kilometers of
mountains. This would require a cooling on the order of 1200°C just to
account for the Alps – and in roughly the same period the formation of the
Himalayas, the Andes and the Rockies had occurred. Moreover, there were
extensive, older periods of orogeny. Further, if the earth were cooling
gradually, one would expect orogeny to be occurring more or less continuously, but it seemed to have occurred in well-defined but widely separated
periods. Finally, if the theory were correct, mountains – like the wrinkles on
a drying apple – should be distributed uniformly over the surface of the
earth, yet this was not the case. Suess's appeal to unequal and spasmodic
contraction was not founded on any reliable information concerning the

interior of the earth; it was tantamount to explaining that which had to be explained by that which had to be explained!

The two chief difficulties for Contractionists arose from developments in physics and geophysics. First, there was a conflict with isostasy. Suess himself opposed the notion of isostasy as either a local or general phenomena. Prior to the early twentieth century, this was not unreasonable. As isostasy was increasingly attested by gravity surveys, the conceptual problems became more grave. If the continents did differ in composition and density from ocean floors such that the continents floated like icebergs in the sima, how could the sialic landbridges and connecting continents used to explain species distribution sink into the denser sima? Was not any interchange between ocean floors and continents physically impossible? Allied to this was the question of where the water had gone if all these land links had existed and had sunk: this would have displaced enough water to drown all the continents and wipe out all terrestrial life. Had the oceanic crust also subsided by just the required amount? If so, this was a remarkable coincidence. These problems were not ignored: possible solutions were put forward. For example, some pursued the line that over time water had been created through volcanic action and other subterranean processes. Others suggested that bridges might have been sunk by accumulations of denser material.

More severe problems came with the discovery of radioactivity. Physicists found that radioactive materials produced heat during decay and that they were widely distributed in the earth's crust. In 1909 the Irish physicist John Joly spelled out the geological implications clearly and unequivocally. Making certain assumptions about the distribution of radioactive materials and drawing upon his calculations as to the amount of heat the radioactive decay of these materials would produce, Joly argued – convincingly to many – that one could no longer assume that the earth gradually cooled by simple loss of internal heat. It was likely that it had been able to maintain a stable heat balance for much of geological history and indeed the earth could be getting warmer (Joly, 1909)! This conclusion was open to challenge but even for some Contractionists it was tantamount to a refutation of the central tenet of Suess's version of the program. The motor of the Suessian synthesis had overheated. The rival tradition of Permanentism was, as we have seen, able to adapt fairly easily to this incursion of physics. The simple theory of secular contraction through heat loss was, however, effectively eliminated as a global theory.[9]

In 1912 there was no program which received global assent. Even within the two research programs we have sketched, there was a multiplicity of variations and modifications. The demise of the Suessian synthesis, the retreat from global theories in Britain and America, the problems facing Permanentism – all these set the stage for a bewildering variety of novel theories and permutations of parts of different theories, including theories of Drift. Greene (1982: 289) aptly characterizes the circumstances in which Wegener's theory was launched as follows:

> In 1912 it [Wegener's version of Drift] was a legitimate but very tentative deduction from a great body of geological and geophysical evidence assembled in the last quarter of the 19th century, one of many different hypotheses created from the same materials. It had no particular claim to predominance . . .

Drift before Wegener

Wegener was not the first to notice a 'fit' between the opposing coasts of the Atlantic nor to postulate a theory of large-scale displacement of continents. The theories of some of Wegener's 'predecessors' sank immediately into obscurity from which they were rescued only when Wegener's version began to command attention. These include the highly idiosyncratic views of Evan Hopkins, Richard Owen and Antonio Snider-Pellegrini. Others devised Drift theories to solve empirical and conceptual problems facing the major research programs near the turn of the century. This group included Osmond Fisher, W.H. Pickering, H.B. Baker and, especially, F.B. Taylor. Their theories are of interest not because of any anticipation of nor influence on Wegener, for Wegener and most other geologists were ignorant of them, but as indications of the theoretical pluralism in geology prior to Wegener, a pluralism which, if anything, was even more marked after Wegener's proposal of Drift.

Hopkins put forward a theory of large-scale continental motion in 1844 (Hopkins, 1844). He was a practical mining engineer, a fellow of the Geological Society of London, who worked in Europe, North America and Australia and had a special interest in the formation and location of gold deposits. He believed that his global theory would be of value in mining operations. It purported that the driving force of continental motion was the earth's magnetism. There was a continuous circulation of magnetic fluid, toward the North Pole, down the polar axis to the South Pole then encircling the globe, being electrically conducted through the oceans and thence back to the North Pole. The oceans contained various substances in solution. At the South Pole these were crystallized by the magneto-electrical flow, forming the continents which were swept to the north. The continents were flexible crystalline compounds floating on and moving through a denser fluid. To support his contention of a northerly movement of land he gave comparisons of ancient and modern latitude determinations, referred to myths and legends and noted the concentration of land in the northern hemisphere. He also discussed tropical remains found in the far North and glacial remains in landmasses now near the tropics. These, he claimed, could not be explained in terms of a general cooling of the earth but fitted with his scheme. The congruence of opposing coastlines was of little interest, given his notions on the origins and movements of the continents. His theory was decidedly non-Lyellian but he was insistent that it agreed better than other global theories with physics and that it was uniformitarian (1844: 68). His discussion of continental motion and the evidence for it was, however, ancillary to his major concern, the provision of a theory of ore deposits.[10]

Owen was a medical doctor and Professor of Geology and Chemistry at the University of Nashville in Tennessee and a fellow of the Geological Society of America. In 1857 he published a book modestly entitled *Key to the Geology of the Globe*. In brief, he argued that the earth was originally of tetrahedral rather than spherical shape.[11] In this configuration Australia lay on top of the Arabian Peninsula and snuggled next to South America which lay on top of Africa. In a great cataclysm, the earth had expanded, the crustal blocks had broken apart and the moon may have been ripped out of what is now the Mediterranean basin. It was a frankly catastrophic theory which had no appeal to American Permanentists and virtually no influence.

Snider-Pellegrini, an American living in Paris, put forward a cataclysmic theory which bore little resemblance to that of Owen and which was tied to theology. He regarded his theory of the earth as a series of deductions from Scripture. His theory, published in 1858 as *The Creation and its Mysteries Unveiled*, was based on a rapidly shrinking earth and various catastrophes. On the Sixth Day, volcanic vapours vented through the fissure between the New World and the Old, blasted the continents apart, and caused a sudden shrinking of the earth by one-quarter of its diameter. This was followed by massive flooding except in the Americas whose inhabitants (the Atlanteans) escaped and were therefore direct descendants of Adam rather than of Noah (Snider-Pellegrini, 1858). His theory attracted little attention though his maps depicting the joined continents were reprinted in a popular English introduction to geology (Pepper, 1861).

O. Fisher (1882), a geophysicist, questioned the permanence of the ocean basins. He suggested that they were recent features, whereas the continents were ancient ones though their locations might have changed. He proposed that the moon had been ripped out of what then became the Pacific Ocean, that this had left a huge depression, and that the continents had moved toward this depression, resulting in a rifting of the Americas from the Old World. W.H. Pickering (1907) supported this idea by pointing to the correspondences between the opposing coasts of the Atlantic and to the marked differences in structure between the Atlantic and Pacific coasts. Howard B. Baker in a series of obscure articles between 1912 and 1914 further developed the theory and sought to answer critics of Fisher and Pickering (Baker, 1912–1914; cf. Baker, 1932). These Drifters were more interested in the formation of the oceans than the continents. They relied upon a great catastrophe, the 'fissipartition' of the moon, as the putative mechanism. They did not delve in any detail into the geological consequences of or evidence for the idea of continental rifting other than the congruence of the Atlantic coastlines. Rather than seeing Fisher, Pickering and Baker as precursors of Wegener, we should instead view them as pursuing a particular line of research which took its importance and meaning from the Permanentist program.

Frank Bursley Taylor in a paper presented to the Geological Society of America (Taylor, 1910) gave the most complete theory of Drift prior to Wegener. Taylor had studied geology and astronomy at Harvard University and had gained a post with the US Geological Survey where he worked under the supervision of Chamberlin. The trigger for Taylor was neither Moses nor the fit between the New and Old Worlds. It was the distribution and character of mountains. Taylor was working in the continental tradition of geology but perceived major problems in Suess's system. He rejected a contracting earth and took the view, perhaps influenced by Joly's work on

Figure 2.1. Snider-Pelligrini's depiction of the continents before (left) and after (right) separation (from Snider-Pelligrini, 1858).

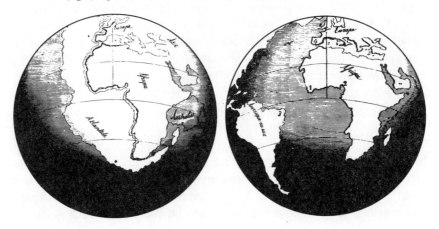

radioactivity, that the earth's heat was essentially stable. More particularly, he argued that Suess's theory systematically misrepresented the direction and nature of the forces which had produced such Tertiary mountain chains as the Alps. Specifically, he claimed that the Alps and the Himalayas indicated that the thrust came from the north rather than from the south. He postulated that the earth had originally been spherical with landmasses at the poles. It had suddenly changed to its current shape of an oblate spheroid (in later papers he was to suggest that this was due to the capture of the moon) and the landmasses by gravitation gradually 'creeped' toward the equator. North America had broken off from the northern landmass. South America and Africa from the southern and then these two had rifted apart, leaving the Mid Atlantic Ridge as a scar. The formation of Tertiary mountains and island arcs was explained by this movement as were the 'many bonds of union which show that Africa and South America were formerly

Figure 2.2. Taylor's diagram showing crustal 'creep' and the dispersion of continents from the North Pole. The heavier arrows near the Pole indicate the extent of displacement; the lighter arrows, the direction of 'creep' and thrust. Note the formation of mountain chains and island arcs around the periphery of the Northern Hemisphere (after Taylor 1910: 209).

united.' Although there are similarities between Taylor's theory and that of Wegener, Taylor ultimately had recourse to a unique, catastrophic cause and gave little geological evidence to support his theory; for example, he did not discuss the nature of the 'bonds of union' between South America and Africa. Whatever the reasons, Taylor's theory was only discovered after Wegener had proposed his version of Drift. The case of Taylor illustrates the ferment in early twentieth-century geology and the plethora of theories stimulated by the collapse of the Suessian synthesis (cf. R. Laudan, 1985).

Voice-Over

The 'old' history of geology provided the basis for and gave plausibility to some of the Kuhnian analyses advanced by geologists in the 1960s and 1970s. Most of the British and American histories of geology written before the acceptance of plate tectonics – and most of the Kuhnian accounts written in the early 1970s – were written by geologists. History can be used to defend or to attack established views, depending on whether the historian agrees or disagrees with them. There is a temptation in dealing with the history of science to regard what we accept today to be the appropriate measuring stick for past scientific beliefs and to write that history as progress from past error to present truth.[12] Scientists writing histories of their own disciplines are especially open to this temptation. Their histories – as in the cases of Lyell and Geikie – express and justify the authors' theoretical convictions which in most cases correspond to the prevailing orthodoxy. In their crudest form practitioner histories are celebrations, not explorations. Kuhn (1970: 1–2) remarks apropos of historical sections in science textbooks that the view of science which they purvey is that of piecemeal accumulation of current orthodoxy. The 'pre-revolutionary' geologists' histories began with the assumption that there was consensus among geologists and that this consensus had come about through the elimination of error by rational argument and empirical results. If there were but one 'correct' geology, then such unorthodox views as catastrophism (or Drift) could be either ignored or criticized as unscientific and as barriers to progress. Even some of the more scholarly studies produced in the 1960s; e.g., that of Chorley, Dunn and Beckinsale (1964) on geomorphology, which contained a wealth of detailed information and which took a more balanced and subtle approach than that of Geikie, emphasized the triumph of truth over error. They maintained the Lyellian legend and sustained the currently-accepted orthodoxy even as that orthodoxy was being over-turned.[13] This apparent theoretical consensus lent itself to characterization as a Kuhnian paradigm.

Practitioner historians writing in the immediate aftermath of the 'revolution'; e.g., Hallam (1973) and Marvin (1973), had a complex task: to give some understanding and justification of the new orthodoxy and to account for why their colleagues had initially neglected Drift in favor of other views but had changed their minds. A simple story of truth triumphing over error would not be very palatable to those who had once held to 'error'. Kuhn's vision of scientific change offered considerable appeal. The former orthodoxy could be labeled the 'old paradigm' and of course its proponents – since they could not stand outside the paradigm and critically examine it – could not be accused of being unscientific for rejecting or ignoring Drift. The period from about 1840 to at least the 1920s or even the late 1950s could be interpreted as a period of 'normal science' practised under a Lyellian paradigm. Wegener could be interpreted as having advanced a candidate paradigm at a time when no 'crisis state' existed for most geologists. The rejection of Drift was therefore a foregone conclusion. The crisis came in the late 1950s and early 1960s and was resolved by a revamped and revitalized theory of Drift. Practitioner historians could also make a place for the importance of fieldwork and facts in this story: new palaeomagnetic data and techniques and interpretations, new knowledge of the ocean floors, new developments in seismology, additional data on the geology of the continents had all been incorporated into geology since Wegener's day and had been instrumental in the development and final acceptance of a new version of Drift. Wegener in this account was the 'Aristarchus'[14] of the modern revolution in geology: his paradigm was proposed at a time when geologists did not have good reason to undergo a revolution in their world view.

'Revisionists' historians of late nineteenth- and early twentieth-century geology acknowledge the existence of and competition among programs and the extraordinarily diverse range of theories within those programs. This history seems far less amenable to a straightforward Kuhnian gloss. We might argue that there was something of a Kuhnian crisis near the turn of the century. Geologists were much concerned with the fundamentals of their field, there was a proliferation both of modifications to the main theories and also of some quite novel theories, the rules of the game seemed to have been loosened considerably, and there was a general recognition that major, perhaps insuperable problems loomed for the established views. So far, so good. But, this crisis state existed not for one paradigm which governed the whole geological community, but for rival paradigms which experienced crisis states simultaneously. Permanentism could be described as a more or less unified approach followed in the English-speaking world;

Contractionism, in Europe. However, not only were they inconsistent with one another, within each of these paradigms there were numerous variant theories before, during and after the crisis. Finally, for the rival paradigms the crisis eventually passed and something like Kuhnian 'normal science' ensued for a few decades *but* 'normal science' practised under at least three paradigms: Permanentism, Contractionism and Drift. Revisionist history depicts a science in which the normal state of affairs was conflict, competition and theoretical pluralism rather than adherence to one rigid paradigm.

Views of science which hold the existence of alternative research programs or theories to be the norm rather than the exception seem more suitable to the 'new history' of geology prior to Wegener. These would stress the continuing rivalry between Permanentism and Contractionism, of the evolution over time of different versions of these two broad sets of guiding assumptions and of the waxing and waning of these approaches and the variants within them, with respect to their usefulness in making sense of the problems with which geologists were concerned. Proponents of the 'internal struggle' view would point out how competing social and technical interests of geologists in the same or in different specialties and in the same or different academic and institutional milieus would sustain this competition and conflict at the theoretical level.

This last point is worth underlining: European geologists were more concerned with orogeny than their Anglo-American counterparts and orogeny was a central feature of the European Contractionist program. This suggests that geologists' decisions to accept theory or program might have been based less on judgment of their overall merit than on their value in contributing to the solution of particular problems on which those geologists place a premium. This might be due to their technical interests, their social interests within the field, their specialties or even the general geology of the region with which they were most concerned. Nonetheless, geologists were not oblivious to the larger setting of their specific research agendas, as we have seen in the 'crisis' of the early twentieth century. This crisis was not mainly due to the empirical inadequacies of the rival global programs; i.e., to the existence of empirical problems which had resisted repeated attempts at solution. It was more due to conflicts between what were thought to be solidly established geological beliefs and seemingly well-attested beliefs in other fields. For both Permanentists and Contractionists a cooling, shrinking earth came into conflict with the discovery in physics of radioactivity as a source of heat. For Contractionists a free interchange between seafloor and dry land conflicted with isostasy. For Permanentists relatively permanent extensive oceans seemed to conflict with evidence from biogeography and thus with Darwinian biology. The importance both of specialist concerns and of conceptual problems in the competition between rival theories and programs will be recurring refrains in the development of geology after this 'crisis'.

Notes

1 Less myopic studies of the Society include Rudwick (1963; 1982b) and R. Laudan (1977).

2 See Gould (1983: 94–106) for a lively sketch of Cuvier, an influential but much-maligned catastrophist.

3 The most comprehensive 'new' history of geology which treats European developments is Greene (1982). Greene concentrates on tectonics, especially orogeny. Rudwick (1972) gives an excellent overview of palaeontology. Rudwick (1985) and Secord (1986) make perceptive comments about the character of British geology in the 1830s–50s and stress the central place occupied by stratigraphy. R. Laudan (1987a) draws upon both British and continental materials to distinguish between and trace the early history of causal and historical theories of geology. Wood (1985: 3–60) draws on Greene and other sources in his discussion of the background to the 'modern revolution'. For other criticisms of and alternatives to the 'old history', see Ospovat (1976), R. Laudan (1982a, b; 1983) and Greene (1984; 1985).

4 For some discussions of the protean nature of uniformitarianism, see Cannon (1960), Gould (1965; 1977: 147–52) and Hooykaas (1959; 1970).

5 We are not concerned with what Lyell himself wrote but the way his successors interpreted his writings. Ironies abound. Some of his contemporaries thought him too much of a theorizer (cf. Rudwick, 1985: 38) yet later geologists were to make him a champion of empiricism and Baconian induction. His rigid uniformitarianism was partly a polemical device aimed at his opponents, yet for some later geologists it became a holy word. Within the Permanentist tradition from his time into the twentieth century there was tension between the aim of geological description and classification and the aim of some form of causal explanations using physical forces and processes (R. Laudan, 1982b; cf. Rudwick, 1982a). The first aim was probably more popularly shared but the second should not be neglected.

6 Rudwick (1982a: 229–32; 1985: 432) and Secord (1986: 24–9) rightly claim that even such seemingly theory-neutral field data as observations of strata are not only 'theory laden' but can also be 'controversy laden'.

7 In North America the fragmentation of a global geology into a geology composed of semi-autonomous specialties was facilitated by the implementation of a mistaken conception of the German academic model. Geology as a discipline was situated in a department organized as a collection of separate but more or less equal specialties represented by more or less equal specialists (see Ben David, 1971). This institutional pattern fitted well with the renewed emphasis upon data collection and description. There were, of course, exceptions; e.g., T.C. Chamberlin and the 'Chicago School'.

8 These estimates were highly contentious. Some drew a distinction between superficial folding, which affected only the uppermost surface of the earth and

which might be brought about by lateral motion of the crust, and folding due to an actual decrease in the size of the earth.

9 World War I may have been almost as devastating as radioactivity to the Contractionist program. The heartland of the Suessian synthesis was the Austro-Hungarian Empire. Its political disintegration in the aftermath of the war resulted in the collapse of the academic and institutional milieu which had nurtured the Central European tradition of geology of which Suess's work was the culmination (Greene, 1985).

10 I thank Robert Stafford for drawing Hopkin's work to my attention. How many other pre-Wegener Drift theories would be unearthed by a systematic search?

11 This is the reverse of the later, more widely known views of William Lowthian Green who argued in the 1870s and 1880s that the earth was contracting and collapsing from a sphere into a tetrahedron.

12 This is sometimes termed 'Whig history' from the school of British historians who represented British history as a struggle between conservatives and (the ultimately victorious) progressives.

13 Davies's (1969) study, which covers much the same ground as Chorley *et al.* (1964) is far less a history of progressive heroes and obstructive villains but it retains vestiges of the 'old history' framework.

14 Aristarchus had proposed over a millenium before Copernicus that the earth revolved about a stationary sun but this had been rejected in favor of the more orthodox system in which all revolved about a stationary earth.

3

Wegener and his theory: antecedents and arguments

Wegener was not the first to propose large-scale 'displacement' of the continents. His version of Drift was rejected in his lifetime. Those few who did take his theory seriously in the two decades after his death were often critical of his presentation and accepted not his version but one of several modifications. His career was not distinguished. He was not a recognized authority in the then-central specialties in geology. His modest reputation was grounded in meteorology, palaeoclimatology and polar exploration. He did not hold a prestigious chair at a major university, attract throngs of admiring graduate students, or win significant prizes or awards. Were we to be writing a history of modern geology, in 1950 or even in 1960 Wegener would receive no more than a footnote. No extracts from his writings would be included among classic papers in the discipline (e.g., Mather, 1967). Yet, today if we were to tell the story of the modern revolution in terms of heroes and villains, Wegener would feature as the heroic but neglected genius.

A drifting career

Alfred Wegener was born in Berlin in 1880.[1] He acquired a taste for science as well as outdoor sports and pursued studies in the former at the University of Berlin as well as spending some terms at Heidelberg and Innsbruck. He concentrated on physics, meteorology and astronomy but also found time for skiing and mountaineering. He capped his studies at Berlin in 1905 with a Ph.D. in Astronomy. The thesis topic, the conversion of a thirteenth-century set of astronomical tables from sexagesimal into decimal notation, precipitated his decision to abandon a career in astronomy. Instead, he joined the Prussian Aeronautical Observatory at Lindenberg. From 1906 to 1908 he served as a meteorologist in a Danish expedition to Greenland. The real attraction was probably adventure rather than science and the major concern was survival rather than research. Nonetheless, he gathered material sufficient for several publications.

On his return he lectured on meteorology, geodesy and astronomy at the

University of Marburg. In three years there he published papers on meteorology and a well-regarded textbook on physical meteorology, and first thought of Drift. He also became friends with Wladimir Köppen, an eminent meteorologist. Shortly after the unveiling, in January 1912, of his Drift theory, Wegener set out on another Greenland expedition. This time he gathered not only meteorological data but also observations relating to the possible westward drift of Greenland. On his return to Marburg in 1913 he married Else Köppen and resumed his research in meteorology. His career was interrupted by World War I. He saw action on the Western Front and was wounded twice. During convalescent leave Wegener set down a fuller version of Drift in *The Origin of Continents and Oceans* (1915), a slim volume of 94 pages.

After the war he lived with his in-laws in Hamburg. With the aid of Köppen he joined the newly-founded University of Hamburg where he organized the geophysics colloquium. He published a short monograph on lunar craters in which he argued that these were impact craters – like other combat veterans, he had first hand knowledge of impact craters – rather than volcanic. This had relevance to the moon's origin, evolution and structure and, consequently, to some geophysical theories. He also continued work on Drift. New editions of the *Origin*, each larger than the previous one, appeared in 1920 and 1922. Köppen initially opposed Drift and cautioned Wegener about the dangers posed to his academic career by the espousal of such a speculative theory but soon became a valuable ally. He was not only a distinguished meteorologist but also a walking reference library of the Central European tradition in geology and geophysics. Wegener no doubt read many of the recent meteorological and geophysical publications himself to prepare his lectures and his monograph on meteorology, but he also likely made free use of Köppen's knowledge and expertise. The citations in the early editions of the *Origin* show a familiarity with a wide range of European publications.[2] He and Köppen also collaborated on a treatise on palaeoclimates, *Die Klimate der geologischen Vorzeit* (1924) which used Drift theory.

In 1924 Wegener was named to the chair of meteorology and geophysics at the University of Graz. By the early 1920s his theory had attracted notice outside Germany; in 1924 French and English translations of the 1922 edition of the *Origin* were published. His appointment to a chair indicates that within the European academic community, his qualifications, contributions and status were above reproach. His teaching duties were light and he had few students. He prepared a fourth edition of the *Origin* (1929) which was revised and expanded from the third and which responded to his

American and British critics. It was his final word on the subject.[3] In 1930 he headed an expedition to Greenland. Its objectives included geophysical measurements bearing on Drift. From the outset the expedition was marked by mishaps. On 1 November 1930, with supplies running desperately low at his inland station, Wegener departed for the coast. His body was found the following May and was buried on the ice-cap. Drift had lost its most famous champion.

Building the theory

Wegener recounted that the stimulus for his theory was a chance observation made in 1910. A colleague at Marburg was given a world atlas.[4] Wegener looked at it and noticed a coincidence: 'Does not the east coast of South America fit precisely with the west coast of Africa, as though they had formerly been joined together?' This coincidence might be easily dismissed, for a fit of the shorelines would be affected by fluctuations in the depth of the oceans. What was striking was that the fit was even better 'if one compares not the actual shorelines of the continents, but the contours of the continental shelves' (letter to Else Köppen quoted in Schwarzbach, 1985t: 68). He commented, however, that he set aside the possibility that these continents had really been once joined together as improbable (Wegener, 1915: 8–9, 1924t: 5). In the autumn of 1911 he came across a summary of palaeontological similarities between Brazil and Africa given as evidence of a former landbridge linking the two. 'This induced me,' he wrote, 'to undertake a cursory examination of relevant research of the results of research in the fields of geology and palaeontology, and this provided immediately such a weighty corroboration that a conviction of the fundamental soundness of the idea took root in my mind' (Wegener, 1915: 8–9, 1924t: 5).

What had excited him? The jigsaw fit of the continents was not a novel observation. Moreover, in Europe the postulation of former land connections between now-separated continents was commonplace: any of several standard works could have provided him with lists of fossil correspondences across oceans, sketches of drowned landbridges and drawings in which the mountain chains of the Old World were continued into the New. This became grist for his Drift mill but it does not indicate why he concluded in favor of Drift instead of the European orthodoxy of Contractionism with its concomitant emergence and foundering of landmasses. We may conjecture that this was because the 'fit' and the palaeontological evidence came to his attention at a time when he was reading widely in geophysics and physics. This may have made him keenly aware of the problems confronting Permanentism and Contractionism. This would likely have been reinforced by his subsequent discussions with Köppen who was also well-read in the relevant field. If there were strong evidence of previous connections between now-separated continents and if the Contractionist theory which purported to explain these connections were unacceptable in the light of recent geophysics, then the way was open to develop a theory which harmonized geophysics and palaeontology. The hint provided by his earlier observation of the fit had become a working hypothesis. By 1912 he had scavenged enough evidence from the literature to state his views publicly. He believed that his hypothesis was well-supported by the various branches of geology and that it solved the outstanding problems peculiar to Permanentism and Contractionism and also some problems common to both.

Wegener opened his formal case for Drift – in the two papers read and published in 1912 and in the 1915 edition of the *Origin* – with a sketch of his views presented as a working hypothesis. The earth was composed of concentric shells of increasing density from the crust to the core. There were discontinuities in properties and behaviour between the shells due to differences in composition and increasing temperatures and pressures.[5] The outermost shell – the continents – was not continuous and as Suess claimed was formed of blocks of sial extending into and floating in the simatic shell. Between the sialic blocks and resting on the sima were the oceans. The ocean floors, underneath accumulations of sediment, were formed of sima. All the continents had once been joined in the super-continent of Pangaea which, for unspecified reasons, had begun to break up in the Cretaceous (about 100 MYA). Since that time the continents had been propelled through the ocean floors, sometimes moving apart, as in the case of South America and Africa, sometimes colliding, as in the case of India and Asia. The mountains along the Pacific coast of the Americas were thrust up by the ploughing of these continents through the sima; the Himalayas, by collision. Some bits were left behind; this explained the formation of non-volcanic islands and island chains. He seems to have been reluctant to discuss the motor which propelled the continents. He initially invoked two forces. The first was a 'Polflucht' (pole-flight) force due to the earth's rotation and which was directed from the poles toward the equator. Wegener quickly admitted that this was grossly insufficient to account for orogeny (e.g., 1929t: 171–4). The second was a tidal force due to the sun and moon and was directed east–west. In later editions of the *Origin* he suggested that convection currents in the sima might also power the continents (e.g., Wegener 1929t: 178–9). He was also reluctant to speculate about what the earth was like in the Pre-Cambrian: there were no fossils and no means of dating rocks that ancient. However, he speculated on ancient positions of the poles in publications with Köppen on palaeoclimatology.

He placed his working hypothesis – which in later formulations and with the inclusion of additional evidence hardened into a theory – squarely in the context of the difficulties facing Permanentism and Contractionism. He astutely borrowed the arguments of each against the other, claimed that these were answered by his theory and that his theory incorporated the major valid parts of each. With Permanentists he accepted isostasy and the essential permanence of continents; with Contractionists, the need for large-scale intercontinental connections to explain species distribution. As for radioactive heat, it posed no threat. A hotter earth meant a lower viscosity of the sima, enabling the continents to move through it more easily. The largest section of the *Origin* from first to last edition was by far the catalogue of evidence he had gathered. This was set out under the headings of the specialties from which it was drawn. Wegener was a researcher in the Suessian mould: he took his evidence from the publications of others, not from his own fieldwork. This was condoned in the European tradition but it was to be a sore point with some Anglo-American critics. These critics also objected to his use of the literature. It seemed that he pillaged others' publications for material favorable to his views and ignored that which was opposed. However pragmatic this procedure might be for a revolutionary, it did not conform to the Anglo-American stereotype of an unbiased, open-minded, neutral seeker after factual data who avoids speculation and theorization.

Wegener's evidence

What were the 'weighty corroborations' which he garnered from the literature? He led off with geophysics. The continents were lighter blocks of sial floating on a denser substratum of sima. He cited seismic waves and other evidence which indicated that the earth reacted to sudden, sharp forces like a bell being struck; it acted like a solid object. However, it reacted like a fluid to smaller, long-term forces. For example, its oblate spheroidal shape was produced by the effects of gravity and rotation; as another example, Scandinavia was isostatically 'rebounding' after the melting of its last glacial ice-cap. Since the continents move up and down, as most agreed, he argued that they could move sideways as well, provided this was a gradual motion produced by some small but long-acting force or forces.

He drew heavily on geological pattern-matching. There was the jigsaw fit. This was not, however, as straightforward as it might seem. He argued for a matching of continental shelves. To get the fit he sought, he bent the continents from their present shape, claiming that some distortion had been produced by their movement through the ocean floor. He also imaginatively 'unfolded' the Tertiary mountains such as the Alps, Andes and Himalayas – which consisted of crust that had been folded since the breakup of Pangaea. This naturally further changed the present outlines of the continents. After these adjustments, he found a pretty good fit.

The mere fact of fit had not been compelling for Wegener; a fit of distorted continents would be even less so for his contemporaries. Of greater weight were similarities in geological features between continents now separated. He called attention to the Sierras of Buenos Aires which seemed to be a continuation of the Cape Mountains of South Africa; the gneiss plateau of Brazil and the gneiss plateau of Africa; the correspondence of kimberlite (a distinctive diamond-bearing mineral) deposits in Brazil and South Africa; the coal fields of North America centered on Pennsylvania and those of Europe in Belgium and northern France; gneiss mountains in Labrador and those in the Hebrides and northern Scotland; and, the Canadian Appalachians and Caledonian mountains in northern Britain and Scandinavia. These features had all been formed *before* Pangaea had broken up. He did not suggest, however, how earlier episodes of orogeny had occurred: his theory dealt only with orogeny produced by Drift since the breakup. Structures formed after the breakup would not correspond in detail if the continents were rejoined: older structures should. He cleverly argued that this sort of geological pattern-matching was of enormous significance: if we had several bits of paper that we were able to reassemble into a page, that was no guarantee that the pieces had formed such a sheet and had been torn apart. The fit by itself was inconclusive. If, however, there were printing on the pieces and if, after assembling the pieces according to shape, the lines of type ran across the page and formed a coherent text, then this was proof that they were the pieces of such a sheet. These geological features were equivalent to lines of type; that they matched up when the continents were fitted together removed Drift from speculation to demonstrated truth at least, as he put it with some hyperbole, that the odds of his theory being wrong were 'one in a million' (Wegener, 1929t: 77).

The third sort of evidence he brought forward was the palaeontological data others had used to argue for intercontinental connections. These proposed connections and the evidence for them Wegener could subsume under his theory. He especially emphasized fossil and living species for which landbridges did not seem feasible; e.g., garden snails and earthworms of North America seemed to be related to those across the Atlantic. But, snails and earthworms were unlikely to dash madly across landbridges. Similarly, freshwater perch in eastern North America seemed identical to those in Europe: landbridges must therefore have been traversed by rivers, surely an unlikely happenstance. Manatee were found in the mouths of tropical rivers of the Americas and Africa. The fossil remains of Mesosaurus (a small Permian reptile shoreline scavenger) had been found only in South Africa

Figure 3.1. Wegener's reconstruction of Pangaea (top) and the movement of the continents after its breakup. Compare the distortion of North and South America in his Pangaea with the later reconstruction of Bullard in Illustration 9.3. (Reproduced from Wegener, 1922: 4, by permission of Friedrich Vieweg & Sohn).

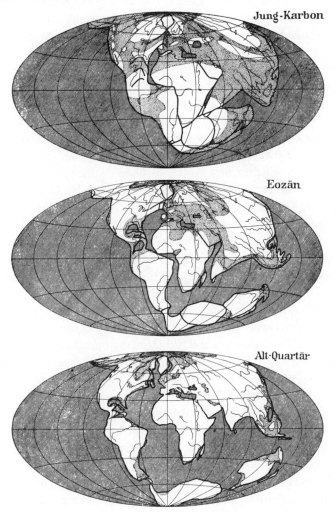

and Brazil. These correspondences, like the geological correspondences, were in his opinion explicable not in terms of drowned landbridges which were a geophysical impossibility but in terms of the former contiguity of the continents.

The fourth sort of evidence came from Köppen's specialty of palaeoclimatology. Even those dubious of Drift were to admit that this was among the most persuasive. At a simple level, the climates of the past differed dramatically from those of the present: a subtropical Spitzbergen; a glaciated Australia. What complicated matters for Permanentists or Contractionists was that such conditions occurred at about the same time; one could not claim simply that the earth as a whole was once hotter or cooler than now. Wegener contended that this could be explained only in terms of a shifting of the geographical poles or a shift of the continents relative to the poles. The single most impressive line of evidence for Drift prior to the 1950s was Wegener's elegant solution to the problem of the great Permo-Carboniferous glaciation (about 275 MYA). This involved both polar wandering and drifting continents. For this period there was extensive evidence of glaciation, especially tillites (deposits of pulverized rock and of boulders pushed along by ice) in some cases resting on top of striated – scored – rock pavements. These in turn were often associated with coal deposits containing fossils of the distinctive *Glossopteris* flora (extinct seed ferns).[6] These deposits had been found in southern South America, Antarctica, the Falkland Islands, Australia, New Zealand, South Africa and parts of India. A world-wide glacial epoch was not a satisfactory explanation because the northern parts of Europe and North America were at roughly the same time enjoying tropical or subtropical climates. This data posed a huge problem, he claimed, for both Permanentism and Contractionism. It was easily solved by Drift. During the Permian – before the breakup of Pangaea – all these landmasses had been joined together. If the South Pole be relocated in this reassembled Pangaea so that the pole lies in the center of the glaciated region, all falls neatly into place. Portions of the more southerly continents would be near the pole and have a polar climate; the more northerly continents would lie near the equator.

Finally, Wegener appealed to geodetic evidence. He claimed that not only had continents moved in the past, they are moving now. Accurate measurements of points on two landmasses supposed to be drifting apart should show gradual changes in longitude and therefore a growing separation. The Atlantic, for example, should be widening. He focussed particularly on Greenland on the grounds that it had broken away from Scandinavia only about 50 000–100 000 years ago and was still moving rapidly. He cited apparent changes in longitude measurements made between 1823 and 1907 on Sabine Island off the Greenland coast. These gave an apparent cumulative westward shift of 950 meters or about 11 meters/year. He admitted that this was based on inexact measurements and that there was doubt that the different sets of measurements were made from the same spot. However, the creative mind cannot be nervous about details. What was important to Wegener was not the absolute precision and accuracy of the data but that the data, flawed though they might be, supported his theory. For North America and Europe, the date of separation was much older. The apparent drift indicated by measurements made between 1866 and 1892 was on the order of 4 meters/year (well within the range of experimental error). Goedetic measurements loomed larger and larger in Wegener's successive presentations of Drift. If he could clearly establish that the

Figure 3.2. Wegener's solution for the great southern glaciation. Parts of present continents (top) with glacial remains (shaded areas) are brought into proximity to one another and to the South Pole in a reconstructed Pangaea (bottom). Such cooler regions of the present continents as the northeastern US were near Pangaea's equator (after Wegener, 1922: 69, by permission of Friedrich Vieweg & Sohn).

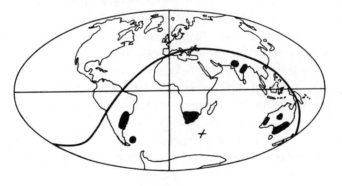

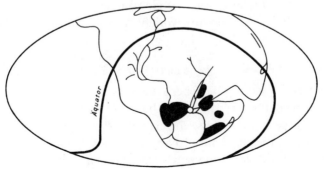

continents were moving now, it would be a great plus for his theory. Geological evidence was problematic: more than one cause could have produced the effects he assigned to Drift. Geologists engaged not in prediction but in retrodiction (predicting what had already happened). His theory offered a genuine prediction which could be tested and this, as he expostulated, set it apart from all others. This would also counter arguments by strict uniformitarians that he needed a force or forces hitherto unknown to break up and propel the continents. He could appeal to processes currently in operation and thereby conform to the dictum that the present should serve as a key to the past. If the continents were moving now, the onus of devising a suitable mechanism to explain this motion would not fall on him.

Wegener and his predecessors

Wegener claimed to be ignorant of earlier Drift theories when he first formulated his own. He referred to Taylor in the first published account of his theory but remarked that he became aware of Taylor's work only after working out 'the main framework of drift theory' (Wegener: 1929t: 4). To the later editions of his *Origin* he added an introduction in which he mentioned some predecessors, particularly Taylor, but deplored the suggestion that Drift should be called the 'Taylor–Wegener theory' (Wegener, 1929t: 4) – and for good reason. Like several of his predecessors, his starting point was congruence and like Taylor he was keenly aware of a number of the problems facing geology in the first decade of the century and saw his theory as answering them. Unlike his predecessors, Wegener put forward detailed geophysical arguments and assembled a large body of supporting evidence from many specialties. Furthermore, while his predecessors generally tossed out their ideas and watched them sink without a trace, Wegener stubbornly battled for his theory in person and in print for 18 years. Finally, he avoided reference to 'catastrophes' and seems to have worked very hard to make his theory acceptable not only to geologists in the European tradition but also to those in the more uniformitarian Anglo-American tradition. Wegener converted Drift from a vague idea to a working hypothesis to a research program taken up by others.

Voice-Over

To examine each facet of Wegener's life with a view to explaining why he became a 'revolutionary' and how he came to propose Drift would be laborious and perhaps fruitless. However, simply to say 'the time was ripe' or 'the idea was in the air' is unsatisfying and probably incorrect. Kuhn rules out the process by which the new theory first emerges in the mind of its inventor as 'inscrutable' but suggests some characteristics of revolutionaries. They are often young or newcomers to the field and therefore have not been indoctrinated into the prevailing paradigm. Further, the revolutionary is someone who is 'deeply immersed' in the 'crisis' besetting that paradigm (Kuhn, 1970: 90). Wegener was young and an outsider but otherwise there are difficulties in matching him with this profile. There was not one paradigm: geologists followed several different and incompatible programs. It is debatable whether or not they perceived the discipline to be in crisis: the programs all faced empirical and conceptual problems but these might be seen as challenges to be overcome or problems to be solved rather than cause for calling one's preferred program into question. Wegener could hardly be described as a scientist 'deeply immersed' in crisis when he noted a congruence of continental shelves or when he happened across a synoptic survey of palaeontological evidence of continental connections. In Wegener's own account the genesis of his theory was a causal observation – not a prolonged struggle with or reflection upon geological anomalies. Finally, he presented his theory not as a totally new world-view but as a compromise which retained the valid parts of both Permanentism and Contractionism while eliminating their major difficulties.

The rationalist theories of Lakatos and Laudan do not deal explicitly with the origin of research programs nor with their originators. This may be an arational part of science not subject to rational analysis and, in any case, they are more concerned with appraising extant programs (cf. Brannigan, 1981: 1–45). For Lakatos, contra Kuhn, empirical anomalies would be an unlikely source of innovation since his scientists can happily work in a Programme awash with anomalies. For Laudan, 'empirical' anomalies would be a source of concern but we may suppose that 'conceptual' anomalies and problems would be crucial as would divergent methodologies or aims.

Social analysts might give an account of why young scientists or newcomers would become revolutionaries. They have only a small stake of authority in the field and are risking little. If their revolutionary ideas be accepted, the return is great. If they be rejected, the loss may be small: these youthful enthusiasms can be written off as 'sowing wild oats'. A more sophisticated analysis might take the following form. In a mature scientific field, there is a hierarchy of authority and authorities. A newcomer may slowly amass authority within the existing organization of that field. Alternatively one may attempt to redefine the field in terms of one's own research interests and expertise or fractionate it by introducing a new specialty or subspecialty (one's own) and immediately rise to the top of a new hierarchy. Either short cut may be achieved by introducing new theories, new methods, new standards of practice, new aims and so on. In Wegener's case we could point to his introduction of a new global theory, a reordering of the hierarchy of specialties and the importation from physics and meteorology of standards, methods and types of argument.[7] If his views became authoritative, geophysics would become the core of the discipline, Drift the master theory, and the other specialties would be of interest only as they related to Drift. Wegener would naturally be the chief authority in this redefined, reordered discipline. This form of analysis may be illuminating but it is an analysis which we retrospectively impose. There is no unambiguous evidence that Wegener was motivated by such considerations. Moreover, this analysis is incomplete: to gain acceptance for his ideas and a position of pre-eminence required convincing other scientists, many of whom would have a vested social interest in opposing such a change. He appealed to 'the facts of the matter', not to presumed social interests.

This approach is plausible and not altogether inconsistent with several theories of scientific change. It addresses a sometimes neglected aspect of the organization of science into disciplines, specialties and problem-fields: the ever-changing cognitive and social boundaries and interactions between them; the creation of new ones; and their growth and decline, their disappearance or redefinition. Scientific communities – at least modern ones – are neither homogeneous nor impermeable. They may be broken down into disciplines, specialties or small groups of scientists centered upon small sets of interrelated problems. Memberships are not fixed. 'Invasions' from other disciplines or specialties may have profound consequences: the 'invasion' of geology by the physicist Kelvin in the 1860s created severe problems not only for British geologists but also British evolutionists (Burchfield, 1975); the 'invasion' of biology by physicists and chemists in the 1940s and 1950s transformed important parts of that discipline. Disciplines and smaller divisions do not exist in splendid isolation.

The academic milieu which Wegener experienced may have hampered his own career but may have facilitated his construction of Drift. German universities were tightly controlled by the professoriate. In the sciences there was usually one professor for each science. Often the chair had a research institute attached to it; professors dispensed patronage in both teaching and research. Within a university the professor exercised great power. Specialties were represented by the teaching staff in the institute, but in terms of funding, courses of study, university governance, academic and social status and extra-mural matters the professor formally and actually exercised control over and was the authority for the whole science (Ringer,

1969; Ben David, 1971; McClelland, 1980). Professors naturally resisted the evolution of new interdisciplinary studies or the splitting up of established disciplines: either would mean a reduction in a professor's institutional and cognitive authority. When vacancies occurred, usually through death and not the internal dynamics of the discipline, the new professor was chosen by the professoriate. Universities were slow to create chairs and thereby slow to legitimate thereby new disciplines or new conceptions of disciplines. Aspirants to a chair might have to wait a long time for their transmogrification. An aspirant with unorthodox ideas would likely encounter still greater difficulties in gaining admission to the select company: this could be an obstacle to the pursuit or advocacy of new theories especially if they were opposed to established views. The wonder is not that Wegener had to wait so long for a chair but that he was able to secure one at all.

A professor supposedly had competence in an entire science. He might concentrate on a specialty or two in research but was expected to have a working knowledge of the others. The organization of academic science could therefore encourage the creation and the balanced evaluation of broad synthetic theories covering a whole discipline. German professors were better placed than most of their more specialized counterparts in America and Britain to judge such a theory on its overall merits rather than with respect to just a single specialty. There was in Central Europe, especially in the countries forming the old Austro-Hungarian empire, a strong tradition in geology of theorizing on a global scale, of producing syntheses which could be used as 'working hypotheses' to guide research and to bring order out of chaos of detailed observations. This was furthered by the 'habilitation', an examination which an intending academic had to pass before being eligible for appointment as a lecturer. In it the candidate was expected to demonstrate some degree of originality and synthetic ability and, above all, a familiarity with the whole discipline; i.e., to exhibit those qualities expected of future professors. It is likely that some of the plethora of general theories of geology which were proposed in Central Europe in the early decades of the century had their origin in habilitations.[8] Wegener's own habilitation dealt with meteorology, not Drift. Nonetheless, his version of Drift – its originality and boldness of conception, its synthesis of previously unrelated theories and data, its integration of specialties – would be a form not at all unusual and actually fostered within the German tradition and the Central European geological tradition.

We could also point to a much more diffuse and highly speculative set of social and cognitive influences which might have affected Wegener and those responding to his theory: on the scientific side, the promulgation of relativity theory, quantum theory and indeterminancy; on the social side, the catastrophe of World War I and the subsequent social and economic dislocations in Germany. When the fundamentals of traditional science and the foundations of the familiar social order were dissolving, was the notion of moving continents rather than stable, permanent ones more appealing and less heretical? An analogy could even be drawn between the breakup of Pangaea and the breakup of Germany's overseas empire. Possible connections of this kind have been traced out for some aspects of German physics (Forman, 1971) but as yet geology has not been analyzed in this way. None of these broader or more specific considerations, separately or in combination, explains how and why Wegener proposed Drift nor do they severally or together explain why that theory had the form and content that it did. Rather, they were circumstances which may have helped create the possibility of such a theory.

Wegener's stimulus may well have been a chance observation. His decision to pursue a vague idea which he at first dismissed as improbable, and the manner in which he developed it into a working hypothesis and eventually into a research program were not owing to chance. In part they were shaped by his prior knowledge of several disciplines and of a number of specialties within geology, a consequence of his participation in the German academic tradition and in the Central European tradition of global theorizing. It is incorrect to label Wegener a meteorologist.[9] The content of Drift was partly shaped by Wegener's single-minded scouring of the literature, once he had the idea, for empirical and theoretical support. Drift was a global theory: Wegener aimed at nothing less than a history of the earth from the Cambrian and a causal account of much of that history. A lifetime of fieldwork would be unlikely to produce evidence of Drift convincing to others. A detailed, secure, first-hand knowledge of a whole country or even a whole continent would not, for example, be sufficient to press home his pattern matching claims. This later formed an obstacle to the acceptance of Drift. The very nature of the research program, were it to be taken up, required global co-operation and fieldwork. This ran counter to the practice of many geologists who tended to be rather parochial in their research; it would also be an expensive undertaking at a time when research funds were miniscule. Wegener in constructing his theory had before him the example of Suess. For Wegener, as for others trained in the European tradition, it was quite legitimate to develop and document a working hypothesis or a theory through a literature survey as opposed to fieldwork. This literature search, together with his prior knowledge, provided much of the building material for Drift.

A final and crucial influence on his decision to pursue his vague idea of Drift, and on the version of Drift which resulted, was the importance he placed upon conceptual problems. He presented his theory in the context of the conceptual problems which he believed faced Contractionism and Permanentism. He insisted on the possibility, indeed, the necessity of resolving those through Drift. We might conjecture that, compared with empirical problems, many conceptual problems by their nature may be less easily defused or avoided by *ad hoc* adjustments, are less likely to be confined to a single specialty or even a single discipline, may therefore draw in a richer range of resources – both scientific and social, and are more likely to engage theoreticians than experimentalists. Since they may involve clashes with beliefs about other theories or methods or aims or world views, we might also suppose that proposed solutions might involve fundamental reorientations of theory and practice. Conceptual problems, perhaps more than empirical ones, may stimulate innovations in theory and act as constraints on or determinants of those innovations.[10] Cognitively, Wegener's acceptance of continental permanence, isostasy, radioactive heating, past intercontinental connections, the value of hypothesis and global synthesis and so on constrained possible solutions and pointed toward a radical alternative to existing approaches.

Drift is indissolubly linked in the geological and historical literature with Wegener. He is credited with the 'discovery' of Drift. What is implied in describing Wegener as a 'discoverer' and Drift as a discovery?[11] Brannigan (1981) gives an analysis of discovery which is in general consistent with several theories of scientific development and which can be applied to Wegener's discovery of Drift. It is not sufficient for scientists simply to claim they have made a discovery. It involves judgment by other scientists. What are retrospectively deemed discoveries are claims which other scientists believe possess specific explicit or implicit elements (Brannigan, 1981: 77). First, the knowledge claim must be unprecedented: in some sense the claim must be novel or at least independent of and different in important respects from earlier similar claims by others. Second, it must be the product of a scientific investigation using appropriate methods. Third, it must be plausible. It must be almost 'expected' given the current state of the field and accord with other generally accepted theories. Fourth, it must be perceived to be valid in terms of the theories and data on which it is based and in terms of the results of subsequent research. To these four can be added a fifth: it must be perceived to be fruitful, to contribute to the solution of existing problems, to define new ones or to suggest new lines of research.

Let us take Wegener's knowledge claim to be that the continents have moved and are moving and that this has enormous implications for geology. From his point of view this claim was unprecedented, was the product of a proper investigation, was plausible with respect to isostasy and radioactivity and so on, was valid and was actually and potentially highly fruitful. As we shall see, geologists disagreed on each of these five elements of discovery. There was no mechanical 'ticking of the box' process whereby geologists could reach consensus on whether or not Wegener had 'discovered' Drift. Each of these aspects was negotiable. Those who refused to accept Drift did not accord to it the status of a discovery or deem Wegener a discoverer: he 'proposed', or 'put forward' or 'advocated' or 'presented' Drift, he did not 'discover' it. The bestowal on Wegener of the title discoverer could only come with the acceptance of Drift. Wegener was very careful to state that he arrived at his theory in ignorance of all the authors before him who had suggested some version of Drift. Moreover, in his 'Historical Introduction' to the *Origin* he pointed out how earlier versions differed from and were inferior to his. To agree with this assessment was to assent to one aspect of Wegener as a discoverer and Drift as a 'discovery'. To suggest, as some did, that Wegener was not original, or that the theory should be called the Taylor–Wegener theory, or even to suggest that the idea of Drift had been floating around for a long time but had never made any headway, was to undermine Wegener's claim of originality and to undermine the case for his theory as a legitimate 'discovery'.

Wegener's claim of originality is one element of his case for Drift. We should not assume that his monograph was a dispassionate recital of his theory and of the facts which bore upon it. He was guilty, as some charged, of being an advocate arguing a case, a lawyer for Drift. The *Origin* was his brief. In structure, content and style it was rhetorical: he aimed to persuade his readers in terms not only of technical arguments and evidence but also of his own credibility. Space does not permit a comprehensive analysis of the text of the *Origin* nor of the later attacks upon and defences of it.[12] We have already noted some elements of Wegener's presentation that are conformable to such an interpretation: his glossing-over of the problem of a mechanism; his hyperbole about geological pattern-matching; his portrayal of his theory as an almost obvious compromise between Permanentism and Contractionism; his extensive references and quotations to buttress his arguments; and his careful selection of data which supported his views while ignoring data which did not. We shall observe further instances of

Wegener's special pleading for Drift, both in his initial case and in his cross-examination and rebuttal of his opponents. Wegener was not here following an exceptional or exceptionable procedure. He was rather more open in his advocacy for his research program than some of his opponents and competitors – particularly in Britain and North America – were for theirs. This may reflect the tradition of global theorizing in which he was steeped and the conventions which were associated with it; e.g., it is probably no accident that he cast his presentation in a hypothetico-deductive form rather than an inductive one. He may also have believed that to call attention to a global theory it was better to shout 'Eureka!' than to speak softly; to emphasize its strengths and play down its weaknesses.

As we shall see, many of his British and American counterparts followed a different approach, one which was more in keeping with their research program and conventions of publication. They often assumed an attitude of impartiality: the histrionics and courtroom antics of Wegener were censured and shunned despite their prevalence in the previous century (cf. Rudwick, 1985; Secord, 1986). Facts alone should and could decide for or against Drift. But facts and theory were not neatly separable. His opponents, even while espousing inductivism, also selected and arranged their facts and discarded or attacked his in attempting to shore up their own programs and to undermine his. Their statements against Drift and in favor of orthodoxy, though encased in a thicker coating of neutrality and objectivity than Wegener's, were as much rhetorical constructs as his. This is not to say that there were no agreed 'facts of the matter' or that 'facts' did not matter, for as we shall also see, appeals to empirical data could be persuasive.

Notes

1 Convenient sources for biographical information on Wegener are Schwarzbach (1985t) and Greene (1984).

2 This may have created difficulties for Wegener's theory in America and Britain. Scientists there attempting to assess the argument in the *Origin* were probably unfamiliar with and unimpressed by many of his sources and the portions of his argument they were intended to sustain. The work of the Hungarian physicist, Roland Eötvös, on his eponymous pole-fleeing force was, for example, not widely known to the Anglo-American scientific community until the mid-1920s (Mott T. Greene, private communication).

3 Wegener admitted that even he could no longer keep up with the details of the 'snowballing literature of drift theory' (1929t: Preface) and he wished to do research on other topics. A posthumous fifth edition, prepared by his brother, was published in 1936.

4 In Germany at this time geography and cartography were undergoing rapid development as university subjects and as fields of research. Dickinson (1969) gives an introduction to European developments. The boundary between geology and

parts of the 'new' geography, especially geomorphology, was hazy. Wegener's theory was the subject of comment by geographers as well as geologists. We shall see that such visual materials as maps, charts and diagrams were ubiquitous and sometimes influential elements in the Drift story.

5 He extended this to the atmosphere which was composed of distinct layers with differing properties and composition (Mott T. Greene, private communication).

6 See M. White (1986: 99–133) for superb illustrations of *Glossopteris*.

7 Wegener was no advocate of the mathematization of geology: he once wrote that he distrusted books in which the logic of the argument could not be made whole without recourse to equations; that is in which arbitrary values could be assigned to constants and variables within equations in order for the argument to hold (Greene, 1984: 760). Wegener admitted that he was skilled in neither mathematics nor mathematical physics.

8 I owe this suggestion to Mott Greene (personal communication).

9 His knowledge of the recent geophysical literature surpassed that of many of his opponents. Moreover, he had a good grasp of the geological literature in so far as it supported his theory. His opponents criticized him for not being an 'outsider', a 'meteorologist' writing on geology, but for being selective: for failing to be familiar with, or, worse, for refusing to acknowledge data which they thought told against Drift. Leuba's remarks (supra: p. 10) do not conflict with this: Leuba was a proponent of Drift who gave a 'social account' for why Wegener might have met resistance from other geologists even though his ideas were (to Leuba) 'correct'.

10 L. Laudan (1977: 55–57) seems to suggest that intra-scientific conceptual problems arise from conflicts between beliefs held in different disciplines. We might amend this to include conflicts between beliefs held in different specialties within the same discipline.

11 My use of the term 'discovery' may be misleading. I do not take the view that a 'discovery' is an unproblematic uncovering of a piece of pre-existing reality or pre-existing order. I would more accurately describe Drift as an 'invention' or a 'construct' in line with the view that scientists in some sense impose order on the natural world.

12 Yearley (1981) is an example of how one might undertake a study of the rhetorical character of publications. See Rudwick (1970) for an analysis of Lyell's *Principles* as a carefully structured polemic. *Cf.* Rudwick (1985: 430–5). The rhetorical function of scientific publications is integral to Feyerabend's (1975) model of science: the arguments of scientists, especially those propounding novel theories, are replete with 'propaganda' and 'psychological tricks'. These create breathing space for new theories. Since his ideal is the proliferation of theories, rhetoric is both legitimate and desirable. For Kuhn, too, rhetoric is indispensable in effecting a change of paradigms. Since the facts, theories, standards, methods, problems, solutions and indeed 'the world' is different in different paradigms, no logical proof or simple appeal to the facts can bring about a 'conversion' to a different paradigm. Instead, recourse must be had to persuasion (Kuhn, 1970: 148–59) which includes claims of having solved the anomalies of the old paradigm, simplicity, elegance, fruitfulness and so on.

4

Debate around the earth

Wegener's theory sparked a lively debate. It did not, however, dominate the geological literature. Only a small fraction of the whole community discussed Drift in their writings. Of those who did enter the dispute, Europeans tended to differ from Britons and especially Americans in their response. In Europe the sort of 'big theory' which Wegener espoused was neither uncommon nor unwelcome; causal accounts of geological structures were regarded as highly desirable. His use of a wide range of evidence culled from the writings of others rather than from his own fieldwork was to most Europeans a legitimate practice. His attention to geophysics and tectonics was a well-established part of the European tradition, as was his use of the 'method of hypothesis'. His theory was accorded a tolerant though neither uncritical nor enthusiastic reception in Europe. It was otherwise among British and American geologists.

A skeptic's guide to Drift

Before sketching the early fortunes of Drift, we should try to put ourselves in the shoes of a skeptical geologist examining Wegener's theory and evidence. If we focus singlemindedly on the 'pluses' of Drift, which is what Wegener and some of his partisans did, then the theory seems powerful and the evidence all but overwhelming. Most geologists, however, were very much underwhelmed. How did they see the matter?[1]

Geophysically, there was nothing compelling in the theory. If isostasy was accepted, it applied to the vertical motion of the continents, not lateral motion. The earth did behave like a fluid in some respects, but no one was proposing that the ocean floors were in fact liquid: they were composed of dense, basaltic rocks. How could the continents move laterally through such floors without crumbling to bits? What enormous force not only moved the continents but had crumpled them up to form the Alps, Rockies, Andes and Himalayas? The forces which Wegener invoked did exist but they were far too weak. A force nearly 1 000 000 times stronger was needed, and if it did exist, it would surely have been noticed by physicists.

What about the 'jigsaw fit' and geological pattern-matching? The 'fit' was simply not as good as Wegener claimed. If the Americas were simply shifted over until North America touched Europe, there was a gap of hundreds of kilometers in the south. Wegener got *his* fit, as represented in his famous maps, by massaging the continents. He had distorted the shapes and unfolded mountains in an arbitrary fashion. Wegener was able to match some patterns of mountains and strata, but with bent and distorted continents, and, though he had produced some evidence for matching, there were other geological features he did not mention which did not seem to have analogues on the opposite side of the Atlantic. Further, some of his data were erroneous or taken out of context or misinterpreted. There were some similarities between the Canadian Appalachians and the British and Scandinavian Caledonians but were there not marked differences which Wegener had failed to mention? Those patterns which did exist could be explained as minor coincidences or perhaps by similarities in geological evolution or, by the landbridgers, as the two ends of mountain chains or strata which had extended across a now-sunken landmass.

Biogeographical evidence was ambiguous. If Wegener was right, should there not be many more similarities of species than had been found, since there would have been no ocean barriers to migration? Contractionists could explain the few similarities which had been found by landbridges; Permanentists, by such chance events as rafting. All geologists acknowledged the imperfections and gaps in the fossil record. Little reliance could be placed on it as an accurate complete source of information about the biogeography of the past.

Permanentists were skeptical of arguments and evidence from palaeoclimatology. It was a highly speculative enterprise and the notion of some sort of relative motion between the poles and the continents – which had been proposed by others and rejected – seemed quite bizarre (Barrell, 1914).[2] Wegener's reconstruction of the Permo-Carboniferous glaciation was ingenious, but based on a thin body of facts. Moreover, he had not cited all the relevant evidence. There was undisputed evidence of glaciation near Boston, Massachusetts about the same time that, according to Wegener, Boston should have been a tropical paradise. The *Glossopteris* flora, thought by Wegener to be associated uniquely with the southern continents, had reportedly been found also in Siberia and European Russia!

Geodetic evidence could be very powerful but Wegener's was ambiguous. The techniques on which it rested were inaccurate and the results he had seized upon were well within the limits of experimental error.[3] Moreover, other data which Wegener had not mentioned indicated that the continents had not moved appreciably over the last hundred years or so – certainly not at the rate which Wegener proposed. At best, all one could do was wait for further observations. It was to be a long wait!

There were challenges to the evidence which Wegener had plucked from each specialty. Could not a Drifter reply that what matters is the way that all is drawn together and interconnected by Drift? Should we worry about matters of detail? Drift stitches these pieces of evidence together, so maybe the whole is greater than the sum of its parts. This argument may sound odd, but Wegener and some Drifters advanced it. Du Toit (1937: 4) wrote that Drift 'represents the holistic outlook in geology' and that it should be judged as a groping toward a general theory of the earth rather than on a piecemeal basis. Wegener claimed that his theory should be judged as a whole rather than by any one specialty or matter of detail. Critics scoffed at such a notion.

Finally, there were fierce criticisms of Wegener's methods, especially from naive inductivists. It is improper, it is unscientific, to rummage through the publications of others looking only for that which fits a preconceived theory, blind to anything which seems to contradict it! The scientific approach consisted of years of extensive, meticulous fieldwork at the end of which one might be so bold as to offer a generalization or two. As we shall see, these attitudes and criticisms appeared again and again in the reactions of geologists to Wegener and his theory. Verdicts on Drift were not simply a matter of 'looking at the facts'.

The view from Europe

The European reviews of Wegener's production were decidely mixed. A few swiftly hailed Drift for providing solutions to problems on which they were working. These supporters, usually geophysicists or those concerned with orogeny, species distribution or palaeoclimatology, were outnumbered by those who criticized Wegener's evidence, his methods, his theory, or all three. Karl Andrée (1914), a German palaeontologist who had a special interest in mountain structure, criticized Wegener's explanation of the formation of folded mountains on the continental margins. He was also skeptical about the forces which supposedly powered the sialic blocks through the sima but was otherwise moderate in his comments. Edgar Dacqué (1915), a German palaeontologist who worked on the Alps, believed that Drift solved the major problems confronting Permanentism and in his textbook on palaeogeography championed a Wegenerian approach against that of Suess. Carl Diener was severely critical. He was a stratigrapher and palaeontologist at the University of Vienna and a senior member of the European geological community. He rejected Drift, listing detailed objections frequently cited by later authors.[4] Wegener was perhaps a corrective to over-enthusiastic bridge-builders but, Diener claimed, isostasy did not rule out the sinking of small landmasses. He apparently regarded 'negative' facts as a sufficient rebuttal since 'Wegener's hypothesis, as impressive as it may seem at first glance . . . is after all only a playing with possibilities. It lacks a foundation of positive evidence . . .' (Diener, 1915: 352).

Wegener's 1915 *Origin* drew others into the fray. Wolfgang Soergel, a young palaeontologist who later gained a post at Tübingen, propounded a meta-theoretic yardstick by which to judge the new theory: to be acceptable, it had not only to provide solutions to the problems facing older theories, but also to provide satisfactory explanations of all that which the older theories had successfully explained (Soergel, 1916: 201).[5] Drift did not measure up. Citing Diener, Soergel sought to demonstrate that Drift was contradicted by a detailed comparison of the two sides of the Atlantic 'split'. Like Diener, he rested his case on a marshalling of data. Drift did not solve the problems it purported to solve and 'unsolved' many old problems. Andrée (1917), in a second publication was more favorably inclined to Drift. He discounted Diener's criticisms for being based on too static a view of the earth. This, he asserted, illustrated the dependency of historical reconstructions on initial assumptions. The available evidence could be interpreted in more than one way.[6] Wegener's version of Drift had merit and was a stimulus to research even though it could not be accepted in all of its details (Andrée, 1917: 79). Theodor Arldt, who published extensively on palaeogeography, took a similar tack, adding that even if a present westward drift of Greenland were confirmed, 'it is not admissible to extrapolate to a similar movement hundreds of thousands or even millions of years into the past' (Arldt, 1918: 7). Both Andrée (1917: 80) and Arldt (1918: 10) concluded that there was a 'valid kernel' in Wegener's theory.

Max Semper's scathing rejection of Drift stands apart from other continental reactions in its vitriolic tone and its substantial use of 'method talk'.[7] He opened his paper entitled 'What is a Working Hypothesis (Arbeitshypothese)' with the comment that Wegener's theory was 'wonderfully suited to explain the character of a working hypothesis – by counterexample' (Semper, 1917: 146). According to Semper the citations and sources used by Wegener were problematic, unrepresentative, erroneous, misleading or otherwise inadequate as evidence or argument. Wegener's misrepresentations and distortions accounted for the erroneous character of his theory and were attributable to his poor methods and

ignorance in various fields of geology (Semper, 1917: 157). Semper rhetori-
cally asked if Drift should not be pursued by some geologists as a 'working
hypothesis'? It would then be legitimate to set aside for the moment the
theoretical difficulties and inconsistencies facing Drift in order to concentrate on empirical investigations suggested by it – an approach, he noted,
that is consonant with 'the English way of thought' (Semper, 1917: 160–1)[8]
Absolutely not! Wegener's hypothesis was a synthetic one. If it be used as
though it were an analytic working hypothesis, 'One truly sews a rotten bag
with good thread' (Semper, 1917: 161). Semper excused himself for going
on at such length about a patently absurd theory on the grounds that some
reputable geologists had actually endorsed aspects of Drift. (Semper, 1917:
157–8)!

A flurry of articles and reviews and a symposium greeted the second
(1920) edition of the *Origin*. The Berlin Geological Society organized the
symposium early in 1921. Wegener (1921) summarized his theory. This
was followed by criticism from Wilhelm Schweydar, Franz Kossmat and
Albrecht Penck. Schweydar, a geophysicist who was soon to join the
Geodetic Institute in Potsdam, stated that Drift was geophysically possible
but cautioned that the movement of Greenland had yet to be confirmed and
that, in any case, 'geographical and geological findings must first decide if
continental drift has actually occurred' (Schweydar, 1921: 122). Kossmat,
an Austrian structural geologist and palaeogeographer, admitted that
orogeny was better explained by large-scale horizontal movement of rigid
continental blocks than by Contractionism (Kossmat, 1921). Nonetheless,
he opposed a general drifting of the continents. Penck adamantly rejected
Drift; his criticisms for the most part amounted to nit-picking. He was
professor at the University of Berlin, and a leading geomorphologist, and had
gained some recognition for his own theory of landforms which was based
on sedimentation and the action of gravity. He cited the earlier papers of
Diener and Soergel in his rejection of Wegener's evidential base, particularly
pattern-matching and the Greenland data. His conclusion (Kossmat, 1921:
119) was one which was reached by others: Drift, though extremely
'seductive',[9] is built up not on conclusive proofs but rather mostly on
circumstantial evidence. . . .' Wegener (1921) closed by expressing opti-
mism about the future of Drift: geophysicists, he claimed, were generally
well-disposed toward it and geologists were slowly coming around.

The comments of other German geologists underline the division of
opinion over Drift. Friedrich Burmeister (1921), who had recently completed a Ph.D. in geophysics, gave Wegener's geodetic argument a thorough
raking but added that this negative judgment did not mean that he opposed

the theory as a whole. Bruno Schulz (1921), a young oceanographer at Hamburg, and Leopold von Ubisch (1921) both gave qualified endorsement of Drift. Bruno Kubart (1926) suggested that a combination of ideas drawn from the older theories and from Wegener could form a suitable basis for further work. For these geologists, *contra* Semper, Drift was a promising hypothesis despite some reservations. Strident opposition came from Friedrich Nölke and Erich Jaworski. Nölke (1922) rebutted the geophysical arguments and criticized the forces employed by Wegener. He concluded with a methodological sally directed at Wegener's claim that even if the physical foundations as advanced be unsound the theory as a whole would not be jeopardized. For Nölke, conceptual problems generated by possible conflicts with other disciplines were of greater importance than empirical problems: further research can eliminate empirical problems but faults in the use of physical theory are irremediable.[10] This implied claim for the primacy of consistency with physics in evaluating geological theory and for the importance of conceptual as opposed to empirical problems did not play a large part in the German reception of Wegener's theory. It loomed larger, as we shall see, in Britain and America. Jaworski (1922), a young palaeontologist at Bonn, drew upon objections previously stated by others to attack pattern-matching. He admitted that Drift did solve some problems but claimed that these were more than offset by the new ones it created.

The German and Austrian reaction was, in summary, not univocal. A few endorsed Drift; a few were openly hostile; others saw merit in it but noted the problems it raised and cautioned that much further research was needed, but most ignored it publicly. Wegener's 'hypothesis' might yield attractive explanations of gross features of the earth's surface but there was much disagreement over whether or not it did – or could – explain those features in detail. Most believed that 'facts' would decide the issue. Critics, with few exceptions, did not deploy 'method weapons' and, excepting Semper, did not attribute the construction and defense of this (erroneous) hypothesis to improper methodology. This toleration of Drift as a legitimate though probably incorrect program continued into the 1930s until that scientific community became preoccupied with other matters.[11]

The response to Drift of other European geologists was variable.[12] Several Dutch geologists with extensive field experience in Indonesia were enthusiastic about the major tenets of Drift though critical of details (Molengraaff, 1916; Wing Easton, 1921a, b; Smit Sibinga, 1927). Felix Vening-Meinesz, a pioneer in marine geophysics, was more reserved. He suggested that horizontal crustal movements might be powered by deep-seated convection currents but maintained an agnosticism into the 1960s (Vening-Meinesz,

1934b). In Switzerland Drift was soon pursued by most Alpine geologists. Suess's southern subsidence as the 'wave machine' was replaced by the northward movement of Africa, crushing against southern Europe which had then itself moved northward thereby opening the Mediterranean. Emile Argand, founder of the Geological Institute in Neuchâtel, introduced his compatriots to Drift in 1916. He later applied it to the tectonics structure of Asia (Argand, 1924). Rudolf Staub, a rising star in Alpine tectonics, had misgivings about many aspects of Drift but saw its value for analyzing the Alps (Staub, 1924). There was opposition, but among Swiss geologists concerned with the Alps, crustal mobility had become orthodoxy by the 1930s.[13]

The French geological establishment, centered in Paris and dominated by palaeontologists and stratigraphers, were skeptical into the 1960s (cf. Buffetaut, 1985; Carozzi, 1985: 126–9). The first formal discussion came in 1923 in the Geological Society of France. Palaeontologists and biogeographers made sallies against details of the theory, hurled vague generalizations at it as a whole and retreated to the safety and security of landbridges (Joleaud, 1923). Indeed, landbridges were built by French palaeontologists in the 1950s (e.g., Termier & Termier, 1952; 1958). The two geophysicists who spoke at the 1923 meeting pointed out that Drift was consistent with accepted geophysical theory but adopted a position of neutrality pending further evidence (Joleaud, 1923: 264–70). Over the next four decades this pattern persisted. Most who mentioned Drift rejected it or where highly skeptical of it. Those who pursued it or wrote approvingly of it were usually geophysicists or geologists who had a special interest in orogeny or were on the periphery rather than in Paris (e.g., Leuba, 1925; Russo, 1930; Choubert, 1935; Roubault, 1949).

Across the Channel

British opinion was divided. There were more outspoken opponents than proponents but most expressed no opinion publicly. Frequent recourse was had to criticisms of Wegener's methods. This likely reflected the generally inductivist flavor of Permanentism. Wegener's views were virtually unknown in Britain until the early 1920s. This was probably due to a concerted boycott of German science by most of the Allies during and immediately after World War I.[14] By 1922, however, British geologists were introduced to Drift. Early in 1922 Frederick Weiss (1922), a botanist at Manchester, penned a glowing notice of Drift in *The Guardian*. Shortly after, he and William Wright gave favorable accounts of the theory at a meeting of the Manchester Literary and Philosophical Society.[15] Awareness of Drift

spread swiftly. *Discovery* published a summary by Wegener (1922) in
which he chose as a 'single striking example' of the palaeontological
evidence the explanation Drift afforded of the Australian fauna. Reviews of
his publications together with a translation of the *Origins* (1924t) promoted
further discussion and debate. Many of those who spoke up deprecated
Wegener's methods. Philip Lake, a geologist and co-author of a widely used
textbook (Lake & Rastall, 1910; 1920; 1927), reviewed and harshly criticized Drift in several articles. Of Wegener's methods, he wrote (Lake, 1922:
338):

> Wegener himself does not assist his reader to form an impartial judgment.
> Whatever his own attitude may have been originally, in his book he is not
> seeking truth; he is advocating a cause, and is blind to every fact and argument
> that tells against it.[16]

Harry Fielding Reid, an American geologist, reviewed Drift for a British
journal and reached a similar but more explicit conclusion (Reid, 1922:
674):

> There have been many attempts to deduce the characteristics of the earth from a
> hypothesis; but they have all failed. . . . This is another of the same type. Science
> has developed by the painstaking comparison of observations and, through
> close induction, by taking one short step backward to their cause; not by first
> guessing at the cause and then deducing the phenomena.

Both Lake and Reid supplemented their criticism of Wegener's method
with specific objections to data and generalizations employed or ignored or
mis-stated by Wegener. These were often borrowed from German publications critical of Wegener. Both Lake and Reid implicitly assumed that
because Wegener must be wrong, the German literature critical of Wegener
was reliable; the literature supporting him, unreliable. This combination of
methodological and factual criticism contributes not only to a demonstration that Drift is false, but also to an explanation of why Wegener could
believe it to be true. Wegener advocated an erroneous theory because his
facts were wrong; this was because he failed and continued to fail to adhere
to the canons of the proper scientific method as it was understood in the
British tradition. He had sinned methodologically and – irredeemably – he
refused to repent. Charges that Wegener was deductive, speculative, and
highly selective with regard to the total array of data were not without
foundation, given Wegener's own description of how he came to his theory
and how he built up evidence to support it. But, the import of the remarks
similar to those of Lake and Reid should not be underestimated. For most
Anglo-American geologists the labelling of a theory as deductive, specula-

tive and selective was tantamount to labelling it as worthless and liable to lead to serious errors.[17]

Wegener had some support from British scientists. Drift was discussed at a meeting of the Geological Section of the British Association and at the Royal Geographical Society. In the former, most argued against Wegener's geophysics and his geodetic evidence – which one member described as 'so much waste paper'. However, several expressed a more cautious 'wait and see' attitude. A palaeoclimatologist said that 'the theory was a wonderful one from the meteorological point of view', particularly the explanation of the southern glaciation (W.B. Wright *et al.*, 1923). The discussion at the Geographical Society was occasioned by a furious attack on Drift and on Wegener by Lake (1923a, b). Lake ridiculed the 'jigsaw fit' (1923b: 183):

> It is easy to fit the pieces of a puzzle together if you distort their shapes, but when you have done so, your success is no proof that you have placed them in their original positions. It is not even a proof that the pieces belong to the same puzzle, or that all the pieces are present.

He also attacked Wegener's use of the *Glossopteris* flora in the reconstruction of Pangaea. This, Lake admitted, was a strong point for Drift but he cited several papers in which were reported contrary observations; for example, reports of *Glossopteris* in Russia. Lake here indulged in the same tactic for which he criticized Wegener, relying upon the published literature instead of fieldwork.[18] Four of the six geologists and geophysicists who spoke in the ensuing discussion expressed support for the idea of large-scale lateral motion of continents even though Wegener's version might be erroneous. C.S. Wright had done fieldwork in Antarctica and had earlier judged Drift very promising with respect to the problems of Antarctic palaeobotany and palaeoclimatology which had then concerned him (C.S. Wright & Priestley, 1922: 466). He disagreed with the character of Lake's criticisms: Drift should be evaluated as a whole, not on points of detail (Lake, 1923b: 193). The sixth participant, Harold Jeffreys, who was soon to revitalize Contractionism (Jeffreys, 1924), in this meeting rejected Wegener's mechanism for Drift but suggested other forces which might suffice (Lake, 1923b: 191–2).

Skirmishes between proponents and opponents continued sporadically over the next two decades but few actually published explicit comments on Drift. Some strongly opposed Drift, for a variety of reasons. Some openly favored the general idea of Drift but had reservations about Wegener's version. The most significant point of contention prior to the outbreak of World War II, which diverted attention from geological battles, seems to have been geophysics, particularly the issue of the mechanism of Drift. Although few made explicit use of Drift in their research (Bailey, 1929;

Holmes, 1929; 1930; Seward, 1933; Holland, 1941; 1943; Zeuner, 1946; Hills, 1947), it did receive favorable comments in several widely used textbooks (Bailey, 1935; Wooldridge & Morgan, 1937; Rastall, 1941; Holmes, 1944) and was featured in a number of popularizations (Steers, 1932; Gheyselinck, 1939; Seward, 1943; Read, 1949). Drift fared more poorly on the other side of the Atlantic.

Drift under fire: The great debate in North America

North Americans were generally negative in their assessments of Drift. This was exemplified at the formal symposium held in New York in 1926 under the auspices of the American Association of Petroleum Geologists. The symposium was held when the controversy over Drift elsewhere was at its peak and when Drift had won endorsement from some English-speaking geologists as the preferable global theory or at least as a legitimate working hypothesis. The edition of Wegener's book recently translated into English (1924t) reported new Greenland observations made by the more accurate method of radio time signals which supposedly showed that Greenland was drifting west by as much as 20 meters per year. Drift was advocated only by a small but vociferous minority but it was perhaps no longer easy to dismiss it as a temporary aberration. It was in the context of the growing challenge of Drift to conventional theories and programs, particularly to Permanentism, that several leading Permanentists confronted Drift in New York. The arguments mounted and the positions taken in that debate helped set the pattern for the North American response for the next four decades. Drift was to have few friends and many enemies but, despite the hopes of some of the latter, it was not to disappear completely as an alternative program.

The symposium was chaired by van Waterschoot van der Gracht, vice-president of the Morland Oil Company and a pro-Drifter. Joly, Taylor, and ten other geologists were featured speakers. Wegener did not attend but sent a short paper. Several participants were antagonistic toward Drift and accused Wegener of being unscientific. R.T. Chamberlin, Professor of Geology at the University of Chicago, caustically asked (1928: 83), 'Can we call geology a science when there exists such difference of opinion on fundamental matters at to make it possible for such a theory as this [Drift] to run wild?' After a shot at 'Wegener's own dogmatism' (Chamberlin, 1928: 83), Chamberlin listed specific geological objections to the theory (often taken from German critics). More broadly, it did not seem to be part of a comprehensive geological philosophy nor a general theory of earth behaviour – such as his father T.C. Chamberlin had put forward – but rather an *ad hoc* 'explanation' of unrelated facts. He concluded as follows (Chamberlin, 1928: 87):

> Wegener's hypothesis is in general of the foot-loose type, in that it takes considerable liberty with our globe, and is less bound by restrictions or tied down by awkward, ugly facts than most of its rival theories. Its appeal seems to lie in the fact that it plays a game in which there are few restrictive rules and no sharply drawn code of conduct.

Bailey Willis (1928: 82) commented in the same vein that Wegener's book 'leaves the impression that it has been written by an advocate rather than by an impartial investigator.'

Charles Schuchert, the dean of American palaeontologists and one of the few American proponents of landbridges, disputed evidence alleged to substantiate Drift. For example, he argued that the continents do not fit unless they are distorted from their modern shape. This is evidence against Wegener. But, if they did fit, this too would be evidence against Wegener because no one could believe that the continents ploughed undistorted through the ocean floor. There is a lack of parity of reasoning in Schuchert's argument: Wegener is damned if they fit and damned if they don't! He also attacked Wegener's methods (Schuchert, 1928: 139):

> the whole trouble in Wegener's hypothesis and in his methods is, as we have said, that he generalizes too easily from other generalizations. . . . It is not, as he says, that the detailed worker cannot see the forest because of the many different trees, or that the palaeontologists need a geophysicist to show them the road on which they should travel. Facts are facts, and it is from facts that we make our generalizations, from the little to the great, and it is wrong for a stranger to the facts he handles to generalize from them to other generalizations.

Thus, Schuchert argues that good inductions are based on an adequate set of properly understood 'facts'; bad inductions, on an inadequate or poorly understood set. The correct view of geology as a science divided into specialties coupled with the correct use of the proper method will yield valid results. He concluded that limited lateral crust movement might occur but not the large-scale movement postulated by Wegener. Edward Berry, Professor of Palaeontology at Johns Hopkins, also censured Wegener's approach (Berry, 1928: 194):

> My principal objection to the Wegener hypothesis rests on the author's method. This, in my opinion, is not scientific, but takes the familiar course of an initial idea, a selective search through the literature for corroborative evidence, ignoring most of the facts that are opposed to the idea, and ending in a state of auto-intoxication in which the subjective idea comes to be considered as an objective fact.

He excused his brief, dogmatic remarks by explaining that he felt 'that it is as futile to discuss the interior of the earth until we have more facts, as it is to attempt a "scientific" proof of a future life, or the divine inspiration of the Pentateuch' (Berry, 1928: 196). Wegener was not following an acceptable scientific method and therefore his theory did not deserve the attention of true geologists.

The passion and direction of some of the rhetoric hurled at Wegener suggests that some of his opponents saw more at stake than the question of which global theory was best. Clues to these issues are provided by passages already cited; others, by reading between the lines. If Drift be accepted, much of geological theory would become outmoded, the established authorities cast down and the 'empirical base' reconstituted. These are obvious reasons for anxiety and for the stridency of some of the rhetoric. What was seen to be the unscientific character of Wegener's methods certainly aroused much hostility. Another source of anxiety may have been a putative connection between Drift and geological catastrophism, an issue raised directly and indirectly in the American geological community and at the AAPG symposium.[19]

Drift was thought by some to pose a threat to uniformitarianism, whatever that term might have meant to different geologists. The myth of Lyell valiantly combatting forces of ignorance and religious bias had passed into the folk-lore of Anglo-American geologists. This was reinforced by the association of uniformitarianism with biological evolution and recent challenges to evolution and uniformitarianism in America. Whether Wegener really was or was not a uniformitarian is for us not at issue: some did associate Drift with catastrophism. Given the broad compass of how different geologists of the day defined uniformitarianism, some opponents may sincerely believed him to have transgressed this metaphysical boundary of acceptable geological theories. It is also possible that more opportunistic opponents saw advantage in claiming such a transgression of an ill-defined boundary. It is not surprising that the odour of catastrophism might be sniffed in Drift. Owen, Snider-Pellegrini, Fisher and Pickering had postulated some type of catastrophe as a cause; at the 1926 symposium Taylor put forward a 'one-off' event, the capture of the moon (Taylor, 1928: 175)!

Arthur Keith of the US Geological Survey had reviewed Wegener's theory in 1923 and drawn an explicit parallel between Drift and the catastrophism presumably vanquished by Lyell (Keith, 1923: 361, emphasis added): 'The hypothesis apparently must, therefore, fall back on a cause which is of the order of the *special convulsion of nature appealed to in the early stages of geologic work*.' At the AAPG symposium similar worries were voiced. Schuchert implied that the acceptance of Wegener's theory would amount to a surrender of the precious principle of uniformity (Schuchert, 1928: 140):

> We are on safe ground only so long as we follow the teachings of the law of uniformity in the operation of nature's laws. The battle over the permanency of the earth's greater features introduced by James D. Dana has been fought and won by Americans long ago. In Europe, however, this battle is not yet fought to a conclusion. . . .

Schuchert's comment must, of course, be read not only in the context of the Drift controversy but also in the context of the rivalry between Permanentism and Contractionism. The issue of catastrophism was more often raised indirectly. Berry's quip about 'a "scientific" proof of a future life, or the divine inspiration of the Pentateuch' has a different complexion in this light. David White (1928) of the National Research Council devoted much of his paper to asking why the continents broke up during a relatively peaceful and geologically recent epoch and not earlier. Others pursued this same point at and after the symposium (Chamberlin, 1928: 83; Nevin, 1936: 307). Had some cataclysmic event precipitated the break-up? Had Drift only operated within relatively recent geological time?[20]

Wegener's own paper (1928) addressed two lines of evidence for Drift: palaeoclimatology and recent geodetic measurements in Greenland. He attempted to answer criticism based on the 'Squantum tillite'. This deposit, found near Boston, had been classified as a tillite and therefore as glacial in origin.[21] It was dated to the Permo-Carboniferous period. Wegener had argued that his explanation of the great southern glaciation during this period was strong evidence for his reconstruction of Pangaea. His reconstruction placed Boston near the equator. How could this region be both tropical and glaciated? Opponents charged that the Squantum Tillite flatly contradicted Wegener's reconstruction: the theory should be abandoned. Wegener responded by questioning the facts.[22] The so-called Squantum Tillite was not a tillite, other deposits in the region supported his claim on behalf of a tropical climate, and therefore his theory was secure. More generally, isolated facts 'should not be used as an objection against the theory of continental drift' (Wegener, 1928: 100). His attitude, which sounded suspiciously as though facts should yield to theories rather than vice versa, was unlikely to mollify his empiricist critics. It was tantamount to admitting Chamberlin's charge that Drift was not 'tied down by awkward, ugly facts'.

In this confrontation by proxy, his major positive argument concerned the Greenland drift (Wegener, 1928: 100–3). His use of geodetics could be construed as a partial response to those who sought to paste the label of catastrophism on Drift. A reliance on geodetic evidence had, of course, been a prominent feature of Wegener's case from the outset. At the AAPG he claimed that measurements of actual drift would be conclusive proof of Drift and that the recent 1922 measurements were favorable to his theory, though he admitted that many more measurements over a period of years were needed (Wegener, 1928). The importance of such a crucial experiment was conceded even by his critics (e.g., Gregory, 1925: 256; 1928: 96).[23] After the symposium Wegener gave geodetic arguments a much more prominent place in the final edition of his work, moving this section to the head of his chapters of evidence (1929t: 23): 'The majority of scientists will probably consider this to be the most precise and reliable test of the theory.' His later emphasis upon the measurement of present drift owed much to longitude determinations conducted in 1927 which showed Greenland drifting at 36 meters per year (Wegener, 1929t: 29–30) but its possible value in warding off methodological criticisms should not be underestimated. Questions of methods and metaphysical commitments may have played in this sense a 'constitutive' role in his articulation of the theory. The *post hoc* analytical dividing lines between theory, method, and presentation – never sharply drawn – were erased by Wegener's attempt to deflect criticism by altering his presentation and implied research agenda.[24] Wegener's contribution to the symposium fell far short of a thorough defense against the barrage of brickbats hurled by most of the establishment geologists there. Perhaps he thought there was no percentage in preaching to the inconvertible.

Drift faced formidable opposition at the symposium but not all participants were hostile. Even critics were not wholly negative. Several agreed that Drift did have some good points. Chester Longwell (1928) did not favour Drift but adopted a wait-and-see attitude. J.W. Gregory, an English participant who had worked extensively in Australia, remarked that he had toyed with the idea of lateral motion of the continents before he had heard of Wegener and that more accurate measurements of longitude would soon prove or disprove the drift of Greenland. However, he endorsed the nearly universal judgment that Wegener's version of Drift was unacceptable (Gregory, 1928). Taylor (1928) pushed his own version of Drift. Most of the other speakers who had had extensive overseas training or experience also favored large-scale horizontal motion of the continents. Van der Gracht defended Drift, though not all the details of Wegener's version, and gave a long rebuttal of criticisms of Drift in the published record of the symposium (van der Gracht, 1928b, c). Joly (1928) maintained a fairly neutral position, but on geophysical grounds accepted the possibility of some form of Drift.[25]

Molengraaff (1928), a Dutch geophysicist, opposed Wegener but accepted Drift.

There was no clear victory in the 'Great Debate'. Upholders of conventional views were unable to deliver a knock-out blow to Drift. They did, however, land damaging punches. Although Drift received some legitimation as an alternative to the two rival 'big theories', much further work was needed before it was likely to be attractive to many American geologists. Over the next four decades only a handful of research papers which invoked Drift appeared in the American literature, although it received brief mentions in some textbooks.

Voice-Over

Scientists may argue at many levels. They may disagree over facts, theories, methods, standards and aims. If scientists dispute empirical results, the 'facts', they may try to settle their disagreement by appealing to some theory which they share or, perhaps, by invoking shared methods or broad method-ological principles or standards. If they argue over theories they may try to resolve their argument by referring to the 'facts' or to shared methods, standards or aims. If they disagree over methods, for example, inductions based on field-work as opposed to hypotheses based on published literature, they might appeal to the shared aims of the discipline or to accepted theories and facts. What if scientists take opposing views of the very aims of their discipline; for example, geology as description versus geology as explana-tion? Disputes at this level may be extraordinarily difficult to resolve: they may involve calling on the history of the discipline or perhaps hoping that the future success or otherwise of theories and methods reflecting these different aims will produce agreement. Disputes may also arise between different disciplines, between physics and geology for instance. Conflict may occur between science and other belief systems; for example, the geological belief that the earth is billions of years old and the religious belief that the earth was created a few thousand years ago. It is more remarkable, perhaps, that scientists can reach consensus on so much rather than that they may often be embroiled in disputes (*cf.* L. Laudan, 1984). All the more is this the case if social analysts be correct in their view that in these conflicts social interests are intermingled with technical ones.

Geologists argued over Drift at most of these levels. Drift was not a simple, isolated theory which could be judged simply by asking whether or not it fitted some well-defined set of geological 'facts'. It was judged in relation to 'facts' but also in relation to rival research programs. Drift was a 'package'. One of Kuhn's keener insights was to propose that neat divisions between facts, theories, methodological rules and principles, standards and aims were not observed in the actual practice of science. Instead, these elements are combined into larger structures, paradigms, whether we mean by paradigms narrower 'exemplars' or broader 'world views' (Kuhn, 1970: 181–91). This notion of 'theories' as part of more complex structures is also captured in many of the post-Kuhnian theories of scientific development. Whether we call Drift a paradigm, a research program, a research tradition or a web of belief, Drift as it was presented by Wegener and as it was developed by others integrated facts, theories, methods, standards and aims into a research program.[26] As we have seen, the debate over Drift was not simply a debate over whether or not Drift fitted the 'facts'. There were disputes and conflicts over what were the valid and relevant 'facts', how global theories should be evaluated and what standards should be applied, whether Drift was or was not in accord with physical theory, what were the appropriate methods and techniques of geology, whether geologists should aim for accurate description or plausible explanations or both and so on. Responses of geologists to Drift were colored by their perceptions of the extent of the relative agreement of Drift and its rivals with their preferred views on each of these matters and with the 'evidence' with which they were familiar. Responses of some geologists seem especially to have depended on whether or not they thought Drift to be useful in their own special research and areas of expertise. Finally much of the Drift literature, both pro and con, was forensic or rhetorical or persuasive in tone, structure and content: it did not consist of detached, disinterested comparisons of theory with fact.

Consider one difference between European and Anglo-American controversies over Drift. Most Europeans were sympathetic toward or tolerant of the method of hypothesis. Wegener presented his theory in the style of a hypothetic–deductive argument and this alone did not send his European opponents away to sharpen their axes. Rather, his hypothesis seems to have been evaluated according to the method of hypothesis; e.g., the requirement of independent or collateral support, broad explanatory scope, a wide range of confirming instances, the progressive character of Drift as compared with its rivals and predecessors, possible 'crucial experiments', and so forth. This attitude was manifested in the concern of both Wegener and his critics with the Greenland measurements as an *experimentum crucis*, the disputes over pattern-matching as constituting confirming or disconfirming instances and the thrust of Semper's, Soergel's and Nölke's 'method talk', directed as it was not against hypothesizing *per se* but at the inconsistencies and failures – as they perceived them – of Drift to follow the method of hypothesis. If there was a measure of agreement between European proponents and opponents of Drift on the aims of geology, on the relevance of beliefs held in other sciences, on methods and standards, then disputes could be confined to the relative adequacy of Drift and its competitors with respect to the available evidence and to their fruitfulness. It was otherwise with many Anglo-American geologists whose objections to Drift ran the gamut of possibilities.

The remarks of Lake and Reid and of several of the opponents of Drift at the AAPG debate reveal a very different perception of the relation between theories and what were regarded as appropriate methods of generating and evaluating them. There was a correlation between opposition to Drift and professions of naive inductivism, as we saw with Chamberlin, Willis, Schuchert, and Berry. Wegener's theory was damned because it failed to conform to methodological criteria. Conversely, support for or at least tolerance of Drift as a legitimate piece of science by some American geologists was often coupled with an endorsement of the method of hypothesis (e.g., Daly, 1926). At the AAPG Joseph T. Singewald, Jr., Professor of Economic Geology at Johns Hopkins, gave an even-handed summary. He did not oppose the method of hypothesis and had no quarrel with the way that Wegener had arrived at his theory. He was, however, critical of Wegener's presentation, which smacked of advocacy, 'devious reasoning', 'special pleading', and evidence which 'can carry weight only with one who already accepts the theory'. So much is this the case, he claims (Singewald, 1928: 190):

> If one entered the discussion in the rôle of an advocate in rebuttal, it would be easy to go through Wegener's book chapter by chapter and apparently over-whelm his case by the multitude of instances in which one can overthrow arguments advanced in support of his theory, in which his conclusion are *non sequitur*, or in which his assumptions are contrary to the best available evidence.

This is precisely what Wegener's inductivist critics had done. Singewald took a different tack. The hypothesis should be tested on the basis of its worth for guiding research. This should be measured in terms of the balance of probabilities: the evidential base of geology is incomplete and problematic so broad agreement rather than exact fit with specific data is sufficient. Despite his scathing remarks about Wegener's tactics, he concluded that Drift 'possesses a great degree of probability and is supported by considerable evidence' (Singewald, 1928: 193).

The relation between Drift as an over-arching theory of geology and the specialties into which that science was divided was also a sore point with some geologists in the Anglo-American community and may have contri-buted to their attacks on Wegener's methods. For many geologists increasing specialization was probably a sign of the growing sophistication and maturity of their science. Introductory textbooks usually reeled off lists of subdivision and specialties within geology. Gregory remarked with a trace of pride on the difficulties of keeping up with the literature even in one's own corner of the discipline (Gregory, 1915: vii). Poincaré neatly put a difference between the European and English traditions in physics as follows (Poincaré 1905: 215):

> The English scientist does not try to erect a unique, definitive, and well-arranged building; he seems to raise rather a large number of provisional and independent constructions between which communication is difficult and sometimes impossible.

This description is apt for geology, particularly at a time when there seems to have been a retreat from global theorizing into specialist fortresses with the confounding of simple contractionism by the discovery of radioactive heat. Wegener stressed that one of Drift's great virtues was that it brought into relation data, generalizations and hypotheses previously thought to be unconnected. He stressed plausible correlations rather than precise fitting of theory with data and accepted the conclusions of others in their specialist areas as grist for his global mill. English-speaking geologists were more typically concerned with the solid rock of data and description. Arthur Hinks, an expert in surveying and mapping and editor of the *Geographical Journal*, at a meeting of the Geography section of the BAAS referred to 'specious theories of continental drift and wandering of the poles' which had been 'built on slight foundations, largely of convenience' and condemned 'geologists who do not scruple to play fast and loose with the principles of every other science than their own' (Hinks *et al.*, 1931: 435–6). Wegener and his supporters had violated the quintessentially English virtue of playing by 'the Rules of the Game', for the cardinal rule in geology was that no player 'may adopt a principle contrary to the best opinion in another man's subject'. Wegener's specialist critics, each of whom was in his own area more knowledgeable than Wegener, had no difficulty in finding him guilty of sins of omission or commission. Each could then conclude that since Drift was unsatisfactory in their specialty it should be flatly rejected (*cf.* Frankel, 1976). This could be expressed in 'method talk'. A specialist might be incensed at specific inadequacies of Drift but to go outside a small fiefdom to do battle with a general theory could present problems. Schuchert pointed to specific flaws in Wegener's palaeontological case but went on to assign those flaws to methodological failures. The method weapon thus forged could then be used to hack away at Wegener's whole theoretical edifice

(Schuchert, 1928: 134, 139). In other words, specific flaws might give rise to general methodological attacks.

Some of Wegener's critics were inconsistent in their preaching and practice. Semper attacked Wegener for selectively quoting from and copiously citing other authors. Yet he resorted to the same tactic. He apparently assumed that any source mentioned by Wegener was open to suspicion since Wegener's theory was wrong, whereas any source or authority invoked by himself must be self-evidently true if it opposed Wegener. This common phenomenon in the Drift debate may be generalized in the following form: A, in building a case against B, may cite the work of C, but typically provides no argument for why he trusts C rather than B, except (inferentially) that C agrees with A. This is another illustration of accounting for one's favored views in 'scientific' terms but asymmetrically accounting for the (erroneous) views held by others in 'social' terms (cf. Yearley, 1981; Mulkay & Gilbert, 1982a, b). Parts of Lake's attack, as we saw, followed this same pattern. Schuchert attributed Wegener's errors to his penchant for generalizing from the generalizations of specialists but, at the level of empirical rebuttal he concatenated arguments against Drift drawn from the writings of others and cited their disapproval of Drift as support for his own negative judgment.

It is difficult to estimate the weight that objections to Drift based on methodological matters might have had. L. Laudan has argued (1981b: 9) that methodologies are often parasitic upon scientific practice and, more to the point here:

> That the reception of methodological doctrines . . . has been determined more by the capacity of those doctrines to legitimate a preferred scientific theory than by their strictly philosophical merits. Correlatively, when methodologies are rejected it is often because of their inability to rationalize what is regarded intuitively as exemplary scientific practice.

It this view be correct, then methodological criticisms have force if there be other grounds for believing that a new theory does not meet some criteria of exemplary scientific practice. This was certainly the case with Drift. On this view, had the evidence for Drift been thought to be overwhelming, not only would methodological criticism have been muffled but it is conceivable that Anglophonic geologists would have scrapped naive inductivism in favor of the method of hypothesis. This casts into somewhat different relief arguments over the merits and demerits of Drift couched in terms of geological and geophysical evidence: those arguments were not just about facts but also theories, methods and aims.[27]

American opponents of Drift made a noticeably greater appeal to uniformitarianism than did European Contractionists or even British Permanentists. Moderate shadings of catastrophism were common in global theories espoused by many European geologists and a developmental view of the earth was at the core of Contractionism. Even Permanentists did not always defend an uncompromising uniformitarianism. How are we to account for the anti-catastrophic rhetoric of the American participants in the AAPG debate? Geological uniformitarianism was linked through Darwin with biological evolution; catastrophism, with special creation. There was a conflict at the conceptual level between uniformitarianism and Darwinism on the one hand and, on the other, beliefs founded on literalist readings of the Bible. This conflict had in Britain been defused through a tacit acceptance by the Anglican church of evolution, symbolized by the appointment in 1896 of Frederick Temple as the Archbishop of Canterbury (*cf.* Moore, 1986). In America the conflict was more bitter and still continues (*cf.* Numbers, 1986). None of the American participants would have forgotten that in 1925 John Scopes had been tried for breaking the Tennessee law prohibiting the teachings of evolution and, by inference, uniformitarian geology. In this context Wegener's theory, with its overtones of catastrophism, may have torn the scabs off recent wounds. For others, uniformitarianism may have been a handy club to beat down a theory which threatened their positions in the geological community.

Wegener's nationality does not seem to have been a significant issue, even in countries which had fought against Germany in the recent war. During the war German science and scientists were harshly criticized for their presumed ambition to dominate the scientific world and their role in the German attempt to conquer the political world. After the war German academic and political leaders regarded science as a major means to regain national prestige. Any theory of a German scientist might be taken by some German and non-German scientists as part of the struggle for national reassertion. There is no direct evidence that such attitudes influenced the response to Drift. All that can safely be said is that the spread of Wegener's ideas was delayed because his writings were in German. If his nationality contributed to negative reactions to his theory, this reaction was already 'overdetermined' by prevailing geological doctrines (*cf.* Ringer, 1969; Forman, 1973; Schroeder-Gudehus, 1978; Le Grand, 1986a).

The presumed 'conservatism' of the scientific community, the possible effect of such social factors as Wegener's nationality or perceptions that he lacked credentials as a geologist, and the threat to established geological authority and authorities which his theory may have entailed, all may have raised the height of the hurdles his theory would have to clear to be accepted. If so, this may have been conducive to the growth of scientific knowledge. Whatever social or cognitive criteria might be used to evaluate knowledge claims, these would be stringently applied to Wegener's version of Drift. This may have meant a widespread rejection or agnosticism toward Drift in the short term but it may also have stimulated theoretical and empirical work by proponents and opponents. For it to be widely accepted, there would have to be some consensus that it broke new ground, provided solutions to problems deemed important, opened up new lines of research, and possessed significant advantages over its rivals.

Specialist interests were important in reactions to Drift, as we saw in the generally more favorable or neutral comments of geophysicists as opposed to the more negative attitude of stratigraphers and other rock-based geologists. We also noticed examples of the importance of 'specialization' in a different sense: a focus upon regional problems. For example, the response to Drift of Dutch geologists in Indonesia was couched in terms of its merit in solving such problems as the arcuate shape of island chains; for the Swiss, its value in explicating the Alps. We might conjecture that contrarily, one reason why Drift was poorly received among American geologists working on the Appalachians or the Great Plains was that it had little obvious application to their concerns. Disciplinary, specialist, and 'local' aspects of the debates over Drift are discussed more extensively in the next chapter.

In the debate there is little evidence of 'incommensurability'. This is a many-splendored term. Kuhn has been taken to mean that proponents of different paradigms cannot communicate with each other and that debates between paradigms cannot be resolved by rational argument, appeal to facts, and so on: resolution is an irrational process which involved undergoing a 'conversion experience', a 'Gestalt switch'. Kuhn does provide plentiful ammunition for this 'strong' interpretation of incommensurability (e.g., 1970: 111–35) in his talk of scientists living in different worlds and having different perceptions in those worlds. Kuhn subsequently either changed his mind or clarified his position on incommensurability. He now takes the view that two paradigms may share much common 'language' but that there are a few key concepts in each paradigm which cannot be accurately 'translated' into the other. Part of the language of each is untranslatable. Though 'interpretations' can be constructed, this hampers communication between proponents of rival views and their understanding of the arguments, theories, methods, evidence and so on; it does not make it impossible (Kuhn, 1983). Feyerabend has taken a more consistent position. For him, incommensurability exists because all of the observational and theoretical terms within a theory (paradigm, research program, research tradition, web of belief) take their meaning from that theory. Though 'translation' is impossible between different theories, it is possible for scientists to speak more than one 'language' and thus to communicate among and understand different theories. Laudan is more concerned about incommensurability at the level of aims and values. Is it possible for a scientist to make a rational choice between two 'theories', each of which is fully consistent with a different set of aims: in the case of geology, with a Permanentist-type theory which serves the aim of description and a Drift-type theory which serves the aim of causal explanation? There were in the debates numerous instances of disagreement, dispute and circular argument. These are compatible with Kuhn's (recent) and Feyerabend's notions of incommensurability. There were also disagreements over the aims of geology, and aims were used to defend or attack Drift. This illustrates Laudan's point. We shall see that disagreements persisted and their resolution seems to have come through a prolonged period of competition among and development of the rivals, not through a sudden, massive Gestalt switch.

Finally, we must remember that only a small proportion of the geological community participated directly and publicly in the disputes over Drift. Kuhn portrays science as a community activity and there is surely merit in this view. It is not clear, however, what one is to understand by descriptions of communities in crisis or decisions of communities to abandon one general view and adopt another or even the notion of paradigms as something shared by communities. It is improbable that some form of head-counting, even were it practical, would give us much insight into these matters or that the process of scientific change itself is better understood in terms of some sort of 'mass' movement as opposed to the activities and beliefs of relatively small 'elites'. Compiling lists of geologists who preferred Permanentism to Drift in 1930, whether we gained this information from citations or explicit declarations, surveys or some other means, would tell us about the relative popularity of them but perhaps little else. In any case, there would be immense practical difficulties in assessing the beliefs of geologists who gave no clear indication of their preferences; by far the largest category would be 'no opinion' or 'undecided' or even 'don't care'. More important, 'majority rules' is simply inappropriate. As we shall see, what was crucial about the early response was that some geologists were sufficiently impressed by Drift to follow that program, to use it in their research even if but to explain their results in its terms and to seek to improve Drift to meet criticisms regarded as legitimate. And, some geologists who did not accept Drift were challenged by it to rework and improve their preferred view.

Notes

1 My summary is a composite. The counter-examples and counter-arguments which I mention were rehearsed in symposia on Drift, in the journal literature and in textbooks and monographs. This summary should offset the favorable impression of Drift the reader may have gained from the previous chapter.

2 Wegener probably did not assist his case by linking in 1924 his and Köppen's palaeoclimatic and palaeogeographical reconstructions with the highly controversial theory of Milankovich in which ice ages were periodic phenomena produced by changes in the earth's orbit and poles of rotation (*cf.* Imbrie & Imbrie, 1979).

3 The Greenland data he cited in his first publications were based upon observations of the moon. He later admitted (Wegener, 1929t: 27) that this method was subject to considerable error as were the specific observations.

4 Diener had been a consultant on palaeontology to the Geological Survey of India. Through the Survey, which was headed by British geologists, Diener had personal and professional links with the British community. His catalogue of objections was cited by some British critics of Drift.

5 Soergel seems to be arguing for a completely cumulative view of the growth of scientific knowledge, a position held by many of his contemporaries and by Lakatos.

6 These points have bearing on Kitts (1981: 233) arguments concerning retrodictive uncertainty in geology.

7 Semper, who was in his late 30s, was on the margin of the German academic community; he held a post at the Technical High School in Aachen.

8 Semper probably had in mind the 'method of hypothesis' employed by English physicists or perhaps such a more theoretically-inclined geophysicist as Joly (*cf.* Greene, 1982: 276–9). The typical geologist working for the Geological Survey could hardly be described as thinking in what Semper characterizes as 'the English way'!

9 He may have meant by this that it was too good to be true or that it might entice geologists away from the rigors of fieldwork.

10 There are parallels between this aspect of the Drift controversy and the earlier controversy in Britain between physicists and geologists over the age of the earth (see Burchfield, 1975).

11 I have not made a systematic survey of the German literature in the 1930s and 1940s but see the 1939 Frankfurt symposium on 'The Origin of the Atlantic' (Kirsch *et al.*, 1939); a summary is given in Hume (1948).

12 Carozzi (1985) gives a digest of the relevant writings of many European geologists.

13 Some of the notable Swiss proponents included Albert Heim after 1928 (Carozzi, 1985: 130), Léon Collet (e.g., 1927; 1935), Maurice Lugeon (e.g., 1940; 1941) and G. Tiercy (1945). *Cf.* Brunnschweiler's (1981: 11) remarks on the Swiss geological schools after the 1920s.

14 This also played a part in the lag between Wegener's early publications and discussions of Drift in several European countries. In Switzerland Argand defied a prohibition against reading any material printed in Germany to discuss Wegener's *Origin*. The International Research Council was created which had as one of its

aims the exclusion of German scientists from the international scientific community. Unofficially, there was much opposition to using the German language at conferences and in publications (*cf.* Schroeder-Gudehus, 1978).

15 See Marvin (1985: 139–41). Marvin gives a handy compilation of most of the British publications relating to Drift from 1922 through 1931.

16 Ironically, similar charges had been directed nearly a century earlier at Lyell who, once his 'cause' had become orthodoxy, had been canonized as the patron saint of inductivism.

17 Many of the methodological objections were standard gambits of inductivists against any hypothetico-deductive theory (see L. Laudan, 1981b: 90–4). Relativity theory and quantum theory, roughly contemporaneous with Drift, met similar criticisms (quoted from Williams, 1968: 122–3):

> Both Professor Einstein's theory of Relativity and Professor Planck's theory of Quanta are proclaimed somewhat noisily to be the greatest revolutions in scientific method since the time of Newton. . . . it is in reality a return to the scholastic methods of the Middle Ages. Undoubtedly the German mind is prone to carry a theory to its logical conclusion, even if it leads into unfathomable depths. On the other hand, Anglo-Saxons are apt to demand a practical result, even at the expense of logic.

18 The claimed deposits of *Glossopteris* in Russia, when examined much later, turned out to have been misidentified. In 1967, however, there was a report of *Glossopteris* traces near Vladivostok: this claim does not yet seem to have been resolved.

19 The 'Scablands' debate offers interesting parallels; for a brief discussion see Gould (1980: 194–203).

20 Wegener had limited his theory to recent stages of earth history. His supporters hoisted the flag of inductivism: the evidential base for earlier epochs was simply too fragmentary to allow defensible conclusions to be drawn. As du Toit remarked (1937: 1), 'There is . . . nothing contained in those writings to suggest that such kind of drift would have been restricted to any particular period and would not have operated from an early stage in the earth's history, only such remote episodes are too obscure to be elucidated just yet.' *Cf.* van der Gracht (1928c: 203), Steers (1932: 160–1), Wooldridge & Morgan (1937: 47). Nonetheless, as a causal global theory, Wegener's did not give a *general* theory of orogeny. It 'explained' the Alpine orogeny but not such earlier episodes as the Caledonian orogeny or the formation of the Appalachians, nor did it suggest that such episodes were rhythmic as many geologists believed had been the case (but see Choubert, 1935).

21 Robert Sayles made this classification in ignorance of Drift and on what seemed to be very sure grounds. He listed 15 criteria for tillites; the Squantum deposit matched 14 (Sayles, 1914: 144–5).

22 This is not an unusual move; e.g., Einstein and the early electron mass experiments of Kaufmann or Miller's ether-drift experiments.

23 The argument also had appeal to 'neutrals'; for example, it likely played a role in Evan's efforts to make available an English translation (Evans, 1924: ix–x).

24 His geodetic argument was potentially a persuasive one, but it was severely undercut by 1936 observations which indicated that the figure of 36 meters/year was wildly inaccurate (Nörland, 1937; Longwell, 1944c).

25 Joly had recently sketched a general theory of the earth based on geophysical considerations, especially radioactive heat, in which he sought to explain periodic orogeny and other global phenomena (Joly, 1925). The approach which he followed was much more in the European tradition than the Permanentist in terms of his aims and methods, though Joly rejected a contracting earth. Heat built up under the continents. This resulted in a liquifaction of the sima and an attenuation of the ocean floor. In such circumstances, 'it seems quite possible that the stresses arising out of tidal forces . . . might have sufficed to create differential movements among the continents' (Joly, 1925: 172). He was only lukewarm in his attitude to Wegener's theory.

26 This does not mean that the elements of a research program are not revisable nor that all those pursuing or accepting a program will agree on every element nor that rival programs may not have some elements in common. For example, Drift enlisted not only geologists who espoused the method of hypothesis but also some inductivists.

27 Critics of Drift, even when deploying method rhetoric and method weapons, did not refer to formal methodological literature. When some basis for these criticisms was advanced, it was usually in terms of previous successful applications. This suggests that methodological treatises had little direct influence on prescriptive and normative judgments. Treatises on method may be more important to philosophers than to scientists, though Chamberlin's method of multiple hypotheses might have been an exception. (cf. L. Laudan, 1981: 8–9).

5

Specialties, localism and problem-solving

Geologists who commented on Drift before the 1960s often judged it on the basis of its usefulness in their specialties. Specialists, especially those advocating empiricism, might argue that any general theory, no matter how well supported by evidence drawn from other specialties, should be compatible with all the evidence in their own specialties before it should be entertained, pursued or accepted by them – or by others. What if a new theory or program seemed to create no particular difficulties in a specialty? If it were of no greater use in solving 'local' problems, if it had no marked superiority over the established views, then was there any reason to favor that theory or program over its competitors? Geological 'localism' and the fractionation of geology into a plethora of specialties could create obstacles for any overarching, synthetic theory including Drift.[1] Geologists were often concerned with the detailed study and elucidation of specific geological regions and structures. Drift offered little or no special assistance in this to most North American or British scientists.

Specialization and localism could in some circumstances favor theory change. If a grand synthetic program like Drift did seem to be compatible with the data in a given geological specialty and to give that specialty a measure of importance, authority or cognitive and experimental order and direction that it had previously lacked, then practitioners of that specialty might welcome the theory even if it were being attacked in other specialties.[2] Some specialists in species distribution, for example, were well disposed toward Drift, perhaps because it supplied a theoretical underpinning for and opened up new approaches within their specialty. Others endorsed Drift as being of marked value in solving particular clusters of problems or of providing a coherent account of 'local' data and structures.

Localism was an important aspect of the positive reaction to Drift of scientists in South Africa, Australia and South America. Many of the apparent simplifications and correlations to which Wegener drew attention came from the Southern Hemisphere. The 'jigsaw fit' and palaeontological correspondences and pattern-matching between South America and Africa had been involved in the genesis of Drift. How could faunal similarities between Australia and South America be explained without recourse to landbridges or parallel evolution? Both, together with India and Africa were once joined directly through Antarctica. How is the existence of the 'Wallace Line' to be explained?[3] Australia–New Guinea broke away from Antarctica after Indo-Asia had moved to the north and only recently has the former moved into proximity to the Sunda Shelf (Wegener, 1912: 286). How is the faunal assemblage of Australia to be explained? By 1922 this was for Wegener 'of quite considerable importance to the question of displacement' (Wegener, 1924t: 285). The three-fold division noted first by Wallace is due (Wegener, 1924t: 76–7, 285–9) to the early connections between South Africa and Australia via India (the oldest element); after the breakaway of Africa and India, to the connection between South America and Australia via Antarctica (the marsupial element); and, with the recent approach of Australia–New Guinea to the Sunda Shelf, to the connection between Australia and Southeast Asia via the Sunda Islands and New Guinea (dingos, rodents, and other recent arrivals). In palaeoclimatology, Wegener

Figure 5.1. The Wallace Line marks a separation between a predominantly Asian fauna to the northwest of the 'line' and a predominantly Australian fauna to the southeast. Why was this line not 'blurred' by migration between Asia and Australia via the numerous islands and island chains? (after Whitmore, 1984, by kind permission of T.C. Whitmore).

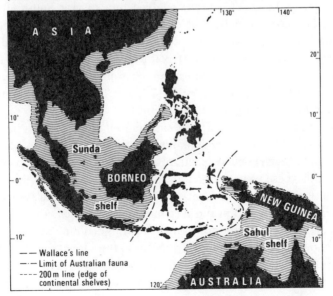

could explain the great southern glaciation; in palaeobiogeography, the distribution of the *Glossopteris* flora (Wegener, 1912: 287; 1924t: 90–111). Scientists interested in these particular problem-fields might well be impressed by the solutions which Drift promised.

A southern partisan

Alexander du Toit was regarded by some of his contemporaries as the 'world's greatest field geologist' (R.A. Daly quoted in Gevers, 1949: 8). From 1903 to 1920 he mapped large tracts of South Africa. His first field expedition involved a study of extensive portions of the Karroo System[4] including sections of the Dwyka Tillite. His work on this formation continued for much of his career and was crucial to his acceptance of Drift. In 1914 he visited Australia to compare deposits there with the Karroo. In 1921 he synthesized his extensive field observations: the heart of the synthesis was Drift, 'revolutionary and heretical as it will appear to orthodox geologists' (du Toit, 1921). Using Drift, the directions of ice flow deduced from the Dwyka Tillite in South Africa and Australia could be harmonized with those reported for India and South America. *Glossopteris* and other fossils could be matched up with those in other southern continents. The Karroo, upon which he had become the world expert, extended throughout the landmasses of Suess's old Gondwanaland which in Wegener's theory included the southern part of Pangaea. Drift offered an elegant solution to problems he had identified, especially those concerned with the great southern glaciation (du Toit, 1921; 1922). In 1923 a grant enabled him to visit South America to compare its geology with that of Africa. He subsequently reported the correlations which he found (especially du Toit, 1927; 1929b) and it was du Toit to whom Wegener gave much of the credit in later editions of the *Origin* for the growing number of pattern-matches on which he relied in arguing for his theory (Wegener, 1929t: 68–72). Du Toit's data – though not his interpretations – were above reproach.

After Wegener's death, du Toit cast himself in the role of champion for Drift, though not necessarily for Wegener's version. In 1937 he issued a polemical but widely-cited defense. Drift was not without problems but he argued that these were not greater than those faced by Contractionism or 'archaic' Permanentism (du Toit, 1937: 7–10). He tried to draw a distinction between the Drift program and Wegener's particular version.[5] An adequate theory of Drift is 'still in the process of being roughed-out' (du Toit, 1937: 12) and it should be judged leniently since 'as in the case of every new hypothesis, the complete and perfect picture cannot straightway be presented' (du Toit, 1937: 1–2). He proposed several modifications; e.g., instead of

Pangaea, there had been northern and southern palaeocontinents (Laurasia and Gondwana). He campaigned for Drift (e.g., du Toit, 1929a; 1939), rebutting criticisms and attacking rivals (du Toit, 1944), until his death in 1948. Other South Africans followed him in explicating Gondwana problems in terms of Drift (Krige, 1926; 1930; Dixey, 1938). His mantle as the southern champion of Drift was eventually assumed by Lester C. King in the 1950s who adopted as well as du Toit's credo that 'the hypothesis of continental drift is not to be proved by idle, armchair theorizing, but stands or falls on one thing – hard work in the field' (L.C. King, 1953: 2174); in du Toit's case, hard work on geological correlations between the fragments of Gondwana.

The boys from Brazil

Geologists in Brazil did not invariably share du Toit's conviction that the correspondences between that region and Africa strongly supported Drift. Alberto Betim of the University of Rio de Janeiro rejected Drift only two years after du Toit's 1927 paper. He admitted that with some fiddling he was able to obtain a 'perfect, let us say even troubling concordance' between the South American and African coastlines (Betim, 1929: 72) and a gross matching of many geological and palaeontological features. However, he stressed differences rather than similarities. The balance was tipped against Drift by the Mid Atlantic Ridge: he interpreted it as a chain of mountains on a drowned continents (Betim, 1929: 85–7). Other Brazilians, Moisés Gicouate (1945), for example, did favor du Toit's conclusions. Ken Caster, a palaeontologist, worked in Brazil from 1945 to 1948 during which time he annotated a Portuguese translation of du Toit's 1927 paper. Later, as Professor at the University of Cincinnati, he argued that Drift provided a plausible explanation of the correlations between Brazil and Africa (Caster, 1952; von Engeln & Caster, 1952: 535–6) although he did not opt for it to the exclusion of other possibilities.

Views from the bottom

Specialization and localism contributed to the skepticism of some Australian scientists (*cf.* Le Grand, 1986b). The debate, such as it was, lacked much of the heat and rhetoric generated elsewhere. Few commented directly on Drift. One geologist recalls that Drift, like other speculative matters, was a topic of tearoom conversation rather than of more formal debate (E.S. Hills, private communication). Theoretical speculation was of minor importance with respect to their major aims. For geologists, these were the mapping of a vast landmass which had not been systematically surveyed and the associated task of locating and evaluating ore and coal deposits (*cf.* Vallance, 1981). An emphasis on fieldwork rather than speculation was probably reinforced by intellectual and social bonds with British geologists.

The empirical, applied orientation of Australian geology – in the classroom, in the office, in the field – was not conducive to enthusiasm over any general theory of geology, much less Wegener's highly speculative, synthetic one. There was no theoretical 'school' in Australian geology; the general attitude could be fairly described as agnostic or even antitheoretic in terms of global geological theories. Academic geologists would have shared with Sherlock Holmes the view that 'It is a capital mistake to theorize before you have all the evidence. It biases the judgment'. The tenor was that the academically-trained geologist should be acquainted with theories but should not take them too seriously. This is borne out by the lecturing style of the Professor of Geology at Melbourne, E.W. Skeats, whose research interest focussed on the Australian state of Victoria. He would give a resume of the various theories on a given topic and then indicate that the data and the state of knowledge were such that is was impossible to decide which – if any – of the conflicting theories was the best (E.S. Hills, personal communication).

This skeptical attitude toward theorization was not confined to the universities. Leonard Keith Ward, Government Geologist for South Australia, in 1927 sketched the major geological features which a general theory should explain and then summarized what he regarded as the strengths and weaknesses of no fewer than eight such theories. Against Drift, which he stated 'has probably some modicum of truth' (Ward, 1927: 18), he leveled seven major criticisms, four of which related to Australian problems (Ward, 1927: 16–7). His conclusion that none of the 'hypotheses' provided 'a logical conception of the evolutionary process that is at once simple and satisfactory for all the phenomena recognized' fits comfortably with his empiricist position that the gathering of more and more data will enable us to 'progress gradually towards the explanation that will satisfy the most exacting and critical intellect' (Ward, 1927: 28). E.C. Andrews, Government Geologist of New South Wales and a specialist in structural geology as applied to ore bodies, dismissed Drift as of no value to orogeny or to locating ore deposits (Andrews, 1926: 452–3). In 1938 he delivered an address on major problems in structural geology which included comments on 'the Wegener Hypothesis, a high-level flight in matters geological, and one of great stimulation in many ways, [but which] nevertheless smacks suspiciously of the waxen wings of Icarus and courts a similar fate when leaving the earth too far below' (Andrews, 1938: v). He argued, *a priori*, that the more ingenious and complicated a theory, the more it should be distrusted, and, on the basis of his own field experience, took a firm stand for Permanentism (Andrews, 1938: xxiv). The overwhelming majority of Australian geologists seem, however, to have concentrated on local problems of mapping and description and did not participate in public evaluations of Drift or its competitors.

Some Australian scientists interested in other problem-fields were more positive. Local and specialist concerns played important roles and institutional factors may also have played some part: geologists at Sydney University, for example, were rather more sympathetic than their Australian counterparts elsewhere. L.A. Cotton, Professor of Geology at Sydney from 1924, specialized in geophysics, especially isostasy, diastrophism (large-scale deformation of the earth's crust), the strength of the earth's crust, and geographical polar wandering. In 1923 he published a discussion of polar wandering in which he referred to a number of German authors on whom Wegener had also drawn. Cotton claimed that polar wandering was consonant with the permanence of continents and ocean basins but urged only that polar wandering 'be merely accepted as a working hypothesis' which 'may stimulate investigations into many fields of research' (Cotton, 1923: 497). By 1928 he had read Wegener. In that year he discussed recent theories of diastrophism. He commented briefly on Drift, even though it was, he felt, peripheral to the topic (Cotton, 1929: 196–7). The forces Wegener had invoked were 'totally inadequate' but Drift was not invalidated. He did not give a full-blooded endorsement of Drift; rather, that he provisionally judged Drift – though not Wegener's version – to be worthy of pursuit or entertainment with respect to the problems which concerned him. T.W. Edgeworth David in the twilight of his career at Sydney seems also to have been well-disposed to Drift (David, 1928). This may have been connected with his long-standing interest in glaciation, particularly glacial deposits in New South Wales.[5] S. Warren Carey was a student at Sydney in the 1930s during the period that Cotton was Professor. Carey recollected that in a 1937 draft of his D.Sc. thesis he had put forward a new version of Drift but that this material 'had to be omitted at the eleventh hour because I realized that these concepts were too radical for acceptance then, and would have cost me my degree' (Carey, 1976: 2–3). His worry was not how such ideas would have been received at Sydney, for he had received encouragement and stimulation from Cotton, but arose because 'I knew I would have senior international examiners, and that in the prevailing climate such a radical concept would survive as long as a snow-man in hell' (S.W. Carey, personal communication).

The only geologist in Australia who was a proponent of Drift was Arthur

Wade, who had been educated in England, had engaged in oil exploration work around the world and had in 1933 emigrated to Australia. He implicitly referred to Drift in an account of his oil explorations in New Guinea and environs. He attributed the crustal deformation of New Guinea to the mashing of New Guinea against Australia and commented that this was of practical value as there was a correlation with the location of 'the most productive of the proved oil-bearing areas' (Wade, 1927: 160–1). Drift offered a key to the structure and history of New Guinea and thereby a tool to solve his problems of locating sites for future oilfields. He discussed theoretical issues more fully in the 1930s (Wade, 1934; 1935). He used a version patterned after Beno Gutenberg's 'continental spreading'[6] in preference to Wegener's. He was careful to describe it as a 'working hypothesis' (Wade, 1934: 77) and stressed its application to the economic geology of those regions in which he had field experience. After emigrating to Australia, he worked on the geology of Western Australia. This formed an important part of his final contribution to the discussion of Drift: an evaluation of the fitting of the southern continents to Antarctica and the matching-up of known geological features. It was only in this paper, it appears, that he fully accepted Drift (Wade, 1941: 34).

If Drift seemed to offer little in the way of solutions to the problems of Australian geologists, it should have been otherwise with Australian biologists. Wegener explicitly offered solutions to a set of interconnected problems in Australian biology: taking account of Australia's relative geographic isolation, how to explain its distinctive flora and fauna, both past and present? Several biologists with special interest in these problems took up Wegener's proferred solutions, perhaps all the more readily because the geological arguments against Drift might have had less force for them than for geologists. More than one geologist grumpily remarked on the prediliction of biogeographers to throw up or demolish landbridges, shift continents sideways, or raise and sink them without regard for due geological process in order to solve problems in their own field. Geologists usually took the view that in such interdisciplinary arguments, geological rather than biogeographical evidence must decide the issue, a view shared by many biogeographers (Frankel, 1981).

At the turn of the century, two approaches were popular among writers on these Australian problems. Landbridges linking Australia with South America through Antarctica (and sometimes other bridges as well) were invoked to explain similarities; the removal of these bridges and consequent isolation, to explain such distinctive fauna as monotremes and marsupials. Or, on a grander scale, similarities were explained in terms of a vast southern continent, Gondwana, which included parts of Australia, Africa, and South America. The subsidence of parts of Gondwana had isolated Australia and since that time divergent evolution had occurred. Both approaches came under severe attack in the early years of the century from Permanentists who cited isostasy and from Joly's work on radioactive heat. William Diller Matthew, an American biogeographer, outlined in 1915 a Permanentist solution: given the permanence of continents and oceans, dispersion of species from the north polar region, aided by isthmian links and chance events such as rafting, could explain all the apparent problems of biogeography (Matthew, 1915: 172–8). A case could be made here for 'localism'. To explain their flora and fauna, North American geologists had before their eyes, so to speak, the 'bridges' or isthmian connections which they needed: the Aleutians and Panama.

Launcelot Harrison's response to Drift underlines the positive role special-ization and localism could play. Harrison, Professor of Zoology at Sydney University, was an outspoken critic of Matthew's theory as applied to Australian problems of biogeography. This led him to favor Drift over its rivals, though not to accept – in a strong sense – Wegener's theory. In 1924 he reviewed the current explanations. Connections between South America and Australia had been rejected on geophysical grounds (Harrison, 1924: 249). The Permanentist solution might be more acceptable to geologists but not, in Harrison's opinion, to Southern Hemisphere zoogeographers. Holarctic dispersion could not explain all cases of distribution. He had built much of his reputation on the use of host–parasite relations to clarify problems of bird phylogeny and had recently extended this same technique to marsupials. As part of his case against Matthew, he applied them to crayfish (Harrison, 1924: 258–62). He concluded there must have been extensive land connections between Australia, South America, and Africa: 'No other hypothesis is adequate to explain some cases of distribution'. The problem was how to establish such connections without completely disregarding geological theory. He hinted that a small elevation of the whole Southern Hemisphere would provide the necessary connections through an extended Antarctica but this – like landbridges – seemed to be forbidden ground. In short, in 1924 he was concerned with detailed problems of species distribution which an acceptance of Permanentism magnified rather than solved.

Harrison returned to the impasse between geologists and biogeographers in 1926 with a new solution which gave importance to the interpretation of data in his own special (hemi-) sphere of competence. He defended land connections and his own position 'as an independent originator of the host–

parasite idea' (Harrison, 1926a: 374) – a method recently impugned by two
of Matthew's supporters (Dunn, 1925; Noble, 1925). In these contexts he
put forward Drift as a solution. He had come across a notice of Drift in
Nature[7] but had 'scoffed light-heartedly' at this 'utterly fantastic' theory
(Harrison, 1926b: 12). His reading of the 1924 translation of Wegener had,
however, led to a reversal of opinion: 'Before this', he confessed, 'I remained
to pray' for it seemed to offer a 'working hypothesis of great value for
zoogeographical problems' which he had begun to apply to 'Australasian
problems' with considerable initial success (Harrison, 1926b: 12–3). He
was aware that many objected to Drift, but commented that the very fact
that many scientists 'have clamoured against it . . . it is a hopeful sign of its
ultimate verity' (Harrison, 1926b: 14–5). He attacked Matthew and other
Permanentists in the pages of their own journal, the *American Naturalist*,
defended the host–parasite technique and supported his contention of a
connection between Australia and Antarctica with evidence provided by
that technique. He did not discuss Drift in detail but concluded that 'the
Wegener hypothesis, if it ever prove acceptable to geologists, would explain
Australian problems admirably' but that 'no hypothesis of northern
dispersal can be reconciled with Australian facts' (1926a: 381–2).[8] It is
plain that his major reason for taking Drift seriously was its perceived value
as a zoogeographical problem-solver. This is borne out by his next and final
paper on the subject.

Later that same year, he gave a formal address on 'The Composition and
Origins of the Australian Fauna, with Special Reference to the Wegener
Hypothesis'. The title accurately describes the orientation of the paper:
Australian zoogeography, not global geology. He set aside geological objections because 'However fantastic this hypothesis may appear at first sight, it
is obvious that, if there be any truth in it, it must have a revolutionary effect
on zoogeography' (Harrison, 1928: 335). At most, it seems he desired to
convince the reader of the plausibility – not the veracity – of Drift: 'I am not
competent to judge the matter from its geological, geophysical, or geodetic
aspects. My aim is simply to show that Wegener's hypothesis, whether it can
be established or not, affords considerable help towards the solution of
Australian biogeographical problems by directing attention to entirely new
modes for northern and southern relations' (Harrison, 1928: 336). Specifically, he remarked that Drift 'supplies the southern landmass which meets
my needs' (Harrison, 1928: 333, emphasis added). He then reconstructed
the history of Australian fauna using Wegener's model. Harrison concluded
by claiming (1928: 393)

that the origins and composition of the Australian fauna are rendered easier of explanation by the adoption of Wegener's hypothesis, and that it, therefore, constitutes a useful working hypothesis, whether it can ultimately be proven or no . . . and that this hypothesis offers the only explanation . . . adequate to explain the divorcement between Asia and Australia.

This is a clear statement of Harrison's reasons – reasons based upon his own specialist knowledge and interests – for pursuing, though perhaps not accepting – Drift. Harrison died before he could further pursue this line of investigation or proselytize among students.

G.E. Nicholls, Professor of Biology at the University of Western Australia,[9] also took a positive stance on the potential value of Drift to Australian zoogeography. Prior to this his few publications had been confined for the most part to identifications and descriptions of Western Australian species. This localism was a feature of his endorsement of Drift, the context of which was an exposition of the zoogeography of Western Australia. For Nicholls as for Harrison, Drift was a *deus ex machina* which gave the extensive land connections needed to solve problems of species distribution. He did not embrace Drift as the only acceptable geological basis for zoogeography but did regard it as superior to its competitors and has having acquired greater probability since Harrison's day (Nicholls, 1933: 133–7). This was Nicholls' only explicit discussion of the theories underlying his descriptive work, and it could be assessed – perhaps uncharitably – as a description of the zoogeography of his state lightly glossed with an uncritical and abbreviated summary of Harrison's views.

The central issue for both Harrison and Nicholls was not geological but biological. It probably does not misrepresent their attitude to say that any theory of geology which provided the land connections they felt essential in Australian biogeography would be regarded with favor. Other Australian biologists evinced little sympathy for Drift (Frankel, 1984a: 152–3). A cursory examination of zoological articles in Australian publications indicates a preponderance of atheoretic descriptions of species and of faunal regions. To this Drift had little to contribute.

. . . and back to Britain

British geologists with field experience in the Southern Hemisphere were not immune to the blandishments of Drift. C.S. Wright, who had been a member of the British 1910–13 expedition to Antarctica, was puzzled by the presence there of extensive fossil deposits of *Glossopteris*. It seemed to him that Antarctica may have been the 'original home' of this flora (C.S. Wright &

Priestley, 1922: 424–66) and that before glaciation must have enjoyed periods of warmth. By the time that he wrote up the results of his study, Drift appeared to offer promise: as he phrased it, 'The theory of Wegener, if tenable, seems to solve both these difficulties . . .'. T.H. Holland drew upon his first-hand knowledge of the geology of India in deciding in favor of Drift. He had worked with the Geological Survey of India for almost twenty years, capping his service with a stint as Director from 1903 to 1909. He had campaigned for the use of distinctive nomenclature and stratigraphic series for the Southern Hemisphere on the grounds that the geological history of these Gondwana countries was markedly different from that of Europe. As he put it, he was among those 'who object to what some of us have called a geological imperialism overriding our craving for self-determination' (Holland, 1933: lxx). He gave two instances. First, those who had worked on Gondwana glacial beds had found extensive correlations among those continents making up Suess's old Gondwana. He cited du Toit's work in this regard. Second, the date of the glaciation was toward the end of the Carboniferous (about 300 MYA) rather than the mid-Permian (about 260 MYA) date assigned by German and American geologists. In 1933 he made only passing reference to Drift. Four years later he spoke strongly for Drift and against its rivals mostly on the basis of its usefulness in making sense of Gondwana correlations but did not unreservedly adopt Wegener's program. (Holland, 1937: 18). He sounded the same note in two papers in the early 1940s (Holland, 1941; 1943).[10] A.C. Seward, a palaeobotanist and Professor of Botany at Cambridge, held 'a favourable and hopeful attitude towards the Wegener hypothesis' (Seward, 1933: 538). Drift afforded accounts of both the intercontinental connections and such past climatic conditions as the glaciation of Gondwana which he needed to solve problems of ancient plant distributions. Although he favored Drift he did not accept it; there were objections lodged by geologists (Seward, 1933: 41–4) and the case for Drift 'for the present must remain *sub judice*' (Seward, 1933: 538).

There were also local interests closer to home. E.B. Bailey inclined to Drift partly on the basis of his extensive fieldwork as District Geologist in Scotland with the Geological Survey. He was an atypical member of the Survey – of which he became Director in 1937 – for he was not only an accomplished field geologist but also had a deep interest in such theoretical matters as tectonics. Among the Scottish problems which occupied him was the unravelling of the complicated folded structure, history and correlation of the Caledonian mountains (about 450 MYA). Bailey was sympathetic to a Suessian global approach and placed his problems in that context. According to a Suessian interpretation, the Caledonians extended north from

Scotland into Scandinavia and to the south, into Wales where they met the east–west Hercynian chain (about 300 MYA). Both chains spanned the Atlantic via a now-submerged landmass. In New England they split apart to form the Appalachian system. In 1929 Bailey dispensed with the presumed sunken landmass: the crossed chains matched neatly when the continents were pushed back together to form Pangaea (Bailey, 1929: 76) as Wegener had suggested (Wegener, 1915: 63–4, 1929t: 75). Bailey stopped short, however, of accepting Drift. Ten years later, in his introductory textbook, he discussed some of the problems which Drift solved and other evidence in its favor, including pattern-matching between Scotland and Canada. The evidence was not yet conclusive: no definite answer could be given to the question of whether or not the continents had really drifted (Bailey, 1939: 7).

As we have seen some scientists in the English-speaking world were sympathetic toward Drift. Their stance was often based on the promise which they perceived Drift to show in solving specific problems which specially concerned them. They were well aware of the many objections to Wegener's version, especially those connected with the actual cause(s) of Drift. Few explicity accepted – in the strong sense – that program. We might conjecture that for these geologists – and for others who remained silent – there was a delicate balance between their specialist interests (both 'technical' and 'social') and their awareness of the conceptual and empirical problems confronting Drift. If the balance were shifted to one side by an elimination of some problems, it might result in an acceptance of Drift; were these to appear insoluble, the balance might tip against Drift and result in its rejection. As we shall see, there were no decisive developments for almost five decades after Wegener proposed his theory.

Voice-Over

Most general theories of science – social, historical, philosophical – assume that scientists when faced with a choice among rival theories, paradigms, programs, or traditions decided for whatever reasons to accept – in a strong sense – one of these. This does not seem to square with the response of many geologists to Drift and its competitors. Most Southern Hemisphere geologists, like many of their northern counterparts, made no public profession of faith in one or another of the rival programs and were instead eclectic. We might conjecture than many were working in problem-fields to which global theories seemed to be irrelevant and which therefore did not require them to make a choice. For these, a shrug of the shoulders might be the rational epistemic response! Some did express preferences but usually hedged with cautions about 'working hypotheses', accepting verdicts from other disciplines, available evidence, and so on. By emphasizing the strong positions taken by a few scientists; e.g., a Schuchert or a du Toit, we could overlook the tentativeness and eclecticism of the theory choices of most workers in a given field.

A simple dichotomy of 'acceptance' (Wade, du Toit, Carey) or 'rejection' (Andrews, Lake and many North American geologists) is inappropriate. Laudan's category of 'pursuit' seems applicable to some. Even this does not capture the many shadings we have seen. One could adopt a five-fold division: 'rejection', 'entertainment', 'employment', 'pursuit' (theoretical or empirical) and 'acceptance'. In this taxonomy, we could say that many scientists (e.g., Ward, Cotton and Wright) 'entertained' Drift; i.e., they implicitly or explicitly referred to it as a respectable alternative but did not actually use it for explanatory purposes or in their theoretical or empirical work (*cf.* R. Laudan, 1986). Harrison, Bailey, Caster, Holland and Sewell went a step further and 'employed' Drift; that is, they used the theory to explain matters familiar to them, though they did not undertake new empirical or theoretical work aimed at or guided by Drift. Wade, Carey and du Toit initially 'pursued' Drift in their research. Wade and du Toit worked at an empirical level to improve the pattern-matching arguments for Drift; Carey, and to a lesser extent du Toit, worked on theoretical articulations of the program. Subsequently, all three 'accepted' Drift. This taxonomy better captures the day-to-day decisions of scientists engaged at the coal-face and is also applicable to the longer term competition among rival programs and theories. Rather than formalizing this 'five-fold way', however, it may be more constructive to regard possible stances as a spectrum ranging from complete and utter rejection to enthusiastic and unqualified acceptance with many shadings in between.

The response to Drift reflected both the character of that program and the nature of geology as a discipline in the early decades of this century. We have stressed the role of specialism and localism in the response of individual scientists to Drift. We often think of specialties within a discipline as an institutionalized division of labor. No one person could master and pursue research in the whole of geology; it is necessary and also more efficient to carve up the subject matter into smaller units. This view has some validity provided there be consensus on the aims of discipline, on preferred theories and programs, on accepted techniques and methods and perhaps even on the hierarchy of specialties within the discipline. The loci of research are the specialties but the meaning and significance attached to that research comes from the central set of shared disciplinary beliefs. If disputes arise, these often revolved around the interpretation and application of core beliefs to specific issues within specialties. Geology in this view is an 'umbrella' under which are gathered a multitude of discrete but complementary subdivisions.[11] This view does not seem appropriate for our study. The 'umbrella' included specialties which sometimes had quite divergent aims: for stratigraphy, description; for geophysics, explanation. There were fundamental disagreements over the 'big theory', the 'umbrella', which should inform geological research.[12] Different research programs could imply different hierarchies of specialties and consequent disputes about their centrality or even relevance to the discipline. In these circumstances, researchers within a specialty might have aims, methods, theories, a preferred program, standards and techniques different from those of researchers within other specialties. Specialties might even be thought of as embodying competing definitions of the discipline. Disagreements among different specialists would represent conflicting cognitive and social interests. Permanentists, for example, often gave pride of pace to stratigraphy, and it occasions no amazement that stratigraphers tended to favor Permanentism. Wegener in proposing Drift in effect also proposed a redefinition of geology and a reconstitution of the disciplinary field which included elements of what were otherwise considered heterogeneous bits of disparate disciplines; e.g., geophysics, biogeography, climatology and palaeobotany, and which moved these from the wings to center stage.

Opponents and proponents of Drift seldom engaged in comprehensive, systematic evaluations of it with respect to its rivals. Rather than taking the global view advocated by Wegener, most were quite parochial. After all, what scientist could claim to have the intimate knowledge of every specialty relevant to judging the overall problem-solving ability of Drift as compared with its rivals or to decide the appropriate weight for the conceptual problems and anomalies that it faced? The ideal of a group of omniscient Dr Smiths judging theories against the total body of available evidence is a myth. Even Wegener confessed to being unable to keep up with the 'snowballing literature on drift theory in the various sciences' (Wegener, 1929t: vi).

How does this bear on some of the general models of science we are examining? Any analysis of the modern transformation of geology should take into account the different levels of specialization within a scientific community: disciplines, specialties, and at the micro-level, research networks, focussed on problem-fields. Rationalistic models usually ignore such divisions, despite their possible cognitive importance. Normally these models focus on the comparative merits of rival theories or programs as judged by what is suggested to be a relatively homogeneous disciplinary community. Further, these models seem to presume that in evaluating theories scientists look or should look beyond the limited horizons of their own specialties and research interests and somehow make computations for the whole domain of the theory (in the case of Drift, for much of the domain of the discipline and beyond into other disciplines). As tools of analysis they seem to be insensitive to the effects of specialization and localism.[13] Wegener (1929t: 5) noted 'that totally different conclusions are drawn about pre-historic conditions on our planet, depending on whether the problem is approached from the biological or the geophysical viewpoint'. This was characteristic of appraisals of Drift. Wegener commented, after du Toit's claim that geophysics should be decisive and von Ihring's call for the primacy of palaeobiogeography, 'it would be easy to add to the list of such opinions, each scientist deeming his own field to be the most competent, or indeed the only one competent, to judge the issue' (Wegener, 1929t: vii). Several social models emphasize the importance of 'technical interests' in the evaluation of competing theories and in the struggle for authority within a scientific field. However, none of these have yet been well-articulated or applied in detail to historical cases (see, however, A. Pickering, 1984; Sapp, 1987).

Judgments of programs and theories do not seem to have been made corporately nor on the basis of overall problem-solving ability. It has been argued that scientists in different specialties (or different disciplines) have, 'so to speak, different access to the system of scientific belief and practice as a whole' (Lugg, 1978: 282) and further that 'we can expect them to differ with respect to what counts as an empirical problem, the sorts of objections they recognize as conceptual problems, the criteria of intelligibility, standards of empirical control, the importance and weight they assign to the problems, and so on' (Lugg, 1978: 290).[14] With a judicious selection of historical material, we could construct a case to the effect that localism and specialization mean that advocates of different programs or theories inevitably disagree to the extent that a resolution in favor of one or another is irrational. This, however, would be misleading. Different specialties did not uniformly reach different judgments about the rival programs. Some biographers favored Permanentism; others, Contractionism; still others, Drift. Even within Australia, not all scientists interested in problems of species distribution chose Drift (*cf.* Frankel, 1984a). Something approaching intra- and inter-specialty consensus did emerge much later, in the late 1960s and early 1970s, but as we shall see it is not necessary to invoke the working of an invisible hand to avoid anarchy or to produce consensus.

For individual scientists, what was usually at issue were not global judgments but judgments with respect to those parts of science in which they had particular expertise, research interests, and authority. For Harrison, his problems were posed conjointly by his acceptance of Darwinian evolution, his empirical knowledge of Southern Hemisphere biogeography, and his awareness that most geologists favored Permanentism. His decision to work on a particular set of problems and the value he attached to their solution preceded his employment of Drift: his localism helped determine rather than was determined by his attitude toward Drift. It seems reasonable for Harrison to have favored Drift as a 'working hypothesis' because it did a good job of solving his problems which its rivals solved in a less satisfactory way, even if other scientists in other specialties or disciplines opposed it.[15] If we accept the views that science is a problem-solving activity and that it is normally characterized by theoretical pluralism, then would we not expect scientists often to select that theory or program which they find of greatest utility with respect to those problems on which they are working. After all, on a problem-solving model we take scientists to be concerned primarily with solving problems, not with the appraisal of research programs or theories. The latter undertaking would be partly contingent on the usefulness of those programs and theories in the former. If it be considered rational to opt for a rapidly progressing, successful problem-solver, then the decisions of many geologists in the 1920s and 1930s to reject Drift or to suspend judgment on it were reasonable and conformable to both cognitive and social interests. If, however, we look to specialties and problem-fields rather than the discipline, it was reasonable for some to employ or pursue Drift as a working hypothesis on the grounds that it offerred solutions to problems in their specific fields of interest; similarly, for most geologists, it offered no advantages over its competitors and was not pursued. From this perspective, specialization and localism can play important roles in the reception of novel theories and the appraisal of rival programs. Further, if one theory be accepted even by a minority of the practitioners in one specialty and another in other specialities, the effect is to spread the risk and perhaps to maximize progress through competition.

This amounts to an extension or refinement of the 'scientific' or 'technical interests' invoked by some social analysts of science and its grafting onto problem-solving models. Most modern theories of scientific change seem to have been conceived with respect to 'theoretical' change. Their extension to scientific change may be improper in so far as they ignore a possible valuable distinction between a scientist's day-to-day research in which he or she uses a particular theory to solve problems of immediate concern, and the long-term judgments – perhaps reached in a semi-collective manner – over the perceived relative adequacy of rival theories of programs with respect to the field as a whole. These two sorts of 'theory–choice' should not be conflated; meta-models address the latter but they fail to take account of its conditional relation to the former. After all, as an anonymous wit has pencilled in a volume in the US Geological Survey library, 'Most of Earth's bigger scars, sutures and dimples etc were more visible on her bottom, and the people down under were quicker to see them'.

The view I have sketched is consonant with theoretical pluralism. It is conducive to progress in empirical and conceptual problem-solving. If one theory is accepted or pursued even by a minority in one specialty or set of 'local' problems, the effect may be progress in depth. If other theories or programs are also pursued by other workers, the effect is to 'spread the risk' and, possibly, to promote progress in breadth. Further, this may provide 'breathing space' for a novel theory, provided of course that some workers judge it promising with respect to some problem-fields. Finally, specialization and localism can produce healthy competition. Suppose that Drift had been totally abandoned in the 1940s and that no scientist today pursued or accepted large-scale horizontal movement of continents. We could still say that the favorable attitude of some scientists toward Drift in the 1920s, 1930s and 1940s had been conducive to progress in that this forced their competitors to sharpen their own solutions and to endeavor to address the problems which Drifters had solved in their program or had created for their colleagues. Indeed, this rationale was often advanced by both friends and foes of Drift who praised the stimulating effect that it had had on theoretical and empirical discussions.

An emphasis on specialization and localism need not lead us to embrace extreme pluralism; i.e., the sort of theoretical anarchy advocated by Feyerabend (1975).[16] Specialists were not utterly blinkered to difficulties outside their areas of special competence and concern. I suggest only that those problems and objections were likely of less weight than for specialists in other fields; e.g., geophysicists. How can consensus arise from fragmentation? The boundaries of problem-fields, sub-specialties, and specialties are not walls. Geologists working on a set of local problems are not isolated; they are at least dimly aware of data, arguments, and theories used in other problem-fields, specialties and disciplines. Moreover, in their education process they are socialized into the belief that specialties are not self-sufficient but are interdependent and are components of a larger enterprise, as a glance at the introductory sections of almost any geology textbook of the time will show. This is reinforced and maintained by academic practices, professional institutions and other aspects of the social organization of the discipline. Consequently, geologists do take into account non-local considerations in choosing and evaluating working hypotheses and they aim at convincing not just fellow specialists but workers in other specialties as well. Consensus can emerge from localism provided a theory or program addresses in useful ways central concerns of several specialties or is at least compatible with them, even though these may vary from problem-field to problem-field or specialty to specialty.

Notes

1 I use the following terminology: *Problem-field*: a group of conceptual and/or empirical problems which may cut across specialties or disciplines, which are perceived to be interconnected, and which form a locus of research. This is similar to the notion of a 'research network'. The problem-field emphasizes the cluster of problems; the research network, the cluster of researchers. *Localism*: this term is used in preference to Frankel's (1984a: 156) term 'provincialism' to avoid pejorative connotations. Localism in geology is a type of specialization. Geologists may specialize in subject matter; e.g., palaeontology, or in the use of particular methods; e.g., geophysics. These may be further subdivided; e.g., seismology as a branch of geophysics. By localism I mean a specialization in time or place. Geologists may focus on problems in a particular geological era (e.g., the causes of the supposed massive extinction at the Cretaceous–Tertiary boundary about 60 MYA) or – especially in specific geographical or geological areas (e.g., the Appalachian orogeny as opposed to orogeny in general, the origin and distribution of Australian species as opposed to palaeobiogeography in general, the mapping of the Old Red Sandstone in England as opposed to establishing world-wide correlations). Localism may be a feature of other historical or field sciences; Latour and Woolgar (1979) suggest that even in laboratory sciences research may be locally contingent upon the particular assemblage and arrangement of equipment.

2 This effect was especially apparent among some British directionalist palaeomagnetic researchers in the 1950s and early 1960s.

3 The Wallace Line is one of the most famous boundaries of species distribution. This imaginary line, proposed by A.R. Wallace in 1858, runs through the middle of the Malay Archipelago (Wallace, 1876). To the west of it the fauna are predominantly Asian in character; to the east, predominantly Australian, yet there are no obvious barriers to the diffusion of species across the 'line'.

4 The Karroo is a series of sediments and lava flow covering a large expanse of southern and central Africa. It includes tillites, coal deposits and evaporites (strata associated with the evaporation of salt water) and contains fossils of many species of reptiles.

5 David & Browne (1950: 688–9) contains the statement that Wegener's theory fails 'to interpret convincingly the known geological facts' but these posthumously-published words were likely those of his collaborator.

6 Gutenberg was a German geophysicist who emigrated to the US in the 1930s. He disagreed with much of the geophysics of Wegener's version. He proposed as an alternative that the earth was once covered with a sheet of sial, that much of this

was ripped away in the birth of the moon, that the remainder was located
asymmetrically about the South Pole, and that it subsequently spread out under
the influence of gravity and the Polflucht force (*cf.* Lake, 1933).

7 Perhaps Evans (1923) or W.B. Wright *et al* (1923).

8 It is tempting to interpret some of the strident tone of Harrison's rhetoric as
   expressing the outrage of the Challis Professor of Biology at the University of
   Sydney at having exceptionable solutions to Australian problems being dictated by
   holarctic interlopers.

9 Nicholls received a D.Sc. from Kings College, London and had been Professor of
   Biology at Agra College, India, before taking up the chair at UWA in 1924. He
   seems to have published only a few short papers after his arrival in Australia but
   from comments in Nicholls (1933) he was an active field researcher.

10 Holland's (1943) paper on Drift read to a joint meeting of the Linnean Society and
   the Zoological Society of London provoked considerable discussion from the
   audience; most of the comments were favorable. Drift was certainly not a taboo
   subject among scientists interested in problems of palaeoclimatology and species
   distribution. R.B.S. Sewell, one of the discussants, had conducted systematic
   surveys of oceanic ridges (Sewell & Wiseman, 1938). He threw out the idea that
   the movement of continents might set up waves in the magma thereby producing
   these ridges (Holland, 1943: 120).

11 Whitley (1976) uses the term 'umbrella discipline' for this type of disciplinary
   structure.

12 Whitley (1976) terms disciplines with these characteristics 'polytheistic'. In such
   disciplines, he contends, 'research often exhibits extreme eclecticism in technical
   approaches under the general heading of a vague positivist ethos' (Whitley, 1976:
   479). This description is apt for the work of many geologists in the period.

13 Kuhn's first version (1962) is subject to these remarks; however he subsequently
   (Kuhn, 1970: 176–8) explicitly defined the communities to which his model applies
   as 'the practitioners of a scientific specialty'. He also noted the possible importance
   of specialization in evaluating contending paradigms (Kuhn, 1970: 177): 'Because
   the attention of different scientific communities [read 'specialties'] is . . . focussed on
   different matters, professional communication across group lines is sometimes
   arduous, often results in misunderstanding, and may, if pursued, evoke significant
   and previously unsuspected disagreement'. There are several objections simply to
   attaching paradigms to specialties. In our case the 'paradigms' transcend the
   boundary of any specialty. Kuhn's own initial (1962) examples, unaltered in his
   later version, are not confined to specialties nor even in some cases to a discipline.
   Further, he does not spell out how differences in specialist concerns may evoke
   communication problems or disagreement. Finally, in our case, even within
   specialties there was no consensus on the relative merits of the competitors. Kuhn's
   amendments are constructive ones but, unfortunately, they are simply 'tacked on';
   he did not rebuild his model as would seem required.

14 L. Laudan (1984: 25, n. 2) rather incorrectly attributes to Lugg the view that if all
   scientists had access to the same data, consensus would result. Lugg, however, as
   indicated in the passage quoted, does not limit access difference to purely empirical
   matters.

15 Apropos of some ideas about science, Harrison expressed no misgivings that Drift might be tainted because it had been, in part, conceived explicity to solve precisely the sort of problems on which he happened to be working. He was not concerned that Southern Hemisphere biogeography was, so to speak, part of the 'problem situation' of Drift. His attitude, from the point of view of problem-solving models, seems quite reasonable: to draw a crude analogy, it is like a carpenter's decision to use a new design of plane because it smooths the wood better than the old design, not because he unexpectedly finds that it can also remote-tune a television set.

16 Feyerabend (1975: 23) is quite right in arguing that the history of science shows that there is not a single methodological rule, however plausible, the violation of which has not been conducive to the growth of scientific knowledge. The anarchist may sometimes be an important figure in the development of science; it does not follow that every scientist should be an anarchist.

6

Interregnum: competition and stalemate

Drift had by 1930 gained adherents in Britain, Europe, North America and elsewhere. Over the next two decades it won few recruits; indeed, it may have lost ground. In the late 1920s the debate entered a stalemate. This could have been broken in several ways. Permanentists and Contractionists sought to solve the problems those programs faced. If they were successful, or able to offer alternative solutions to those of Drift in areas in which it had showed promise or able to mount convincing arguments against any large-scale lateral motion, then Drift might have been 'knocked out' of contention. This did not occur. Drifters sought to meet objections to Wegener's version.[1] The major objection to Drift by friends and foes was the inadequacy of the forces invoked by Wegener. The evidence so far assembled for Drift, though impressive and suggestive, was not convincing. Drift might remain a contender but there was no reason to abandon the more established program in favor of it. If Drifters developed a version which had a more plausible mechanism or if such strong evidence of Drift were amassed that it would be preferable even without a suitable mechanism (after all, geologists accepted past ice ages even though they had no agreed explanation for them), then Drift might yet emerge the victor. Drifters pursued both of these strategies but were unable to sway their opponents. The stalemate continued in the 1940s and early 1950s.

The Permanentist program and species distribution

W.D. Matthew, B. Willis and G.G. Simpson advanced versions of Permanentism aimed at explaining the similarity of fossil or living species on continents separated by oceans. Both Drift and Contractionism had solutions; this may have made it all the more pressing a problem for Permanentists (*cf.* Frankel, 1981; 1984a; 1985). Matthew, a Canadian palaeontologist at the American Museum of Natural History in Chicago, flatly rejected any purported large-scale landbridges or drowned continents: they were contradicted by isostasy (Matthew, 1915: 175). Drift was probably unknown to him. He would have applauded the sentiments of A.P. Coleman, President of the Geological Society of America in 1915, who deplored the 'great recklessness [of palaeontologists] in rearranging land and sea [for] the convenience of a running bird, or of a marsupial afraid to wet its feet' (Coleman, 1916: 190).[2] Matthew, however, agreed with bridge-builders that separate creation or wide-spread parallel evolution were unacceptable solutions to distribution patterns. Darwinian evolution was inviolate (Matthew, 1915: 306–7). His solution relied upon the dispersion of species from the Holarctic region and long-term climatic cycles. Most species, he asserted, originated in the northern circumpolar region. These species were hardier, more active and more adaptable than those inhabiting warmer climes. Dominant races spread from the north, displacing the feebler warm-weather ones.[3] This was aided by climatic changes. He adopted the theory of T.C. Chamberlin according to which cool arid conditions alternated with warm moist ones due to a 'greenhouse effect'. These cycles produced eustatic (world-wide) changes in sea-level: during cool dry periods which promoted glaciation the sea-level fell; during warm moist ones, ice-caps melted and the sea-level rose. In the cooler periods northern species had a double advantage: their adaptations favored them over warm-weather species and the fall of the sea exposed land connections which enabled them to spread to the south. Extant links such as the Panamanian isthmus became broader and new links appeared. He claimed that a fall in sea-level of 100 fathoms would unite all continents – except perhaps Australia and Antarctica – and larger islands into a single landmass ripe for the invasion of hardy northerners (Matthew, 1915: 309). Similarities between South American and African species were due to their common holarctic ancestry.

This theory formed the framework for a systematic treatment of species distribution. His handling of exceptions is revealing. He suggested that where there seemed to be no counterparts for Southern Hemisphere species in the north or, more generally, where there seemed to be a lack of evidence in support of his theory, this was due to imperfections in the fossil record (Matthew, 1915: 190–1). His argument was circular: the evidence was adequate if it fit the theory; if evidence conflicted with the theory, the evidence was inadequate, incomplete, or misinterpreted. To explain other exceptions he invoked rafting. His exposition of the probability of species dispersion through rafting was influential despite obvious mathematical errors (Matthew, 1915: 206–7). He concluded that his theory, especially as it applied to mammals, was well supported. The past and present distribution of species in North America was particularly well explained. The species similarities which he adduced, if landbridges were eliminated on physical grounds, entailed the acceptance of his own theory 'unless we are to abandon all belief in the actuality of evolution and are to treat it as merely a convenient arrangement of successive species and faunas independently created' (Matthew, 1915: 306–7). Darwinian theory was not threatened. Matthew's solution was augmented in the 1930s and 1940s by Willis and Simpson.

Willis emphasized 'isthmian links' for intercontinental species migrations (Willis, 1932: 919). He was encouraged by Schuchert who had abandoned continent-sized 'bridges' in the face of isostasy and who now pursued Permanentism (Schuchert, 1932: 876). Willis placed his isthmian link version in the context of solutions offered by rival programs. He rejected landbridges of the sort erected by Contractionists. As for Drift, which he had previously condemned (Willis, 1928), he coyly remarked that his own explanation of species distribution 'may affect it' (Willis, 1932: 930). Granitic continents 'floated' in isostatic equilibrium in denser, basaltic ocean floors. However, the seafloor could somehow be thrust up to form chains of mountains rising above sea-level. Such a chain had once consti-tuted an 'isthmian link' (in effect, a narrow landbridge) composed of ocean basin material between South America and Africa. Eventually the 'links' would, in accord with isostasy, subside into the ocean floor. Willis, following Matthew, also made use of changes in sea-level to generate other 'links'. He further suggested that the patterns of ocean currents would be so altered by these 'links' that northern regions might experience near-tropical condi-tions at the same time that there was glaciation in India and the Southern Hemisphere (Willis, 1932: 942–52).

G.G. Simpson in the 1940s gave the most comprehensive and influential discussion (*cf.* Frankel, 1985: 165–7). He was an American vertebrate palaeontologist who specialized in Cenozoic (recent) mammals. He did field studies in South America and had expertise in statistical methods. He forcefully stated his case in the context of a four-cornered argument. In three of the 'corners' were Permanentists, proponents of sunken continents and Drifters. In the fourth 'corner' were the purported 'facts' of species distribu-tion (Simpson, 1943). His tactics were admirable. He cast himself as the referee, not a combatant. First, he emphasized the importance of specialist knowledge:[4] summing up the use of species distribution data by Drifters, he bluntly remarked, 'It must be almost unique in scientific history for a group of students admittedly without special competence in a given field thus to reject the all but unanimous verdict of those who do have such competence' (Simpson, 1943). He placed himself above the fight: before inclining to any global theory on the basis of biogeographical data, the accuracy of that data must be assured. He then raised the standards for admissible 'evidence'. He ridiculed previous writers for using improper methods or poor logic, for uncritical copying and the misrepresentation of the writings of others, for the misidentification of species, for ignoring more recent fossil finds, and so on (Simpson, 1943: 10). He discounted the oft-cited similarities between Australian and South American marsupials, claiming that these were few and distant. He proposed instead an Asian origin of Australian marsupials: echoing Matthew, the lack of appropriate fossils in Asia 'is likely to be an accident of non-discovery' (Simpson, 1943: 13). After purging the data, he proposed a statistical method of determining similarities of faunal populations which he regarded as preferable to talking about individual cases. Again, the result told against extensive connections. The purified 'facts' of species distribution were best explained in terms of continental permanence. Exceptions could be accommodated by 'corridors', isthmian links (narrow corridors) or 'sweepstakes routes' (island-hopping, rafting and similar chance circumstances). Not only had he blown up bridges and anchored

Figure 6.1. Willis engineered a link between South America and Africa by lowering the sea-level, temporarily raising a portion of the oceanic crust and using existing oceanic ridges and rises (after Willis, 1932: 930).

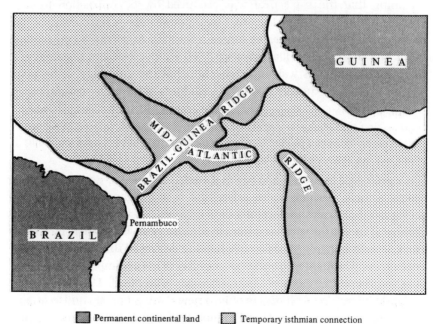

Permanent continental land Temporary isthmian connection

Continental shelf Oceanic basins

drifting continents, he had also given acceptable explanations for the few similarities which remained. The impact of his case would have been all the greater because it appeared in the context of his other highly influential writings on mammal classification and evolution. The following year, he coupled his biogeography with the 'modern evolutionary synthesis' in biology of R.A. Fisher, J.B.S. Haldane and Sewall Wright (*cf*. Mayr & Provine, 1980; Laporte, 1985). He thereby allied his conclusions in geology with the latest theories in biology.

The Contractionist program and radioactivity

The major problems for Contractionists arose from geophysics and physics. Isostasy posed a difficulty for the emergence or sinking of landbridges or continents. The more urgent problem was due to the discovery of heat production in radioactive decay. Both Drifters and Permanentists readily pointed out the anomalous character of these issues for Contractionists. The discovery of radioactive heat removed one conceptual problem faced by more rigidly uniformitarian versions of Permanentism and created one for Contractionists. The Scottish physicist Lord Kelvin had calculated in the 1860s that the earth must be far younger than thought by most geologists or by Darwin: it was perhaps less than a few tens of millions of years old, assuming that the sun's heat was produced by its contraction from a gaseous mass, and by making other assumptions about the cooling of the earth. From these estimates he forged a weapon to defend a contracting earth and to attack Lyellian geology. T.H. Huxley riposted that Kelvin's mathematical physics was impressive but that the soundness of his conclusion depended crucially upon the soundness of his many assumptions:

> Mathematics may be compared to a mill of exquisite workmanship, which grinds you stuff of any degree of fineness; but, nevertheless, what you get out depends on what you put in; . . . pages of formulae will not get a definite result out of loose data. (T.H. Huxley, 1869: 1)

Some geologists in Britain and North America disputed or ignored Kelvin's claims. Others, perhaps more deferential to the senior science, revised geological time scales to accommodate a 'young' earth or even reworked geological measures such as sedimentation rates to bring their estimates into line with Kelvin's.[5] Radioactive heating destroyed the basis of Kelvin's estimates and brought into question a number of the assumptions underpinning the simpler versions of Contractionism (Burchfield, 1975).[6]

Harold Jeffreys used physics to build a new Contractionism and to attack Drift in the 1920s. Jeffreys, later Plumian Professor of Astronomy at Cambridge, imported not only the conclusions of physics but also its form and rhetoric into the debate over global theories. Quantum mechanics and relativity theory were then eclipsing his own more orthodox interests. Perhaps he sensed the opportunity to stake a claim for physics – and himself – as the arbiter of disputes in the obviously inferior – i.e., relatively non-mathematical – science of geology.[7] He admitted the paucity of empirical data on the interior of the earth but argued that using physics he could extrapolate to a model of its structure and processes although it was necessary to introduce 'some ad hoc assumptions' (Jeffreys, 1924: 1). Physics should lay down the boundaries for any acceptable geological theory; primacy should be placed upon quantitative agreement 'of theory with fact' (Jeffreys, 1924: vii). The geophysicist provides a general theory; the geologist fills in the details (Jeffreys, 1935: 168). This implied a new style of discourse about the earth: he remarked that his book might be too mathematical for geologists but expressed the hope that 'if the geologist cannot follow a part of the book . . . he will omit it and go on to the next non-mathematical passage, trusting that someone else will point out any intervening mistakes' (Jeffreys, 1924: viii). His mathematics of the earth led to but one conclusion: contraction explained the gross surface features of the earth.

He contended that radioactive materials were confined to the upper part of the crust. Making this assumption, he calculated that despite the heat produced from their decay the earth as a whole was gradually cooling and contracting. He incorporated isostasy into his model. The rigid crust extended to a depth of 100 or more kilometers and had sufficient strength to hold up Mount Everest and hold down the great ocean deeps (trenches). The crust 'floated' on a substratum which at greater depth and increased heat pressure was more plastic. Mountains were due to horizontal compression resulting from contraction. Carefully choosing appropriate estimates of the crustal shortening represented in various mountain chains, he computed that the amount of cooling and contraction he had calculated was sufficient to explain orogeny. Page after page of mathematical physics, with only occasional forays into geological data, led inexorably – so it appeared – to the conclusion that Contractionism fitted his mathematical physics. Drift, however, was an 'impossible hypothesis' unless 'forces enormously greater than any yet suggested are shown to be available' (Jeffreys, 1924: 260–1). As for problems of palaeoclimatology, he endorsed C.E.P. Brooks's suggestion that these could be explained by past elevation of existing landmasses without recourse to Drift (Jeffreys, 1924: 262–7).[8] Physics (or rather, Jeffreys) had spoken, and with authority and finality.

Jeffreys regarded his approach as the only proper one and his conclusions as unchallengeable. The pose adopted is that of an impartial physicist applying well-honed mathematical tools to remove the undergrowth of wild speculation in a juvenile science not yet schooled in proper scientific procedures. One cannot argue with numbers: they are exact, true and impartial. Physics, the senior science, had spoken, even if not in a language understandable to geologists. Had this approach been followed earlier, 'many wild theories might have remained unpublished or, if published, have received only their fare share of attention . . .'. The adoption of his approach will mean that 'Geophysics is no longer a field for uncontrolled speculation . . .' (Jeffreys, 1924: 4).

His staking of a hegemonic claim for physics met approval from some quarters and opposition from others. Geologists both before and after Jeffreys's foray hesitated to draw firm conclusions about the interior of the earth from the scanty evidence available through study of surface features, bore holes, mines, seismological measurements, gravity irregularities, and similar data (Oldham, 1906; Daly, 1923b: 365; Bull, 1927; van der Gracht, 1928b: 5; Gregory, 1930; Holmes, 1933: 169). This hesitation harmonized with the naive inductivism of many British and American geologists and contrasted sharply with Jeffreys's free use of 'ad hoc assumptions' and mathematical deduction in an argument which in form was analogous to a demonstration in geometry. It was ironic – though the irony was surely lost on geologists – that Jeffreys pushed forward his methods as the methods of physics and therefore as the preferred methods of all the sciences when relativity theory and quantum mechanics could be seen as undermining those methods. Some proponents of Drift regarded his methods as disingenuous or, more kindly, self-delusive. The mathematical clothing of his presentation did not completely disguise his selecting from a range of data and assumptions those which produced the desired result. Huxley's mathematical mill was grinding again. His pronouncements were nonetheless welcomed by many opponents of Drift. This co-optation of Jeffreys into the community of geologists probably owed less to the force of his arguments or to his methodology than to the fact that the conclusions of those arguments were inimical to Drift and favorable to the older programs.

This uncharitable interpretation of the use made of Jeffreys is given some credence by the remarks of some opponents of Drift. Hinks (see above, p. 72) claimed that the best opinion in geophysics – which must not be contravened by those working in other fields, lest they contravene his rules of the game – was Jeffreys's, whose second edition of *The Earth* 'is an impressive canon of the laws of geophysics as they are formulated by the mathematicians' (Hinks *et al.*, 1931: 434). Having set up Jeffreys's physics as the judge of geological theories, Hinks then cited Jeffreys's conclusions as *prima facie* refutations of Joly, Wegener and – inferentially – Arthur Holmes (Hinks *et al.*, 1931: 435–8, 536). His tactics did not go unchallenged. In the discussion which followed, several participants who were sympathetic to Drift rejected Hinks's stand. G.C. Simpson, who had trained in meteorology at Göttingen, worked in India, and who was then Director of the British Meteorological Office, went so far as to say that until mathematicians could provide better answers instead of merely criticizing those that had been offered by Wegener, he proposed to continue 'not playing the game' (Hinks, *et al.*, 1931: 442).[9] This is, however, not to say that Jeffreys had no influence on the course of the debate nor that defenders of Drift closed their minds to his case.

Wegener's counter-moves

Until his death, Wegener spearheaded the campaign for Drift. He incorporated new evidence, responded to criticisms of his mechanism and rebutted the charge that his methods were questionable. In successive editions of his book, he added evidence of pattern-matching found by his supporters. He also stressed geodetic evidence, giving prominence to the Greenland measurements of 1927 which indicated a westerly displacement of 36 meters per year since 1922 (Wegener, 1929t: 29). This, he claimed, '*is therefore proof of a displacement of Greenland that is still in progress*' and that 'the whole discussion of the theory . . . transferred from the question of the basic soundness of the theory to that of the correctness or elaboration of its individual assertions' (Wegener, 1929t: 30, emphasis in the original).

In reply to geophysical criticisms from Jeffreys and others, he downgraded the importance of geophysics to his case. As late as 1927 he had affirmed that 'Were geophysics to come to the conclusion that the drift theory is wrong, the theory would have to be abandoned by the systematic earth sciences as well, in spite of all corroboration . . .' (Wegener, 1927: 103; 1929t: vii). In 1929 he explicitly retracted these words and asserted that no one specialty of the earth sciences had primacy, that evidence from all specialties had to be combined, and that 'the geophysicist easily finds himself on the wrong track if he does not keep constantly in mind the results provided by these branches of science [palaeontology, zoogeography, phytogeography] in order to check his own' (Wegener, 1929t: 97). The apparent failure of his theory in one area should not result in its rejection (Wegener, 1929t: vii–viii). This reformulation of the role of geophysics likely was intended to ward off some of the empirical, theoretical and methodological brickbats hurled at his program. As a second line of defence, he argued in later editions of *The Origin* that to demand a mechanism or, inferentially, to stress the inadequacies of his suggested mechanisms was misguided. The deductive demonstration whereby drift could be derived from a geophysical formula was not yet possible. As he put it, 'The Newton of drift theory has not yet appeared. His absence need cause no anxiety; the theory is still young . . .' (Wegener, 1929t: 167). As for Jeffreys's calculations, he admitted that pole-flight and tidal forces were inadequate but added convection currents as a possible mechanism (Wegener, 1929t: 178–9).

He answered criticisms that he had used improper methods with a bald assertion (Wegener, 1929t: 167):

> The determination and proof of relative continental displacements have proceeded purely empirically, that is, by means of the totality of . . . data without making any assumptions about the origin of these processes. This is the inductive method. . . . The formulation of the laws of falling bodies and of the planetary orbits were first determined purely inductively, by observation; only then did Newton appear and show how to derive these laws deductively from the one formula of universal gravitation. This is the normal scientific procedure, repeated time and again.

This stands in seeming contradiction with his own account of his path of discovery and development: perhaps this was a rhetorical flourish aimed at defending his theory rather than being a seriously intended methodological statement. This passage also links his defense of his methods with his mechanism. He is trying to kill three critics with one stone. First, he argues that he followed the correct scientific methods, the inductive method, in developing his theory. Second, he argues that inductively, he has proven that drift has occurred and that the lack of an explanation of how drift can occur is no reason to reject it. Third, he tries to give his theory respectability by drawing parallels with the highly successful work of Galileo, Kepler, and Newton in astronomy and mechanics. Just as it would have been wrong to dismiss the work of Kepler and Galileo because they could not derive from the force of universal gravitation the inductive laws which they found, so his own work should not be dismissed even though he has no force account.

His counter-moves seem to have been ineffective. Authors who gave Drift an impartial or moderately favorable treatment in the 1930s continued to disparage Wegener's methods. Charles Nevin described Wegener's presentation as 'a type of special pleading whereby he glosses over or completely ignores many of the difficulties and at the same time presents his viewpoint in the best possible light' (Nevin, 1936: 306). Steers, in his relatively even-handed summary, commented that Wegener 'makes some really brilliant suggestions, but frequently spoils them by [errors] . . . or by the omission of evidence of considerable import which tells against his own views' (Steers, 1932: 161–2). Arthur Holmes, who pursued Drift, acknowledged that 'there is a general agreement that Wegener's methods in advocating his particular group of hypotheses are to be condemned. His plausible selection of data, frequently erroneous age determinations, faulty analyses of causes and devious reasoning, have undoubtedly had the effect of weakening his case.' Holmes asked for a careful assessment of the Drift program distinct from Wegener's version (Holmes, 1928b: 431): 'proving Wegener to be wrong is by no means equivalent to disposing of continental drift' (Hinks *et al.*, 1931: 449). Such a policy – had it been adopted – would not have given breathing space for other versions, it would also have put the onus on the critics of Drift to construct a general version to demolish.

The Drift program and mechanisms

The criticisms of geophysicists had a definite impact upon the assessment and development of Drift. Drifters readily acknowledged that Wegener's forces were quantitatively inadequate. Daly, du Toit, Holmes and others postulated alternative mechanisms. Wegener, van der Gracht, Rastall, du Toit, Daly and others advanced the methodological claim that the lack of a known mechanism should not be the cause for rejection of Drift provided there was ample evidence that drift had in fact occurred. Holmes, Rastall and a few others rebutted Jeffreys. None of these strategies proved decisive.

Robert H. Rastall instanced three cases in which physics had proved to be either wrong or unhelpful in geology: Kelvin's determination of the age of the earth in which mathematical theory 'led to completely wrong results'; ice ages, for which it 'failed to produce an explanation'; and the process of nappe formation (complex folding and overfolding of strata as observed in the Alps) with respect to which it 'has not yet tackled one of the fundamental points of tectonic geology' (Rastall, 1929: 447). By analogy, the use of physics against Drift was ill-founded. The latter had to be judged on the basis of geology. He argued for a suspension of 'mathematical judgment' until 'geology and other sciences of observation have shown whether such drift did or did not take place' (Rastall, 1929: 450).

Reginald Aldworth Daly, Professor of Geology at Harvard University, thought that geology could advance only 'through the erection and testing of competing hypotheses; in other words, through speculation. . . . Science progresses by systematic guessing in the good sense of the word' (Daly, 1926: xx). A Canadian by birth, he had taken a Ph.D. at Harvard and then pursued postgraduate studies in Heidelberg and Paris. He read widely in the European geological literature and had worked in South Africa with du Toit prior to du Toit's acceptance of Drift (Menard, 1986: 9). He was impressed by the instability of the earth's surface and was interested in the phenomena of earthquakes, volcanoes, mountain-building and especially the formation of igneous rocks. His earth was a bubbling reservoir of magma beneath a rigid crust no more than 35 kilometers thick (Daly, 1923b: 349–50). Theories of the earth which postulate that the interior is rigidly crystalline to great depth are insupportable: the substratum is hot, basaltic and glassy. This told against Permanentism. The earth is slowly contracting but this is not the cause of orogeny and therefore Contractionism is problematic. Given his attitude toward speculation and his concern with orogeny and surface change, it is not so suprising that Daly was an early – and highly regarded[10] – North American exponent of the lateral motion of continents. He first aired his views in 1922 (Daly, 1923a) and maintained them in several major works (Daly, 1926; 1933). He judged Wegener's version to contain 'highly doubtful and even obviously erroneous assumptions' (Daly, 1923b: 365–6). The chief difficulty was explaining why the continents moved: until this was done, the conservatism of geologists toward Drift was justifiable (Daly, 1926: 263). In the interim Drift should be retained as a 'working hypothesis' because 'it explains so many details and major features of the earth's plan' (Daly, 1923b: 371). He discarded Wegener's version in favor of his own 'downsliding' or 'crust-sliding' version (Daly, 1933: 257). His motor for Drift was gravity. If the crust were distorted so that the continents were piled up in an unstable configuration through the action of contraction (Daly, 1926; 1933: 253) or convection currents in the substratum (Daly, 1933: 255) or the tidal effect of the moon (Daly, 1923c), then the continents might slide 'downhill' over a glassy substratum. His advocacy of continental displacement was muted in his later writings (e.g., Daly, 1942: 47) and his version of Drift, like that of Gutenberg which it resembled, attracted little support.

Du Toit amassed geological evidence for Drift. He believed that the orthodox views could only 'be overcome by the marshalling of the relevant data, their skillful analysis and presentation and the closer study of the evidence, wherever such is in apparent conflict with preconceived ideas' (du Toit, 1937: 4–5). Geophysicists could not referee the dispute (he mentioned Jeffreys only in passing) because their formulae were based on assumptions which need not be granted. The gathering of data was therefore all the more important, for 'under such circumstances we have no option but to conclude that geological evidence almost entirely must decide the probability of the Hypothesis' (du Toit, 1937: 7). He was unable to provide striking new types of evidence, unambiguous evidence that Drift alone could explain. Du Toit also sketched an alternative mechanism but met with little success. His empirical work on behalf of Drift was complemented by that of Arthur Holmes in geophysical theory.

In the 1920s and 1930s Holmes argued against Jeffreys and worked to develop a version of Drift compatible with geophysics and physics. He studied physics at Imperial College, London but became interested in the application of physics to geological problems. He was encouraged by the physicist R.J. Strutt (Fourth Baron Rayleigh) to construct an absolute geological time-scale based on the half-life of radioactive materials. Geologists had had to be content with a relative time-scale based on the fossil content of strata to which were assigned very approximate dates based on assumptions about rates of sedimentation and similar processes. In 1913 Holmes published a technique which won approbation and he became the leading authority on radioactive dating. He continued to refine the technique and steadily pushed the age of the earth back to about 4500 MYA. He had also field experience, first in Mozambique and then from 1920 and 1924 in Burma, with an oil exploration company. He returned from Burma as Professor of Geology at Durham University, England. Holmes entered the battle over global theories with a solid reputation among geologists, geophysicists and physicists. He began as a Contractionist (Holmes, 1925a: 504) but his research on what he termed 'radioactivity and the earth's thermal history' led him to the conclusion that the earth was not cooling. It might undergo cyclical expansion and contraction produced by the build-up and release of radioactive heat, but it was not slowly irreversibly shrinking (Holmes, 1925b). In 1925 he rejected Contractionism and, following Joly's lead, pursued the idea that the lateral motion of the continents might provide a means for discharging heat built up under them through radioactivity and convection currents (1925b: 535).

Holmes, although not yet an advocate for Drift, in 1925 took issue with Jeffreys. Jeffreys had calculated that contraction of the earth would account for estimated compressions of 118 kilometers for the Alps, 100 kilometers for the Himalayas, and 70 kilometers each for the Rockies and the Appalachians. These estimates were selective: they were chosen from the lower end of a series which ranged in the case of the Alps up to 1500 kilometers. Holmes criticized Jeffreys on this point (Holmes, 1925a: 505–6; 1933: 181), one crucial to Jeffreys's claim of quantitative agreement between theory and fact. The 'facts' were rather elastic. He also disputed Jeffreys's assumptions concerning the distribution of radioactive material, a key element of his computations (Holmes, 1925a; 506–15). Holmes used different assumptions to reach his conclusions that the earth was not contracting. His

counterattack, though well aimed, did little to defuse geophysical objections
to Drift. Even after he linked his thermal model of the earth with Drift, he
admitted time and again that the forces invoked by Wegener were 'hope-
lessly inadequate' (e.g., Holmes, 1930: 561). Further, most of Holmes's
objections to Jeffreys's case were couched in terms of application of theory,
data fit, and 'fudging'; i.e., objections perhaps more typical of discourse in
physics than geology. Finally, Holmes's writings usually did not have the
full-dress mathematical physics style of Jeffreys's. A comparison of page after
page of Jeffreys's equations and calculations with Holmes's articles in which
geological discussions are buttressed but not dominated by complex mathematics suggests that Holmes wrote geophysics so that it would be understandable to geologists but Jeffreys deliberately wrote 'over the heads' of
most of those involved in the debate over global theories.

Holmes soon pursued Drift as his preferred global theory. By early 1928
he endorsed the large-scale lateral motion of continents on the basis of the
many converging lines of evidence which Wegener had adduced and
because it provided a mechanism for the release of heat from the interior of
the earth. Radioactive heating and other thermal processes led him to
develop an alternative mechanism for Drift (*cf.* Frankel, 1978): convection
currents in the mantle (Holmes, 1928a). The idea that mantle material near
the core might be heated up, then rise, circulate, cool, and descend – forming
convection currents analogous to those in the atmosphere or in a teapot –
was not novel. European geologists previously had suggested that such
currents, if they existed, might explain orogenesis.

The immediate stimulus for Holmes's interest in Drift and his proposal of
convection currents as the mechanism seems to have been some remarks of
A.J. Bull, a chemist by training, but a keen and respected amateur geologist.
Bull (1921) referred to the convection current theory of orogenesis, but
what seems to have captured Holmes's attention (Holmes, 1928a) was
Bull's (1927) Presidential Address to the Geological Society of London in
which he examined theories of mountain-building. He dismissed
Contractionism (in part, on the basis of the work of Joly and Holmes) and
lauded Drift (Bull, 1927: 154). The major difficulty was that of forces to
propel continents and raise up mountains. He inclined toward convection
currents, perhaps generated by radioactive heat, which might drag apart
and carry along the continents (Bull, 1927: 155–6). Bull had linked
together radioactivity and the inadequacies of Contractionism, two topics
dear to Holmes's heart, with convection currents as a mechanism for Drift.

Holmes gave his most detailed exposition of Drift in 1930 under the title of
'Radioactivity and Earth Movements'. He reviewed the forces which had

been invoked to explain tectonic processes: the crucial factor which most omitted was radioactive heating. He deprecated Jeffreys's Contractionism (Holmes, 1930: 563–4), adding to his list of grievances that 'under the hypothesis continental drift is manifestly impossible'. He introduced Drift obliquely: the theme of the paper, reflecting his specialist interests, was not Drift *per se* but 'a mechanism for discharging the excess heat', produced by radioactivity, 'involving a circulation of the material of the substratum by convection currents, and continental drift operated by such currents'

Figure 6.2. Holmes and other geologists suggested that the successive stages of mountain-building might be connected with the development of convection currents, driven by radioactive heating, in the earth's mantle. (Reproduced from Holmes, 1944: 411, by permission of Van Nostrand Reinhold (UK).)

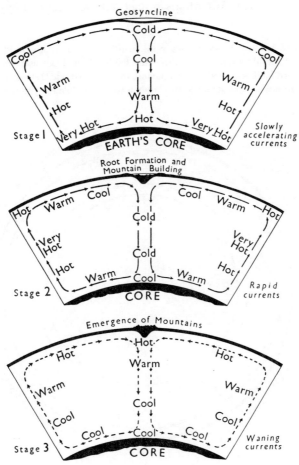

(Holmes, 1930: 565). On the assumption that there is some radioactive material in the deeper layers of the earth, he claimed that heat would build up under the crust (particularly under the continents, for they act as insulating blankets). This gave rise to convection currents in the mantle. The heat must be released: this entailed Drift. The currents, rising and spreading under a large landmass (Laurasia and Gondwana), exerted a 'drag' on the undersurface which could slowly pull the mass apart, perhaps with some subsidence. The continents would be carried 'on the backs of the currents'. Gaps were filled by torn and thinned-out continental crust into and onto which was injected magma which cooled and solidified. This subsided, forming new ocean floor. The motion of the continents would be facilitated by the softening of the ocean floor at their leading edges through the release of heat and by the dragging down into the mantle of the oceanic crust in front of them. Mountains formed at the leading edge by compression

Figure 6.3. The convection cells in the mantle could, according to Holmes, not only build mountains but also cause the breakup of Laurasia, the great northern landmass, and Gondwanaland, the southern one. Tethys was the ancient ocean separating the two landmasses (after Holmes, 1930: 578, 586).

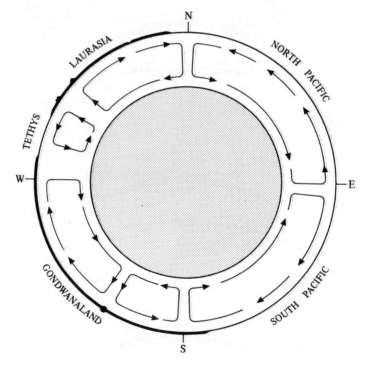

where currents descended (Holmes, 1930: 579–80). He estimated that these currents might move with a velocity of about 5 centimeters per year. This was three orders of magnitude less than Wegener's estimate for Greenland, but it would produce the Atlantic ocean basin in about 100 million years (Holmes, 1930: 583).

Holmes argued for his version throughout the 1940s but to little avail. He devoted a chapter to it in his widely used *Principles of Physical Geology* which included a diagram of how such currents might 'engineer' Drift through crustal 'thinning' and subsidence (Holmes, 1944: 506). Holmes's textbook quickly became a standard work in the field; his version of Drift was less well-received. Proponents of Drift viewed convection currents as a promising mechanism and some opponents admitted that this was more plausible than that of Wegener or others. Nonetheless, there was no marked swing of opinion to Drift. How were geologists to choose between the models of the earth proferred, for example, by Holmes and Jeffreys? Both incorporated radioactivity and isostasy and used similar physical and mathematical techniques but arrived at contradictory conclusions. Jeffrey's earth was crystalline to a considerable depth, was slowly cooling and contracting and did not permit Drift. Holmes's earth had a thin crystalline crust, was being heated from within and entailed Drift. They had begun with different assumptions about the distribution of radioactive materials, about the

Figure 6.4. Holmes's detailed model of how convection currents might provide a mechanism for continental fragmentation and Drift. The essentials of the theory are easily summarized in graphic form. Note the formation of mountains on the leading edges of the new continents as they move through a softened seafloor. New seafloor is formed at the trailing edges by the thinning and subsidence of old continental crust and the filling of gaps by molten material from the mantle (after Holmes, 1944: 506).

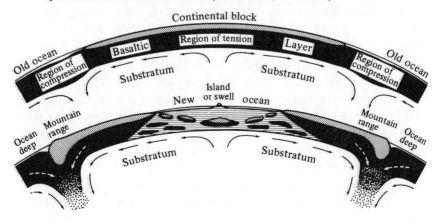

amount of crustal shortening and folding represented in mountains and so on. Each of the models, though inconsistent with the other, was self-consistent. How were geologists to evaluate the conflicting assumptions? The short answer was that they were in no position to do so. As Holmes (1933: 169) himself put it, 'every worker in this field finds an uncharted margin in which the underdetermined conditions governing the earth's thermal history offer a considerable range of personal choice.' He could give no 'independent' evidence on crucial points. Did convection currents exist in the mantle? If they did exist, were they strong enough to break apart the presumed primitive landmasses? If they did exist and were sufficiently strong, were the patterns of convection stable enough over millions of years to account for the lateral displacement of continents as argued from the geological record? If they existed, were strong enough and sufficiently stable, then why did the unique event of the break-up of Laurasia and Gondwana (or Pangaea) occur when it did? One could only appeal to the familiar evidence for and against Drift.

Still, the only research topic directly connected with Drift which did seem to make some headway between 1930 and 1950 was the convection-current hypothesis. The Dutch physical geologist F.A. Vening-Meinesz in the 1920s developed a method for making gravity measurements at sea which could be used aboard a submarine. His observations indicated that isostasy applied to the seafloors as well as the continents. These findings might have put paid to the idea of sunken trans-oceanic landbridges but they do not seem to have been picked up by geologists. He also found unexpectedly large gravity anomalies (i.e., the force of gravity was considerably less than expected) over oceanic 'trenches', the great 'deeps'. He interpreted these trenches and associated gravity anomalies and seismic activity in terms of a downward bulge or down-buckling of the ocean floor produced by the descending flow of converging convection currents (Vening-Meinesz, 1934a: 117–20; 1934b). He also linked them with the more traditional problem of orogeny (Vening-Meinesz, 1934a: 127–33). Vening-Meinesz's research had not gone unnoticed in North America. In 1932 an expedition was organized by the US Navy and Princeton University to carry out measurements over a trench in the Caribbean. Vening-Meinesz co-operated in this research and was joined by a student from Princeton, Harry Hess, with whom he shared his techniques and ideas on crustal downbuckling and convection currents. Hess subsequently continued work on this research with other American geophysicists (H.H. Hess, 1938). David Griggs (a geophysicist at Harvard and later at the UCLA) in a widely cited article drew upon the earlier foreign mathematical and empirical work and combined this with his own experimental and model-building endeavours to argue for thermal convection currents as the most likely mechanism for periodic orogeny (Griggs, 1939: 646–9). In presenting his case, he too drew an oblique connection with Drift. After the war Griggs, Harry Hess, and several emigrés to the United States worked on the hypothesis. In 1950 an international colloquium in Pennsylvania (in which Griggs, Hess, Gutenberg, and Vening-Meinesz participated) devoted considerable attention to the hypothesis. The papers and discussion show, however, that there was still no general agreement even on the existence of such currents, much less on whether or not they could move continents (Gutenberg *et al.*, 1951). Despite these uncertainties there were in many textbooks before and after 1950 references to convection currents as a promising possible solution to a major difficulty for Drift.

Like a rolling stone: Drift in the 1940s and 1950s

Drift in the 1940s and early 1950s has been characterized as stagnating, sinking into obscurity or even eliminated as a research program in North America and Britain. Drift became the butt of jokes: Percy Raymond, Professor of Palaeontology at Harvard, told of finding half of a fossil in Newfoundland, half in Ireland. The two halves matched perfectly, therefore they must have been wrenched apart by the splitting off of North America from Europe! In the next few pages we narrow our focus to the North American geological community. Drift was more stoutly resisted here, where Permanentist or 'Fixist' views were orthodoxy, than elsewhere. It is the 'hard case' for the continued existence of Drift as a research program. A description of the period as one of headlong retreat for Drift has encouraged Kuhnian interpretations of the overall story. By eliminating any but a coincidental connection between the versions of Drift in the 1920s and 1930s and the later seafloor-spreading and plate tectonic ones, the events of the mid- to late 1960s seem all the more dramatic, sudden and revolutionary. This characterization, though not wholly wrong, is misleading. These years were a 'frozen frame' in which there were no breakthroughs in the rival programs nor were there marked shifts of allegiance.

This stasis was perhaps due in large part to World War II and its immediate aftermath. Many North American geologists were employed in the war effort (Schlee, 1973) and the immediate post-war years were not especially conducive to theoretical developments: they were a time of demobilization and transition, of rapid expansion of funding and of dramatic growth in numbers of students. It was also a time of establishing new relationships between the government and university research, of carving out new employment and research niches, of knocking the rust off research projects set aside, of absorbing data gathered during the war, and of starting up new projects in new fields. The results of these activities were only beginning to be felt in the early 1950s.[11] This was the case, for example, in oceanography, which began to attract relatively large numbers of recruits. Much of the work was carried out under the aegis of simple fact-gathering but this did not mean that oceanographers were unaware of Drift and its rivals. With few exceptions the geology of the ocean floors seems to have remained peripheral to 'global' theories until after World War II. Drifters might be thought to have had great interest in submarine geology: perhaps evidence of the 'wakes' of the continents could be found. Yet, they would have been deterred by the impossibility of doing fieldwork on the bottom of the Atlantic. Given the practical difficulties of marine geology, the few geologists who did pursue marine research were reasonable in placing emphasis on the collection and collation of scarce hard-won data. Francis Shepard's 1948 monograph, *Submarine Geology*, was the pioneer work in English in this specialty and remained the standard synthetic treatment over the next decade. Shepard (1948: 1), Professor of Submarine Geology at the Scripps Institution of Oceanography, emphasized the orderly presentation of facts rather than theorizing but included a brief section on the origin of ocean basins. There, much in the style of more general texts, he summarized and criticized various theories including Drift. The possible influence of Shepard's volume should not be underestimated: its ubiquity as a reference work and specialist text meant that many of the rising generation of students in this rapidly expanding field would be exposed to – and not necessarily inoculated against – Drift as one program among several.[12]

Bailey Willis and fairytales

The evidence most celebrated by those who have argued for the death of Drift in the 1940s is Bailey Willis's article 'Continental Drift, Ein Märchen [A Fairytale]' (Willis, 1944).[13] Willis had argued against Drift in the 1926 AAPG symposium and developed the 'isthmian links' version of Permanentism. His hostility to Drift, even as a permissible working hypothesis for other geologists, was unabated in 1944. There was no interrogative in the title of his article. He believed that Drifters unreasoningly clung to that program despite insuperable objections and alternative solutions to problems posed by Drift offered by himself, Simpson, and others (Willis, 1944: 509–13). Drift was a fairytale, it was a hypothesis which had 'conclusive negative evidence' against it and should therefore 'be placed in the discard, since further discussion of it merely incumbers [sic] the literature and befogs the minds of fellow students'. If Willis had spoken here for the Anglo-American or even the North American geological community, then an argument could be made for a lack of continuity between earlier versions of Drift and the emergence of mobilist theories in the late 1950s and early 1960s; Wegener could be cast as the prophet to whom no one listened.

Let us place Willis's attack in context. There was no need to beat a dead horse if it be a decaying corpse.[14] Ergo, there must have been some supporters of Drift around and indeed we find that du Toit had defended Drift theory and sought to rebut Simpson's (1943) attack on Drift in an earlier number of this same volume (du Toit, 1944). What could be overlooked is that Willis was responding not so much to du Toit as to Chester Longwell. Willis's article was bracketed in this volume by three of Longwell's. Longwell was one of the senior members of the American community. He was Professor of Geology at Yale, a former Chairman of the Geology Section of the National Research Council, co-author of a widely used textbook and soon to be President of the Geological Society of America. He was no warm friend of Drift (Longwell, 1928). Nonetheless, his stance toward it differed appreciably from Willis's. Longwell's first paper was a rejoinder to du Toit's.[15] However, he was not 'unalterably opposed to "drift"' and separated his rejection of specific versions of Drift from a condemnation of the whole program (Longwell, 1944a: 220):

> The Wegener hypothesis has been so stimulating and has such fundamental implications in geology as to merit respectful and sympathetic interest from every geologist. Some striking arguments in his favor have been advanced, and it would be foolhardy indeed to reject any concept that offers a possible key to solution of profound problems in the Earth's history.

Cogent arguments for Drift were the explanations it advanced for two sets of data which were problematic for its rivals: the southern glaciation[16] and pattern-matching across the Atlantic. Neither spoke unequivocally for Drift, but both 'establish the "drift" concept among the hypotheses that must serve as equipment' for the tasks ahead (Longwell, 1944a: 225–6). He scolded both Drifters who ignored criticism and anti-Drifters who rejected the theory out of hand. He confessed to being a fence-sitter who refused for the present to adopt any one global theory to the exclusion of others and who believed that geologists should retain the method of multiple hypotheses including the hypothesis of Drift (Longwell, 1944a: 230–1).

Willis explicitly addressed his article to Longwell. Willis's denunciation of Drift was apparently intended to push fence-sitters off their uncomfortable perches. If this was the case, he was unsuccessful. Longwell (1944b: 515) immediately replied that he did not share Willis's view, that he was 'not ready to brand as a fairytale the entire concept of displaced continents', and that though one could refute some arguments for Drift and show others to be inconclusive, the 'hypothesis will have value at least as a stimulant' until an explanation of diastrophism (crustal deformation) was established which excluded the possibility of Drift. At the same time that he dished up more criticism of dogmatic drifters, he chided Willis for failing to take into account the development of Drift since Wegener's day and, especially, for rehearsing the same old objections to Wegener's mechanism and ignoring convection currents. Arguments against earlier versions of Drift might not be valid against more recent ones.

Longwell's third piece (1944c) was a one-page note concerning some 1936 geodetic measurements in Greenland. These indicated no measurable displacement between 1927 and 1936. His conclusion is carefully phrased. He did not suggest that Drift be abandoned, merely that its proponents set aside their claims based on the cruder measurements on which Wegener had relied. For Longwell, geologists should ignore neither theoretical or empirical advances bearing on the rival programs. This was consistent with the position he had earlier taken when he stated that should such measurements not indicate drift was occurring, 'the hypothesis of drift would still be tenable, since movements may have been episodic, or at rates too slow to be detached within a human generation' (Longwell, 1944a: 228).

This contextualization of Willis's oft-quoted remarks makes it clear that his views were not fully shared by his fellow geologists in North America. Ironically, the very last paper in the volume (excepting an obituary notice of a palaeontologist – not of a program!) was entitled 'A New Idea on Continental Drift' in which Joseph Chelikowsky (1944), a geologists at Kansas State University, proposed yet another mechanism for Drift. What we have seen thus far is suggestive but by no means persuasive. One swallow does not a summer make nor does a Longwell or a Chelikowsky establish that Drift remained – and was regarded as – a research program which was not covered with moss.

Programmed texts

Textbooks, according to Kuhn, enshrine and provide an index of the shared beliefs of a scientific community: they display and are designed to communicate 'the basis of the current normal-scientific tradition'. Indeed, in Kuhn's view, the 'domination of a mature science by such texts significantly differentiates its developmental pattern from that of other fields' (Kuhn, 1970: 136–8). This latter claim may be overstated but textbooks may be a useful gauge of shared beliefs. We would expect that a textbook written from a highly idiosyncratic viewpoint would be regarded by publishers and professors alike as unsuitable for use in introductory classes. This is likely to be all the more so for general texts in which the authors attempt to transcend their own research interests to present a synthetic account of the whole discipline. If this be correct and if Drift had been banished from polite society in North America in the 1940s and early 1950s; if this were a period of 'normal science' guided by one paradigm, one would expect to find among texts published in those years a high degree of uniformity on theoretical issues and no mention of Drift.

A study of a selection of these texts fails to substantiate this. Geologists often looked askance at speculation and theorizing and few introductory texts claimed to do much more than lay out 'the facts'. Even so, Drift was discussed. It might sometimes receive only a few sentences, but it was covered in over two-thirds of the 51 texts checked; fewer than one-third made no reference whatsoever to it.[17] Texts which were silent on Drift[18] usually gave a descriptive, encyclopaedic, relatively atheoretic introduction to the discipline and eschewed explicit discussion of any geological theories, including Drift. This probably reflected a natural history tradition which was rather prominent in North America in the first half of the century (R. Laudan, 1982b).

Authors of texts which discussed theories often presented several competing programs or theories, noted their relative merits and demerits (usually in terms of providing alternative solutions to specific problems) and concluded that much more empirical work was required before definitive pronouncements could be made. Few expressed either great enthusiasm or open hostility.[19] Over three-quarters included relatively neutral summaries of variable quality, length and content. As far as one can trust the textbook evidence, the polarization over Drift was rather less than might have been expected from Willis's diatribe. Longwell's own textbooks on physical geology (Longwell, Knopf & Flint, 1941; 1948; Longwell & Flint, 1955) included brief summaries of Drift and other theories of crustal deformation. Against Drift he cited the problem of mechanism but added a mention of convection currents as 'one hypothesis that has received considerable attention' (Longwell *et al.*, 1948: 477–8). Carey Croneis and William Krumbein's[20] general geology text, first published in 1936 had by 1966 gone through 18 printings without revision! It was pitched at a much lower level than many. Even so, they gave a very brief outline of Drift, described as 'stylish in Europe today', complete with a depiction of Wegener's famous maps. A passing reference was made to the 'jigsaw fit', pattern-matching, and species distribution; rather more attention was paid to the problem of mechanism and to geodetic measurements apparently contradicting Drift (Croneis & Krumbein, 1966: 288–9). Despite a generally hostile tone, there was no suggestion that the theory should be abandoned. We generalize from texts in this group[21] that references to Drift usually occurred in relation to specific problems to which it seemed to afford possible solutions: the great southern glaciation, pattern-matching across ocean basins, the fit of opposing coasts and, least frequently, the distribution of fossil and living species. By far the most commonly mentioned objection was the inadequacy of the mechanisms proposed, though several saw promise in convection currents. This textbook material perhaps reflected a widespread sentiment among members of the North American geological community that some version of Drift was worth retaining as a working hypothesis and as a rival to other approaches – despite its admitted difficulties – because of its promise with respect to certain groups of problems and its demonstrated ability to stimulate research on the part of both its opponents and proponents. Charles Nevin (1949: 291), a structural geologist at Cornell University, expressed this attitude concisely:

> Whatever may be the final outcome of the many theories of continental drifting that have been proposed, the fact remains that the conception of large horizontal movements of the crust is valid. In any discussion of the tectonics of the crust no one should ignore the possibility of both inter- and intra-continental drifting.

There were no major theoretical advances or changes in attitude toward Drift in the decade after the war. As Wood (1985: 111) describes it, it was a period of trench warfare. Although the battle-lines might not have shifted in the period, I think it wrong to assume that Drift had receded to the fringes of geological consciousness. If nothing else, the occasional exchanges of volleys would have kept the audience awake. Moreover, the proponents of the contending programs found fresh stores of ammunition provided by recent research though it is true that the initial supplies of bullets seemed to fit the guns of Drift's opponents better.

In 1949 the Society for the Study of Evolution sponsored a symposium on 'The Problem of Land Connections Across the South Atlantic' (Mayr, 1952a). The subject matter was palaeobiogeography; the principal programs under review included Drift. The arguments for and against Drift differed little in kind from those of the 1920s and 1930s. The noticable change was that more species were fired back and forth. There was no unanimity. Of the 14 speakers only one, Ken Caster (1952), strongly favored a Drift interpretation. Two favored land connections between Africa and South America but were vague as to whether or not Drift was required to provide them. Three adopted a position of studied neutrality. Theodore Just,[22] for example, commented that 'modified drift theories continually attract favorable comment from many paleobotanists' (Just, 1952: 198) but added that Drift was no panacea and that much further work was needed before a judgement could be made. Six made no reference to Drift but presented their work within a Permanentist framework. Maurice Ewing, who that same year was setting up the Lamont–Doherty Geological Observatory in New York which was to become a leading institute for marine geology, gave an overview of recent work in submarine geology with the implication that the results told in favor of a rigid ocean floor and thus against Drift. Only two, including Ernst Mayr (1952b),[23] the organizer of the symposium, specifically rejected Drift. Walter Bucher,[24] a Permanentist and an implacable foe of Drift, argued the familiar line that Drift was physically impossible and therefore concluded 'the concept of continental drift cannot be used as a working hypothesis by the student of animal and plant distribution' (Bucher, 1952: 101).

Drifting around

American geologists were, of course, not isolated from the world-wide geological community. Though there may have been few active proponents of Drift in North America, 'foreign' Drifters on occasion visited and spread their curious ideas; Gutenberg and a few others actually took up residence there. Lester C. King came to the US from South Africa as the AAPG Distinguished Lecturer in 1951–1952. As he spoke to various groups of the AAPG around the country, he observed that there was 'much greater interest evinced in the hypothesis of continental drift than had been expected', (King, 1953: 2163). Though he commented that Drift received little attention in the curriculum and that many in his audience professed ignorance of the program, they also expressed 'a desire for information on which they could base a judgment for themselves'. His espousal of Drift seems not to have been greeted with overt hostility, and he remarked (perhaps optimistically) on the 'new spirit' of 'impartial inquiry and broad outlook' (King, 1953: 2163). In the course of his travels he engaged in a debate with Bucher at Columbia and purportedly won support among the students there (Wood, 1985: 106). American journals gave space to the arguments of Drifters, both home-grown and imported. King followed the lead given by du Toit in defending Drift in print as well as in person. He argued in 1953 in the AAPG *Bulletin* for the 'necessity for continental drift' on the basis of extensive pattern-matching evidence, including new correlations between South Africa and Antarctica (King, 1953). 'Foreign' publications which contained discussions of Drift also found their way into the hands of American geologists. Were we to confine ourselves to the Anglophonic world, it is clear that many American geologists continued to be exposed to Drift. Their cousins across the Atlantic were one carrier of the infection, for British geologists continued to be rather less hostile to Drift. S.E. Hollingsworth (Bullard, 1964: 27) remarked that when he began teaching physical geology in England in the late 1940s, the available texts in Britain were very sympathetic to Drift.

Some American geologists, while themselves not advocates of Drift, publicly stated in texts and research papers that it had a place as a working hypothesis and that it offered interesting solutions to some difficult problems. It was, in terms of actual proponents, very much a minority view in the American geological community but it had not been extinguished and some did use it to inform their teaching and research.[25] Drift, relative to its rivals, was neither forgotten nor gathering moss. However, in the absence of new and striking evidence or of new versions of the program, Drift was unlikely to make much headway against its rivals. It was 'revived' in Britain and North America in the 1950s and 1960s. The revival was promoted by new lines of investigation, new types of data unknown to Wegener or his immediate successors and also new versions of Drift. The revival was not simply a matter of the random collection of new facts. It owed much to an awareness of the existence of Drift as an alternative global theory which might co-ordinate, interrelate and give significance to this new material. It is likely that some geologists were not introduced to Drift in their education or in their later professional practice but it would be misleading to exaggerate the novelty of Drift in the late 1950s and 1960s and to suggest that there was no connection between the earlier versions of Drift and later ones which did win acceptance. The 'victory' of Drift in the 1960s was dramatic but it should be seen as the climax of a protracted competition among rival programs.

Voice-Over

Geologists' evaluations, selections and modifications of research programs were not a matter of comparing theories with data.[26] All three of the major programs underwent considerable modification. Not only were 'facts' elastic and changeable, so were the programs themselves. The judgments which geologists made were comparative. Appeals were made to the 'facts' of the matter and to the importance of gathering more 'facts', but the context was what were perceived to be the *relative* merits of competing programs. The major criticisms and problems each program faced were principally conceptual, not empirical. The major changes proposed in each program came in response to these conceptual difficulties and a premium was placed on solving them partly because they had been solved within one of the rival programs.

'Programs' and 'theories'

Despite changes in the programs, each retained coherence and identity. Geologists could distinguish among Wegener's, Daly's du Toit's, Gutenberg's and Holmes's 'theories' but classed them all as lying within the Drift 'program'. The same could be said of Jeffreys's version of Contractionism or of Matthew's, Willis's and Simpson's versions of Permanentism.[27] Kuhn regards as part of the theoretical activity of normal science the articulation and reformulation of the paradigm under which that science is practised. He does not, however, spell out the relation of the paradigm to the specific theories which form part of its structure. Moreover, he appears to conceive of paradigms as relatively inflexible and unchanging, at least at the theoretical level. Articulation and reformulation amount to restatements of the paradigm in an equivalent but more coherent, less equivocal and more aesthetically satisfying form (Kuhn, 1970: 32–3). This limited view does not seem to capture the relationship of, say, Holmes's version to that of Wegener or, as we shall see later, of those to the versions of Drift proposed in the 1960s. Only in the loosest sense could it be claimed that Holmes's Drift theory is equivalent to Wegener's: in each there is lateral motion of the continents, but they differ not only in matters of detail and emphasis but also on more crucial points such as the constitution of the interior of the earth and the actual mechanism of continental motion.

Lakatos and Laudan both specify how theories are related to larger conceptual structures: respectively, 'Research Programmes' and 'Research Traditions'. I have referred to such larger structures as the 'research programs' or 'programs' or 'global theories' of Drift, Permanentism and Contractionism. A Programme or a Tradition is an explicit set of assumptions which is embodied in a succession of specific theories.[28] With reference to the Drift program, the theories of Wegener, Daly, Holmes and so on may each be regarded simultaneously as specific instantiations of that program and proposed changes in it. The development of Drift thus occurred partially through the construction of new theories lying within the program. Some of these were taken up by other Drifters, most were not. These theories were connected with one another and with Drift both conceptually and historically. Conceptually, each presupposed and was justified by the larger context of Drift. Each represented an attempt to improve the program by solving or dissolving problems that the program confronted. Historically, each geologist who proposed one of these versions set it in the context of Drift as the preferred global theory, made reference to the merits and demerits of Drift and its rivals and claimed that their version marked an improvement over others then available.

There is some correspondence between the evolution of Drift and its competitors and the views of Lakatos and Laudan on the nature of programs but there are also discrepancies. Lakatos, for example, argued for a tight logical connection between the successive theories making up a Programme; specifically, that each new version should include all the features of the previous version plus some new ones: there is an accumulation from one version to the next. This does not seem to apply: du Toit's postulation of Laurasia and Gondwana instead of a single Pangaea, Holmes's reliance on convection currents and his dismissal of Wegener's propulsion system for continents, Willis's proposal of temporary isthmian links in a Permanentist program which had resolutely rejected transoceanic land connections[29] – all involved not only the addition of new ideas or data but also the scrapping of portions of older versions of programs. There was addition but there was also elimination as new versions were put forward. This is even more apparent in some of the later versions of Drift. Both Kuhn and Laudan allow for 'losses' as well as 'gains' in scientific development.

Comparing research progams

Most geologists did not join in debates over global theories nor explicitly accept one of those programs. Chester Longwell commented on this as follows (Longwell, 1944a: 220; *cf.* Menard, 1986: 86): '[Most geologists are] emphatically not uniform in their thinking on these problems. They find much in their geologic work to occupy their attention aside from the conflicting hypotheses on the history of continents.' Those who did join in did not make absolute judgments about global theories solely on the basis of empirical data. They reached their decisions to accept or entertain or employ or pursue or reject a program through a comparison of that program with others as well as with the data which they regard as relevant. Willis did not demand the death penalty for Drift purely on the basis of empirical matters but also because in his opinion another program was far superior. Jeffreys not only presented a new version of Contractionism, he also attacked Drift; du Toit not only worked to shore up Drift, he also criticized Permanentism and Contractionism; Holmes's version of Drift was placed in the context of a rebuttal of Contractionism; Simpson in arguing for Permanentism castigated Drift and Contractionism. Drift, even in textbooks, was rarely if ever appraised simply on the basis of empirical data; it was appraised with reference to its rivals. Most geologists, in both textbooks and research literature, seem to have adopted a position of agnosticism. They claimed that the empirical base was simply inadequate thus far to single out one program or theory from all others.

Given the many lacunae in geological knowledge, the ability of a hypothesis to suggest novel lines of research or to suggest connections amongst apparently unrelated phenomena, was sometimes taken into account in judging among rival hypotheses. Hess (1938: 75) epitomized this attitude in endorsing the convection current explanation of trenches. It is not discordant with the method of multiple working hypotheses in which hypotheses serve as compasses to direct research, a dictum to which many geologists still at least tugged their forelocks.[30] I do not suggest that an individual scientist might simultaneously have used competing programs or theories; rather, that different scientists used different, competing programs and theories. However many did entertain several programs. Theoretical pluralism at the operational level is, so to speak, a group characteristic rather than an individual one. The practice followed in many texts of summarizing and critiquing contending research programs no doubt owed much to what were perceived to be huge lacunae in empirical data but it also echoed this dictum. A methodological rule may well be more than a rhetorical flourish; in this case it may have been constitutive of the actual instruction in and practice of a discipline. Evaluations of Drift and its rivals were sometimes guided by specific presuppositions about what criteria theories – or, to use their language, 'working hypotheses' – should meet in order to be acceptable, and occurred most often in the context of comparative evaluations.

Despite obstacles which may have been presented by specialization, regionalism, lack of data and perhaps conflicting technical and social interests, geologists were able to make some near-unanimous decisions. For example, Wegener's version was abandoned by 1940. It was flatly contradicted at both the 'theoretical' level, by physicists and geophysicists who computed that the magnitude of his forces was grossly insufficient, and at the 'empirical' level by the later Greenland measurements carried out under the direction of N.E. Nörland (Director of the British Geodetic Institute) in Greenland and Denmark in 1936 in association with the Danish Geodetic Institute (DGI). The comparison of longitude measured by the DGI for a site in Greenland in 1927 with that for 1936 revealed no appreciable westward movement (Nörland, 1936; summarized in Longwell, 1944c). Nonetheless, the program was seen to have some 'pluses'; i.e., some possible solutions to outstanding problems. This agrees with Laudan's and Kuhn's views that a 'plus' for a program is that it can solve some of the problems facing its rivals

(Kuhn, 1970: 152; L. Laudan, 1977: 18, 27). On the other hand, Lakatos's (1978: 185–6) contention that appraisal should be based only on empirical data anticipated by a program; i.e., predicted by and indeed recognizable only in the context of that program, finds little support in the comments of geologists. Moreover, even opponents of Drift admitted that it was 'seductive' in its non-empirical characteristics: it was simple, clear, self-consistent, elegant, bold, sweeping, and easily visualized. Such comments provide evidence for the importance of what Kuhn (1970: 155–6) terms aesthetics and also with what Laudan describes as consistency, coherence and clarity (L. Laudan, 1977: 49).

If evaluations be comparative and relative, how does this apply to the succession of theories which make up those programs? In evaluating the specific versions, geologists took into account the 'track records' of the programs with which those versions were associated. Holmes's theory was judged on how well it answered objections to Wegener's mechanism, not as an independent global theory. This was the way also in which Holmes presented his version: he did not rehearse all the evidence or arguments bearing on Drift or its competitors, he confined himself to several specific issues. This is in general agreement with Laudan's views (L. Laudan, 1977: 82) and is not incompatible with a Lakastosian or an 'interest' or 'internal struggle' model. Those who were well-disposed or at least neutral to Drift were more sympathetic to Holmes's version than were those who adamantly rejected Drift. After all, if Drift as a general approach were unacceptable, why would this be different for yet another version? Holmes, of course, tried to distance his version from Wegener's and to claim that the former was an improvement on the latter. This carried more weight for scientists who shared Holmes's conviction that Drift offered a promising approach despite the flaws in Wegener's account than for scientists who flatly rejected any large-scale lateral motion of the continents.[31]

In evaluating programs some geologists focussed on the most recent versions of them: Holmes's criticisms of Contractionism were aimed primarily at Jeffrey's version, not the older ones; Lake commented that Gutenberg's version, though still highly problematic, was 'impressive' and 'more consistent than the lawless wanderings imagined by Köppen and Wegener' (Lake, 1933: 121); Jeffreys admitted that Holmes's version was not open to the same objections as Wegener's even though he continued to pour scorn on the latter; and a few geologists who entertained Drift admitted that Holmes's version made Drift more plausible. This seems the reasonable procedure to follow: the most recent versions presumably reflected the 'best case' for the program and met some of the empirical and conceptual criticisms of earlier versions. Indeed, a later version might be quite inconsistent with one or more of the earlier ones: Holmes's was inconsistent in numerous respects with those of Wegener, Daly, and du Toit; Jeffreys's with Suess's; Willis's with Dana's (*cf.* L. Laudan, 1977: 81, 85; Lakatos, 1978: 57). However, this was not a procedure followed universally. In the 1930s, 1940s and 1950s most of those who dismissed Drift referred to the inadequacies of its mechanism as proposed by Wegener and paid little or no heed to the later versions.

The importance of conceptual problems

Conceptual problems were important to geologists not only in comparing the merits of rival programs but also in suggesting lines of research within them. For Wegenerian Drift, there were conflicts with what were taken to be well-founded theories of physics and geophysics; for Contractionism, physics and geophysics; for Permanentism, biology. These conceptual problems could be characterized as inter-disciplinary conflicts. Additionally, for Wegener's version, there were conceptual problems arising from a conflict between his apparent methodology and what many took to be the appropriate methodology of geology (*cf.* Le Grand, 1986a) and between his view of the aims and definition of the discipline and those held by many British and American geologists.

The debate over Drift underscores the inter-relationship of scientific disciplines. It may be that scientists in a discipline can sometimes function in serene isolation from those in others. In the case of Drift, however, the standards, theories, methods and techniques of other disciplines, especially physics, were a resource for opponents and proponents. Wegener drew on meteorology and physics; Joly and Holmes used the standards, methods and theories of physics to call into question accepted views about the earth. Physics could also be used as a weapon against Drift. L. Laudan (1977: 50–7) gives a taxonomy of relationships between theories from different disciplines: entailment, reinforcement, compatibility, implausibility and inconsistency. In the controversy we have sketched there was a fair measure of agreement that Wegener's mechanism seemed at least implausible if not inconsistent with physical theory. If we took Wegener's version of Drift to mean that continents of solid soft rock simply push through ocean floors of solid hard rock, any conceivable mechanism would conflict with physics: no small force acting even over millions of years could thrust a stick of butter through a steel plate to (to exaggerate somewhat: Drifters did not advance such a simple account)!

The seriousness of some conflicts derived in part from the perceived relations of status and authority between physics and geology. For those who postulate a common aim or an ideal method (or both) for all the sciences, physics has often been the exemplar. In the nineteenth century Auguste Comte expressed this view:

> There are two kinds of natural sciences. The abstract and general have as their object the discovery of the laws governing the different classes of phenomena, in every conceivable case; the concrete, particular descriptive, sometimes called the 'natural' sciences apply the said laws to the natural history of existing beings. The first are fundamental . . .; the second, whatever be their intrinsic importance, are truly secondary. . . .

Physics fell in the first class, followed by chemistry and biology. Geology was of the second class. Physicists had need only of mathematics but geologists, to unravel and describe the past history of the earth, had to draw upon and apply not only physics but also chemistry and biology. Physics dealt with abstract phenomena; geology, with the messy complexity of the observable world. Physical theories were predictive and verifiable, those of geology were retrodictive and descriptive. This conception of the sciences has figured and continues to figure in scientific disputes and philosophical analyses of science; for example, in the debates over biological reductionism and in positivist philosophies of science. Hierarchical views of the relationship of the sciences may, of course represent the historical evolution of the different disciplines and our cultural values associated with 'pure' versus 'applied', 'abstract' versus 'concrete', 'explanation' versus 'description' and so on, rather than some intrinsic order within science. Although such disciplinary hierarchies may lack cognitive foundation, this does not mean that they played no role. For scientists who accepted such a view, physics provided the constraints for acceptable geophysical theories and geophysics in turn gave constraints on acceptable geological theories. A conflict between physics and a global geological theory thus posed a very grave objection to the latter. The pronouncements of the 'pure' physicist Jeffreys would perhaps carry more weight – inside and outside geology – than those of the geophysicist Holmes (and Holmes's more than those of the geologist du Toit).

There were and are views on the relationship of the sciences other than hierarchical ones. Disciplines may not be purely intellectual divisions but also social ones which have occurred as a consequence of a struggle for authority within science (e.g., Sapp, 1987). Disciplines could be products of institutionalized patronage (*cf.* Kohler, 1982). Or, we might reject the assumption that there is one aim and one method for all disciplines. After all, this assumption sails perilously close to the reefs of recent history and philosophy of science which have wrecked the old discourse on The Scientific Method. There may be a plurality of aims and methods among disciplines and within a discipline. Axiological and methodological pluralism may be as normal and desirable as theoretical pluralism.

Pantin (1968: 5) suggests that a more appropriate categorization of the sciences may be into those that are 'restricted' in scope and number of variables (e.g., physics) and those that are relatively 'unrestricted' (e.g., geology). The former are more abstract and exact but they exclude 'much of the grand variety of natural phenomena'. The latter are neither purely descriptive nor derivative. Their scope is broader, their objects of study are immensely varied but so too is their range of aims and methods. Geologists, for example, draw upon the methods and theories of other sciences not because geology is somehow 'inferior' to those sciences but because its scope it broader, its objects of study more varied, and its aims different (Pantin, 1968: 9–20). Kitts (1977: 98–9, 117–9) takes a slightly different tack. He leaves aside the question of whether or not geology should defer to physics, but as a matter of description states that geologists have always deferred to physics by accepting physical theory as the unproblematic background which constrains their historical explanations. Geology does not aim at general laws or theories; as he puts it (Kitts, 1977: 99): 'historical knowledge is the goal and theory is the means of achieving it'. Even for scientists who did not accept a disciplinary hierarchy and who may have argued for some autonomy for geology, the conflict between physics and Drift – or between physics and Contractionism – posed a serious conceptual problem.[32]

Conceptual problems were raised again and again in discussion of the rival programs. So too were more narrowly empirical problems but given that the data seem often to have been ambiguous and open to different interpretations these do not appear to have been as decisive as the conceptual ones in forming attitudes to the programs.[33] Specialization and localism may bear on this point. Granting that 'empirical' matters are not theory-free, nonetheless a recognition of empirical problems often depended on a detailed knowledge of a particular region or a grasp of the techniques and data within a particular specialty. The significance assigned by a geologist to specific empirical problems likely depended on specialist interests. In this sense, empirical matters were limited in scope unless, as in the later Greenland measurements, the empirical claims of the theory or program were sharply defined and were 'independently testable'; i.e., did not rely in an obvious way on assumptions of one or another of the programs. Even in this case the 'null result' of the later Greenland measurements was taken to be strong evidence against Wegener's version but not against the Drift program. Conceptual problems, though sometimes reliant on specialized knowledge, were far more accessible to a broader range of geologists. *Prima facie*, most geologists could readily grasp the implications of radioactivity for a contracting earth; would regard the notion of continents crunching through solid rock as improbable; would be acquainted with evolutionary theory and how it bore on biogeography; and so on. Conceptual problems of the sort which exercised geologists in the 1920s, 1930s and 1940s rose above narrowly specialist or regional concerns and seem to have had greater currency – and force – in the discipline. These could be construed as conceptual anomalies for a program which had as yet failed to solve them. As we have seen, some researchers lavished much effort on removing these anomalies. This indicates the importance attached to anomalies in the comparative evaluation of rival programs and as foci for further research. It goes against Lakatos who, in respect of empirical anomalies, takes the view that scientists should freely ignore anomalies (Lakatos, 1978: 36, 50–2) provided their programs are 'progressing'. This was not the case for empirical anomalies (using Laudan's definition of anomaly) or, *a fortiori*, for conceptual problems.

Social and institutional pressures

Scientists are not, of course, disembodied intellects. We could, for example, take the view that Holmes's development of a new version of Drift and his attack on Jeffreys was not only a reflection of his specialist interest in radioactivity and geophysics but also a defense of his own credibility and position in the scientific community. In his case, 'social' and 'scientific' interests merged and contributed to his pursuit of a relatively heterodox program. It is probably more often the case that social and institutional pressures encourage conformity to established views. In North America, for example, where Drift had few supporters, it would be daunting to a young geologist, much less a student, to argue openly for Drift.

We have given for the most part a cognitive account of the apparent reluctance of scientists in North America to employ, pursue or accept the novel program of Drift: after an initial burst of interest in Drift, its proponents seemed to make little progress in meeting the major problems that it faced and, moreover, workers in the rival programs were not standing still. We could go so far as to say that from a North American perspective, not only did Drift seem of little use, but at both theoretical and empirical levels the program seemed to be stagnating. This would presumably make it less attractive, therefore fewer geologists might choose to work within a Drift approach. The upshot could be, unless there were very able workers in the program, that it would stagnate or even wither. This is complemented and reinforced by a social account. Younger geologists might well believe that advocating Drift could jeopardize their chances for academic employment, support for their research and publication of their results. Some proponents of Drift have later claimed that such a situation did exist. Carey reminisced (1976: 9) that in the 1940s and 1950s, 'Although any loose statement denigrating or mocking continental dispersion got easy passage and approval for publication, any who was unwise enough to argue for displacement of continents was cold-shouldered by referees and editors and became the butt for snide comments'. Nonetheless, even in North America, Drift as a research program was not dead in the 1940s and early 1950s, rather it was in a phase of stagnation and, possibly, decline.

Drift theory: a Pretender to the throne

From 1939 through the mid-1950s Drift, at least in broad outline, would have been familiar to many geologists and their students. They would have encountered it in introductory and advanced textbooks, in the odd lecture, and in some of the journal literature. It may not have been advanced with enthusiasm, at least by American authors, but it was advanced on grounds, which by their standards were rational ones, as a possible program which did have a small following. For senior geologists and for recent graduates alike, in the late 1950s and early 1960s Drift was not a rediscovery or a reinvention; it had never been cast aside. One might be fanciful and liken Drift in the 1940s and 1950s to be a pretender to the throne: always lurking in the background, usually able to attract a small but dedicated following, never completely absent from public memory or political calculations, always ready to assume the throne if the occasion should present itself. If this is too fanciful an image, then simply think of Drift as a rolling stone which gathered no moss.

Beginning in 1954 research workers in paleomagnetism began to make active use of Drift in the interpretation of their results. A number of specialists in this area took up Drift as a preferred program and cited the compatibility of the new data with Drift as opposed to permanentist theories. This 'unfroze the frame', so to speak, and the rest of the story up through the late 1960s is almost a 'fast forward'.

Notes

1 There may have been some urgency in this task. As a novel approach Drift may have been judged more leniently than otherwise. It had shown initial promise in such problems as species distribution and the great southern glaciation. If that promise was not fulfilled and if its supporters made no headway in meeting

empirical and conceptual objections to it, the very passage of time could result in a hardening of opinion against Drift.

2 Coleman had field experience in Canada before joining the University of Toronto. He was later to reject 'confidently' Wegener's proposed solution to the great southern glaciation (Coleman, 1925).

3 We might think of this as 'Yankee (Canadian?) aggression' at a biological level! Matthew was not the first to assume the superiority of northern species over tropical ones: Montesquieu had taken this view in the 18th century and it was a staple belief in the 19th century.

4 According to Drifters, the continents had separated long before the Cenozoic. Cenozoic mammals were not an obviously appropriate group to use in evaluating Drift.

5 For a comprehensive account of this episode, see Burchfield (1975); for a brief overview, Hallam (1983: 82–109).

6 Joseph Barrell, Professor of Structural Geology at Yale University, responded by incorporating radioactive heat into the Contractionist program in an effort to solve the problem posed by isostasy (Barrell, 1917; 1927). The lighter crust could, he claimed, be sunk by the accumulation on it of denser matter pushed to the surface from deep within the earth. The process was driven by heat generated from radioactive decay.

7 There are parallels with William Hopkins (1793–1866), a famous mathematics tutor at Cambridge, who sought to extend the rigor and precision of mathematics to geology. He opposed uniformitarianism and favored a contracting developing earth. He pioneered a British tradition of mathematical geophysics which influenced Kelvin and which was carried on at Cambridge by, among others, George Darwin (Plumian Professor of Astronomy until 1912) and Jeffreys (see Brush, 1979). This tradition and its institutional base merit further study.

8 Brooks, a meteorologist with the British government, favored Contractionism. To explain the great southern glaciation, he proposed (Brookes, 1922; 1926; 1949) that the old Gondwana continents had undergone elevations of thousands of feet. That such elevations might produce glaciation was admitted by Drifters (Holmes, 1929: 340) but the problems of how this massive elevation could have been produced and how it might have vanished without trace, were even more stupendous than the palaeoclimatic problem the hypothesis was intended to solve!

9 Simpson had a specialist interest in glaciology; he later proposed that recent ice ages resulted from an increase in solar radiation.

10 His sympathy for Drift did not prevent his election in 1932 as President of the Geological Society of America.

11 The archives of the Scripps Institution of Oceanography would provide the basis for an interesting micro-study. See, e.g., the Roger Revelle papers and the Records, Office of the Director 1943–1955.

12 Revelle in 1951, his first year as Director of Scripps, used Shepard as a text in his marine geology seminar and posed a set of questions on the origin of ocean basins which dealt with both Permanentism and Drift (Revelle Papers: SIO, box 24 folder 34). Kuenen (1950) was also used, often in conjunction with Shepard, and although his text was more openly hostile to Drift it did devote several pages (Kuenen, 1950; 125–9) to it. Zeuner's (1946; 1950; 1952) monograph on

geochronology filled a similar role in that specialty but was more openly favorable to Drift.

13 Wood (1985: 109), for example, refers to Willis's 'small town book-burning zeal' and implies that his categorical rejection marked a thorough-going American hostility to Drift.

14 Wegener's nationality seems to have played no more of a role between 1939 and 1945 than it had in the previous twenty years.

15 Wood (1985) refers only to this paper of the three in discussing the episode and then only to certain of Longwell's remarks critical of Drift.

16 Frankel (1984b) correctly adduces that other solutions for this glaciation were put forward. I disagree with his claim that the importance of Drift's solution was consequently decreased. In both research and textbook literature this problem–solution was frequently mentioned in the 1940s and early 1950s as one of the stronger if not the strongest arguments for Drift, suggesting that the proposed alternatives (e.g., Brooks's) were not held in high regard.

17 I have examined nine texts first published in the period 1935–1939 which were reprinted in the ensuing decades; 31 texts first published or issued in new editions in 1940–55, and, for purposes of comparison, 11 texts from 1956–60. There seem to be no significant variations among the three groups with respect to Drift. These texts are introductory texts in general geology, geomorphology, and physical geology published or co-published in the US and some 'British' texts which were widely distributed in the US; e.g., Holmes (1944) and Jeffreys (1952) even though the latter could scarcely be termed 'introductory'.

18 13 of 38 for 1935–1955; 14 of 49 for 1935–60.

19 One pro-Drift and four anti-Drift in 1935–55; one pro-Drift and three anti-Drift in 1956–60.

20 Croneis and Krumbein were geologists at the University of Chicago. Their rehearsal of various theories fit the 'method of multiple hypotheses' of T.C. Chamberlin, the late founder of the Chicago department, to whom they paid tribute.

21 17 others for 1935–55; four others for 1956–60.

22 Just was an Austrian-born palaeobotanist who had come to the US in 1929. He was a curator at the Chicago Natural History Museum.

23 Mayr had studied zoology in Berlin before emigrating to the US in 1931. Curator of the American Museum of Natural History, he was a noted exponent of the modern evolutionary synthesis. His opposition to Drift probably owed much to G.G. Simpson's work in biogeography.

24 Bucher was an American geologist but had received his Ph.D. at Heidelberg. He was at this time Professor of Structural Geology at Columbia University, an eminent specialist in tectonics and orogeny and President of the American Geophysical Union.

25 Nitecki *et al.* (1978) conducted a survey of 293 North American geologists active in the 1940s through the 1970s. They report that of the 215 respondents, some 7% recalled that they supported Drift prior to 1940; another 22% had come to share this attitude by about 1960. These results should be treated with caution since scientists, no less than historians, may reconstruct the past in light of the present; nonetheless, these results accord with the other lines of evidence and argument we have developed.

26 The data were, as we have previously observed, theory-contingent. Facts took their meaning and significance from a theoretical context. Most modern views of science deny a sharp fact–theory distinction. Lakatos may at times seem to maintain such a distinction. However, he explicitly states that there is no 'natural' demarcation between observation and theory (Lakatos & Musgrave, 1974: 99). The distinction is one adopted by 'convention' (Lakatos & Musgrave, 1974: 107–108).

27 There were many versions suggested other than those discussed above; e.g., for Permanentism, Bucher (1933) and Umbgrove (1942; 1947); for Drift, Grabau (1939), Chelikowsky (1944) and Hills (1947); and for Contractionism, Hobbs (1921), van Bemmelen (1939; 1954) and Stille (1939). There was also an embryonic program based on the expansion of the earth (cf. Carey, 1976: 24–7).

28 Kuhn argues that many of the guiding assumptions of a paradigm are not made explicit (Kuhn, 1970: 42–51) except perhaps during 'crisis'. Many of the geologists engaged in the global debates were explicit about aims, methods, the relationship of theories to evidence, preferred techniques, standards and so on. Kuhnians could argue that this was atypical and indicative of either a 'crisis' or 'preparadigm' state but, if so, one would presumably have to argue that this had been true of geology for most of the time since the late eighteenth century. Lakatos (1978: 47–50) and L. Laudan (1977: 86) claim that guiding assumptions are explicit at the initiation of the program.

29 Willis's version came close to contradicting the central tenet of permanence of continents and ocean basins. He appealed to the same forces which produced mountains to raise up his isthmian links but he did not specify the nature of those forces. The geosynclinal model of mountain-building which involved the compaction and uplift of sediment seems to fit poorly with submarine mountains formed of oceanic crust. Contractionists might find little to cavil at in his version.

30 Pyne (1978); Rachel Laudan (1980b) includes a discussion of how this dictum may have functioned in the research of J. Tuzo Wilson. I would interpret his work as illustrating the 'consecutive' pursuit of several 'hypotheses' rather than a 'simultaneous' pursuit of them.

31 Holmes was impressed by the early promise of Drift but he once confessed to always having had some deep reservations about Drift even though he actively pursued it (Holmes, 1953: 671):

> despite appearances to the contrary, I have never succeeded in freeing myself from a nagging prejudice against continental drift; in my geological bones, so to speak, I feel the hypothesis to be a fantastic one. But this is not science, and in reaction I have been deliberately careful not to ignore the very formidable body of evidence that has seemed to make continental drift an inescapable inference.

32 Following L. Laudan (1977: 56), a conflict between theories in physics and geology should be symmetrical in that it creates a problem for both domains. Kitts would disagree: if there be a conflict, neither geologists nor physicists challenge physical theory; instead, geologists reformulate their historical accounts. Kitts goes so far as to deny that a 'revolution' in a Kuhnian sense took place in the 1960s and 1970s on the grounds that what changed was not a theory but only a descriptive, particular, historical account (Kitts, 1977: 119–21). Physics would have to change for there to be a true revolution. R. Laudan (1980a) gives a persuasive rebuttal of

Kitt's characterization of the attitude of geologists toward physics, of his claim that the only 'theory' in geology is that which is taken from physics and, more specifically, of his view that the absence of a suitable mechanism for Drift was a 'knock-down' argument against it. In our case, the conflict for the participants was not symmetrical in so far as many – but not all – gave primacy to physics over geology.

33 In the case of Drift we might well wonder how much further empirical data, for example, systematic and highly detailed data related to pattern-matching across the Atlantic, would have to be gathered to 'prove' Drift. Even if a perfect geological correspondence between opposing coastlines had been established, would it have been unreasonable for geologists to continue to resist Drift?

7

Palaeomagnetism, Drift and polar wandering

A flood of new data, techniques, problems and theories in the 1950s and 1960s reawakened interest in Drift and posed fresh challenges to the dominant programs. Palaeomagnetic studies (studies of the earth's ancient magnetic field as recorded in geological deposits) made the most immediate impact.[1] Many who took up palaeomagnetism in the 1950s and helped reshape its techniques and aims came to it from geophysics rather than from more traditional geological specialties. Some of the key researchers had been exposed to Drift, read widely in the Drift literature and interpreted their results as evidence for Drift.

Palaeomagnetic phenomena had been noticed in baked clays, lava flows, rocks and even pottery long before Wegener's proposal of Drift. In the nineteenth century, for example, it was observed that lava beds were magnetized roughly in accord with the earth's magnetic field. The presumption was that this magnetization was due to the alignment of iron in the molten lava with the earth's magnetic field and that it was 'locked in' as the lava cooled. A similar mechanism presumably operated for other igneous rocks. This 'thermoremanent magnetization' could be measured in the laboratory with some precision with respect to direction and, to a lesser extent, with respect to inclination.[2] For historic times it seemed to be stable, though samples occasionally had an orientation to the earth's present field.

Paul L. Mercanton, a meteorologist at the University of Lausanne, suggested in 1926 that palaeomagnetic studies might have some bearing on Drift (Mercanton, 1926a, b). He examined basalts from Greenland, including samples gathered by the Danish expedition in which Wegener had participated, and also samples from the Southern Hemisphere. Some of the northern samples had a 'southern' inclination; some of the southern samples, a 'northern' one. This he took as evidence for a reversal of the earth's magnetic field at the time these rocks had been formed (Mercanton, 1926a). If there were a connection between the magnetic poles and the geographical poles – the earth's axis of rotation – this implied that the earth's geographical poles had undergone considerable movement in the past. A world-wide palaeomagnetic survey comparing present and former locations of the earth's poles, he believed, might 'corroborate in an unexpected manner the hypothesis of great displacements of the earth's axis of rotation defended by A. Wegener' (Mercanton, 1926a: 190). His plea for this project to the International Union of Geodesy and Geophysics fell upon deaf ears. Palaeomagnetism was a marginal specialty, any link between the earth's magnetism and its rotation was speculative and Drift was a controversial program. The concatenation of all three on the basis of a handful of measurements did not, it seems, constitute an attractive research agenda. For most geologists before the late 1940s, palaeomagnetism was of little interest.

Researchers in this field confronted many practical and theoretical difficulties. There was no accepted explanation of the earth's magnetic field. The techniques of palaeomagnetism were not well developed; for example, there were no established sampling methods or agreed mathematical procedures for reducing results or indicating limits of error. Igneous rocks (which did not contain fossils) were traditionally of less interest than sedimentary rocks but the remanent magnetism of the latter was weaker, more randomized and possibly less stable. Even for igneous rocks, there was no *prima facie* reason why stability should be presumed for rocks formed millions of years ago and folded, elevated, eroded, and altered in other ways. Were not reversely-magnetized rocks evidence for the instability and unreliability of palaeomagnetism itself? At the outbreak of World War II, researchers in the Department of Terrestrial Magnetism at the Carnegie Institution in Washington developed a 'spinner magnetometer' to measure weak magnetic fields. They applied it to sediments and clays, some of which dated back some 20 000 years. They found some divergences from the current magnetic field; however, they were uncertain of the stability of their samples (Glen, 1982: 104–5).

New currents

In the immediate post-war years, the Carnegie group, which included Ellis A. Johnson, Oscar W. Torreson and John W. Graham, renewed their studies but now with a specific aim: to evaluate conflicting theories of the origins of the earth's magnetic field.[3] The aim was provided by conflicting theories of the origin of the earth's magnetic field. An 'internal' origin had been postulated by the physicist Walter M. Elsasser just before the war but he did not present his detailed case until peace-time (Elsasser, 1946: 106, 202). He suggested that geomagnetism arose from processes inside the earth. The

English geophysicist Edward C. Bullard (Bullard, 1949; Bullard & Gellman, 1954) developed a sophisticated version according to which geomagnetism was produced in a manner analogous to a self-exciting dynamo by the formation and circulation of thermal convection currents in the earth. The English physicist Patrick M.S. Blackett put forward a 'general' or 'fundamental' theory: magnetic fields, including the earth's, are a fundamental property of all rotating matter. Blackett's Nobel prize in physics did not hinder his gaining a hearing. Johnson argued that if the 'fundamental' theory was correct, ancient magnetic poles determined by palaeomagnetic studies should coincide with the present ones, assuming that the earth's axis of rotation had not changed appreciably. His bid for palaeomagnetic studies to be the arbiter between these theories gave his research agenda and his technical skills new direction and importance.

Johnson and his co-workers refined their instrument and tested samples dating back 1 000 000 years. There were only slight variations from the present field. He concluded that the magnetism of his samples was stable and, on their evidence, that the earth's magnetism was generally stable (Johnson, Murphy & Torreson, 1948). His results were consonant with a 'fundamental' theory but he needed to trace the 'Prehistory of the Earth's Magnetic Field' further before drawing firm conclusions: after all, even a million years is an eyeblink in geological history. The farther back these measurements were pushed, the greater their importance for evaluating theories for the origin of the earth's magnetic field but the more concern there might be about the validity of methods, techniques, instruments and results. Palaeomagnetics was far from being a 'black box'; that is, a set of procedures and instrumentation the principles of which are no longer controversial, which can be utilized by semi-skilled technicians and the output of which is treated as unproblematic; e.g., a pH meter, barometer or x-ray machine. Torreson a year later presented data which indicated that there had been no marked change in the orientation of the earth's field for 50 MYA. The bulk of his paper was devoted to a description of sampling techniques, methods of sample preparation and the mathematical treatment of the data generated (Torreson, Murphy & Graham, 1949). A discussion of instruments and methods was characteristic of many of the early papers reporting palaeomagnetic results.

Graham addressed the trustworthiness of palaeomagnetism as a guide to earth history in reporting measurements of sedimentary rocks dating back approximately 200 MYA. He suggested two tests which were quickly accepted. One test, applicable to folded strata, was to compare the palaeomagnetic orientation with the present orientation of the earth's field.

If the rock magnetism be stable, the orientation of the strata should not coincide with the earth's field, but should more or less follow the contours of the folds (see Figure 7.1a). The other test was applicable to deposits which contained pebbles or similar extraneous matter; if the magnetization of the strata be stable, the inclusions, which presumably had been formed earlier and possessed their own magnetization, should be randomly oriented with respect to the overall orientation of the strata (see Figure 7.1b).

Graham also proposed an additional aim for palaeomagnetic studies which soon gave them pride of place in revitalizing Drift. If the stability of palaeomagnetism was accepted and the methods and techniques used to measure and interpret it sufficiently developed, not only would it be possible to trace the history of the geomagnetic field and discriminate between theories of geomagnetism, one could answer 'questions regarding large-scale movements of the crust, such as continental drift and polar wandering' (J. Graham, 1949: 160). However, the Carnegie group and most other American palaeomagnetists tended to focus on rock magnetism itself – its nature, reliability, reversals and self-reversals – rather than its application to Drift.[4] Graham's suggestion was soon taken up in England.

English directionalists and theories of geomagnetism

In contrast to the topics pursued by the Carnegie researchers, a small group of physicists and geophysicists in England by the mid-1950s applied studies of rock magnetism to polar wandering and Drift. Their approach stressed the determination of the direction – the location – of the ancient magnetic poles by measurements on rocks from different geologic periods and the treating of 'normally' and 'reversely' magnetized rocks as equivalent. These 'directionalists', associated with the Universities of Cambridge, London, and Newcastle-upon-Tyne, gained pre-eminence in this area. Their early claims

Figures 7.1.*a,b*. Graham's tests for the stability of palaeomagnetism could be applied to strata which were folded or which contained such material as pebbles. In the figures, the strata 'pass the tests'. In 7.1.*a*, the palaeomagnetic orientation of the strata as indicated by the arrows follows the contours of the folding and has therefore not been acquired since folding: in 7.1.*b*, the palaeomagnetic orientations preserved in the pebbles are random rather than uniform.

(a)   (b)

on behalf of Drift were greeted with skepticism. They then devoted much effort to refining and defending their techniques, gathering additional data and developing corroborating lines of evidence. By the early 1960s palaeomagnetism had pushed Drift back into the limelight. Their starting-point was the same as the Carnegie group's: the competing Blackett and Elsasser–Bullard theories.

Blackett, a physicist at the University of Manchester and from 1953 at the University of London, had before the war conducted research on cosmic rays. During the war he developed a very practical interest in magnetism: he worked for the Admiralty on defences against magnetic mines. In 1947 he was intrigued by a less hazardous and more theoretical problem. Observations of the rapidly rotating star, 78 Virginius, indicated that it had a magnetic field. He speculated that magnetism might be a property of all rotating matter: the geomagnetic field was generated by the earth's rotation (Blackett, 1947). Bullard, who had served under Blackett during the war, championed the rival 'internal' theory. Bullard, like Blackett, had studied physics in the Cavendish Laboratory at Cambridge but he had shifted to geophysics in order to take up a post in the new Department of Geodesy and Geophysics at Cambridge. Prior to the war he had used geophysical methods to investigate not only the continents but also the ocean floors (*cf.* Wood, 1985: 130, 133–4). A friendly rivalry ensued which extended into the field of rock magnetism. Blackett conceived a laboratory test for his theory: if a massive object was rotated, it should generate a small field which could be measured with a sensitive instrument, provided extraneous magnetic fields could be 'damped out' (Blackett, 1947: 665:6). By 1950 he had developed an astatic magnetometer which incorporated an array of magnets to give high sensitivity and 'damped out' background magnetic noise. Alas, his hopes were dashed: no measurable field was generated (Blackett, 1952). His results agreed instead with the views of Bullard and Stanley Keith Runcorn. Runcorn had studied geophysics at Cambridge under Bullard and then had gone to Blackett's department at Manchester as an assistant lecturer. Soon after his arrival, he heard Blackett lecture on his theory (Runcorn 1975: 156). Runcorn preferred Bullard's. He carried out his own experimental program involving measurements in mineshafts of changes in the earth's magnetic field which yielded results opposed to 'fundamental' theories (Runcorn *et al.*, 1951).

Blackett did not regard his negative result as fruitless nor did he give up hope for his theory. The astatic magnetometer was 'admirably suited to the measurement of very weakly magnetized geological specimens, for instance certain sedimentary rocks' and he called attention to the work of the

Carnegie group, which he described as being of 'extreme interest' (Blackett, 1952: 312). His enthusiasm was fired by the application of his device to elucidate the 'pre-history' of the earth's magnetic field. If ancient magnetic poles corresponded to the present ones (and, inferentially, roughly with the earth's axis of rotation), this would be evidence for 'internal' theories: convection currents would presumably be unstable over geologic time and should give rise to apparent 'wandering' of the magnetic poles or, perhaps, non-dipolar fields (i.e., several, simultaneous pairs of North and South magnetic Poles). He was not oblivious to Drift. About 1950 he began to read himself into geology and into the Drift literature. He seems quickly to have become convinced, possibly through discussions with S.W. Carey, that Drift had occurred (Runcorn 1975: 157). Blackett was also impressed by Arthur Holmes's textbook, especially its discussion of the southern glaciation (Blackett, 1965: vii).

Blackett and his group swiftly applied his instrument and techniques to measuring rock magnetism. In 1951 he visited Runcorn, who had returned to Cambridge, to concert a palaeomagnetic survey of Britain (Runcorn 1975: 156). Britain was a cross-section of paradise for field studies: it had exposed strata from the Pre-Cambrian through to the present. Blackett's group, now based in London, had found by 1953 that contrary to previous studies there seemed to be a systematic discrepancy between the modern poles and those calculated from measurements on British sedimentary rocks from Cambrian and more recent periods. This could have been simply explained had Blackett surrendered his 'fundamental' theory. His group instead proposed that England had rotated some 34° and moved toward the North Pole. This did not necessarily entail Drift, they noted, since it could reflect a general shifting of the crust about the poles, but data from other landmasses were needed (Clegg, Almond & Stubbs, 1954a: 596–7). They seem to have anticipated that their unexpected findings and startling conclusion would draw criticism. The astatic magnetometer was a new piece of instrumentation; it fell far short of being a 'black box'. They dwelled on the design and use of the instrument and the reliability and stability of the rock samples. This may often be a feature of the introduction of new techniques and instrumentation in science, especially when these are used in the evaluation of competing theories or programs or in the furtherance of relatively heterdox ones. Considerable time and effort is focussed on the techniques and instruments before the information produced becomes accepted as neither an artefact nor simply 'noise'.[5]

Runcorn and his group, located in the Department of Geodesy and Geophysics at Cambridge and later in the Physics Department at Newcastle-

upon-Tyne, accepted Blackett's techniques and data but rejected his interpretation. Runcorn continued to favor an 'internal' theory of geomagnetism
and in 1954 presented a version which involved convection currents and a
coriolis force (an apparent force caused by the earth's rotation) in the mantle
(Runcorn, 1954). His group built their own magnetometer after Blackett's
design and in 1951 started measuring rock magnetism, no doubt hoping to
find support for an 'internal' theory. They tested samples from major
geological periods going back to the Pre-Cambrian. Their data were consistent with Blackett's. They shared his presumption that the magnetic poles
were in rough alignment with the poles of rotation but there agreement
ended. A northward drift and clockwise rotation of England could not be
ruled out, but they proposed that the earth's poles had moved with respect to
landmasses, not that the landmasses had moved relative to one another, and

Figure 7.2. One of the first polar-wandering paths constructed. The
orientation of samples from Britain dating from the Pre-Cambrian to the
modern period were calculated, plotted and then connected in a 'smooth'
curve. The representation of the data in this form is striking and suggestive
(after Creer, Irving & Runcorn, 1954: 165).

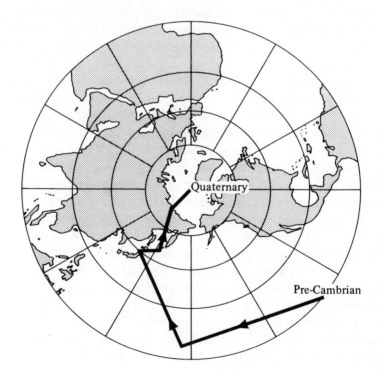

they represented their data in graphic form as a polar-wandering path (Creer, Irving & Runcorn, 1954). Runcorn, like Blackett, felt it essential to defend palaeomagnetic techniques, not just his interpretation of the results they yielded. Both Blackett and Runcorn were in part responding to the contention of Louis Néel (1951) that rock magnetism might be altered by chemical and physical processes. Néel, a French physicist, was to become a Nobel laureate for his research on a variety of magnetic phenomena. His contention stepped on a sore toe: the phenomenon of reversed magnetism. This was problematic for both 'fundamental' and 'internal' theories of the earth's magnetism, especially the former. A reversal of the earth's field implied a reversal of the earth's direction of rotation! If reversals were due to the properties of rocks themselves, as Néel suggested, then how sure a guide was palaeomagnetism to past geography?[6] A Japanese group soon reported spontaneous reversals in pumice from an active volcano. This seemed to fit with one of Néel's mechanisms and thus to undercut an assumption – the stability of palaeomagnetism – central to the directionalists' agenda. Only a few other instances were soon reported but it was not until the 1960s that it was agreed that most reversals could be explained by reversals of the earth's field, not special properties of the samples themselves. In the interim, the phenomenon could constitute an objection to drawing sweeping conclusions from palaeomagnetic studies.[7]

Runcorn in an article in *Endeavour* discussed possible mechanisms for rock magnetism, replied to arguments that it could be due to causes other than the earth's magnetic field, and described Graham's tests for sample stability (Runcorn, 1955a). All this preceded his brief remarks on the interpretation of results. This same pattern appears in a longer, more technical treatment by Runcorn (1955b). He discounted Drift as an explanation though he admitted that it could not yet be excluded (Runcorn, 1955b: 289). He also questioned the value of palaeoclimatic data used as evidence for Drift (Runcorn, 1955a, b: 283–4). He plainly expected further data from other locations to tell in favor of polar wandering. Later in 1955 he published results of studies he had begun in 1953 on North American sedimentary rocks. There was 'rough agreement' between the poles he calculated for these samples and the curve for British rocks of the same approximate age: additional evidence for his interpretation and, inferentially, evidence for his model of geomagnetism and against Blackett's (Runcorn, 1955c). As he put it, 'These results appear to dispose of the possibility that since Pre-Cambrian times the continents of America have drifted any appreciable distance from the continents of Europe and Africa' (Runcorn, 1955c: 506).

English directionalists and Drift

Edward Irving at the Australian National University (ANU) gave support to and charted new directions for the Blackett group. Irving had been an assistant to Runcorn and a postgraduate student at Cambridge before going to the new Department of Geophysics established by J.C. Jaeger at the ANU. Jaeger had astutely decided that his department should specialize in a few research fields which, he believed, would become important. Two of the fields were palaeomagnetism and rock dating. Irving, despite his Cambridge connections, favored Blackett's interpretation. He thought his own studies pointed not only to relative motion between the continents and the magnetic poles but also to relative motion among the continents themselves,; i.e., drift. In a paper submitted in February 1956 Irving clearly and concisely spelled out the major contentions of and the research agenda for palaeomagnetic investigations relating to Drift which were to be pursued by both the Runcorn and Blackett groups over the next few years (Irving, 1956).[8]

Irving assumed a coincidence between the magnetic and rotational axes. If this be correct, the magnetic latitude of a deposit on the earth's surface would correspond to its ancient geographical latitude. The palaeolatitudes so determined did not correspond to modern ones, indicating that large-scale crustal movement had taken place. Using the measurements reported by others as well as his own, he concluded that Europe, North America, India and Australia all seemed to have undergone considerable displacement with respect to the present poles. These movements could be depicted as polar wandering curves by plotting successive locations of the poles with respect to each continent.

Irving developed two additional claims. First, at the suggestion of Jaeger (Irving, 1956: 40), he claimed that palaeoclimatic evidence should agree with palaeomagnetic data; e.g., extensive salt deposits indicating a tropical climate would support a tropical palaeolatitude indicated by palaeomagnetism. A more established specialty could thus be used to help legitimate a new one. For each of the four landmasses considered, there were no major discrepancies between the palaeomagnetic and the available palaeoclimatic data (Irving, 1956: 35–7) – except for the notorious Squantum Tillite which he hinted might be due to a local phenomenon such as a mountain glacier instead of a more general glaciation. His second major claim was that the pole positions for each of these landmasses were inconsistent with those for the others and that this constituted evidence for Drift. If the path of the 'pole' was determined from, for example, European samples, and it was assumed that the crust of the earth had moved as a whole relative to the poles, this clashed with both the paleoclimatic and palaeomagnetic evidence for the other regions. The continents had moved with respect to the magnetic (and rotational) poles but they had not moved as a unit – they had moved relative to one another. Some form of drift had taken place (Irving, 1956: 40).

The London group pursued these same lines. They applied their methods to rocks in India. India was a logical choice: not only was it well-surveyed, there was extensive palaeoclimatological evidence especially in respect of the southern glaciation suggesting that India had moved northwards. Wegener had drawn on this evidence. In light of this, the results were not unexpected: assuming a 'fundamental' theory of geomagnetism, India had apparently moved some 40° of latitude to the north and had rotated about 25°. Moreover, since India's calculated pole positions did not coincide with those of Britain and America, it must have drifted relative to them (Clegg, Deutsch & Griffiths, 1956).

Blackett did not abandon his 'fundamental' theory in response to these results. Instead, he officially shifted the aims of his research. He connected rock magnetism with Drift, citing Graham on this point, and suggested that his own techniques were of value not only for the history of the earth's magnetic field, but also 'to settle the long-debated and highly controversial problems of continental drift and polar wandering' (Blackett, 1956: Preface, 5, 32:4). The interpretations of the measurements in England and India, taken together, eliminated explanations in terms of a simple shifting of the crust as a whole relative to the poles. Moreover, a northward movement of India of some 7000 kilometers, resulting in India's crushing into Asia and throwing up the Himalayas, was in accord with Drift. Finally, the palaeoclimatic arguments previously made for Drift could be turned to advantage (Blackett, 1956: 33–6; Blackett, Clegg & Stubbs, 1960: 104–5). If palaeoclimatic and palaeomagnetic data agreed, they would be mutually supporting, would support Drift, and would not conflict with his 'fundamental' theory or with the more limited claim that the rotational and magnetic poles were always roughly parallel. From the mid-1950s, Drift was the research program pursued by Blackett's group.

Runcorn switched his position in 1956 and accepted Blackett's view that Drift was needed. In a paper published earlier that year, he had reaffirmed his opinion that the discrepancies between polar positions given by American and British rocks were minor and that both sets of data could be accounted for by polar wandering without drift (Runcorn, 1956a). About the same time Walter H. Munk, who was visiting the Cambridge group, re-examined the old arguments against the possibility of polar wandering. His conclusion that the traditional objections (see Barrell, 1914) were flawed in their assumptions and in their mathematical analyses would have given comfort to Runcorn as it removed some objections to polar wandering.[9] Runcorn however rejected polar wandering as a sufficient explanation as a result of rechecking his data and giving greater weight to longitude than to latitude determinations. He found a small but systematic difference between the sets of polar-wandering curves. The North Pole for America seemed to have shifted steadily to the west. Given his assumptions that the earth at any one time had only one set of magnetic poles and that these were aligned with the poles of rotation, the only way he believed he could match up the polar-wandering curves was by postulating (Runcorn, 1956b, c) that North America had been displaced about 24° since the Triassic (about 213 MYA). This, he saw, provided a new and possibly powerful argument for Drift and he warned that 'the geophysical objections to continental drift' – objections which he had himself endorsed only the previous year (Runcorn, 1955b: 286) – 'must be re-examined' (Runcorn, 1956c: 253).[10]

By the mid-1950s both the Blackett and Runcorn groups had redirected their research agendas. Drift now became central. Palaeomagnetic conclusions offered new evidence for large-scale horizontal motion; traditional arguments for Drift such as reconstructions of ancient climates could serve

Figure 7.3. Runcorn's polar-wandering curves for North America (solid and dashed line) and Europe (dotted line). The curves, he argued, should coincide if the earth at any one time only had one pair of poles. They could be made to coincide if it were assumed that the two continents had been joined before the Permian and had since drifted apart (after Runcorn, 1962b: 24).

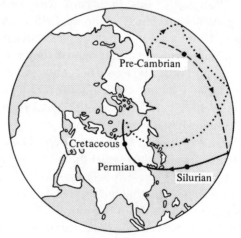

as independent verifications of palaeomagnetic conclusions. The audience was no longer limited to physicists and geophysicists. Runcorn's group swiftly established ascendancy in directionalist studies and by the early 1960s their work formed part of more general geological discussions.

Runcorn and his co-workers tried to dispel skepticism concerning their techniques, results and revised interpretations. The more radical their conclusions, the more solidly argued would have to be each link in the chain from sample to instrumentation to Drift. In a beautifully orchestrated series of six papers, the soundness of each link and its connection with the others was set forth. The first (Collinson *et al.*, 1957) dealt with the design and theory of the astatic magnetometer, the techniques of its operation, and the mathematical treatment of the instrument readings. The second (Irving & Runcorn, 1957) dealt with the theoretical aspects of the stability of rock magnetism, especially the question raised by the phenomenon of reversed polarity. Irving and Runcorn (1957) claimed that mechanisms other than reversals of the geomagnetic field were complex and usually invoked special characteristics of particular sorts of rocks which had not been observed in the field. In contrast, the Runcorn and Blackett view that reversed rock magnetism was caused by reversed geomagnetism when the rocks were formed appealed to precisely the same mechanism as employed for 'normally' magnetized rocks and thus gave a 'simple and general explanation'. The third paper (Irving, 1957) used this preferred mechanism to explain the origin of magnetism of the samples tested and concluded that their measured palaeomagnetic orientation corresponded to that of the earth at the time they were formed. Creer (1957a, b) discussed at an empirical level the problems of stability and reversals in specific samples which had been studied. The sixth and final paper (Creer, Irving & Runcorn, 1957) in the series interpreted the data in terms of multiple polar-wandering curves which coincided if large-scale Drift be accepted. As further evidence for this interpretation, Wegener and du Toit's palaeoclimatic reconstructions were brought forward (Creer, Irving & Runcorn, 1957). The group noted that although palaeoclimatic evidence for the Northern Hemisphere was ambiguous, it was much stronger for the Southern Hemisphere (Creer *et al.*, 1957: 154). This was soon followed by two papers on the drift of Australia (Irving & Green, 1958; Creer, Irving & Runcorn, 1958).[11]

Runcorn addressed a larger more general audience through the pages of *Science*. He built a case for the value and reliability of rock magnetism, sketched the more striking findings of his group, and summarized the support given the Drift interpretation of palaeomagnetic results by palaeoclimatology. His conclusion was unambiguous: drift had occurred.

He also contended that it was occurring now: he cited movement along the San Andreas Fault which was of the same magnitude as demanded by his Drift interpretation of palaeomagnetic results. That there was no mechanism, excepting possibly convection currents, which explains these movements was irrelevant: they had occurred and were occurring even if geologists could not explain them (Runcorn, 1959). Runcorn's clarion call for Drift in the heartland of opposition gained publicity for this new line of evidence.

Palaeomagnetism, Drift and the geological community

The Runcorn and Blackett approach of interpreting palaeomagnetic results in terms of Drift and using Drift arguments to buttress their palaeomagnetic work accelerated from 1956 and drew in more and more workers. In 1956 Blackett's group sponsored an international conference on rock magnetism. Naturally, they campaigned for support from the other participants, who had widely varying interests in the phenomena and applications of rock magnetism. Some support was forthcoming: W.T. Graham and A.L. Hates from South Africa reported on their study of the Karroo, which had figured so prominently in the work of du Toit. They concluded that the polar-wandering curve which they constructed did not fit those of Europe, North America and Australia and that this was most simply explained in terms of the drift of Africa along the lines postulated earlier by du Toit (Graham & Hales, 1957). The conference helped publicize the goals and the current state of play as defined by Blackett and Runcorn.

The stream of publications relating palaeomagnetism to Drift conveyed the impression of a promising and rapidly progressing line of research. It attracted new recruits into the small community of directionalists and drew the attention of palaeomagnetists working on other problems, it also stirred interest outside this sub-specialty. Three symposia, in 1956, 1958 and 1960, illustrate different stages of reaction to the novel palaeomagnetic evidence. Each of the three was explicitly concerned with a re-examination of Drift, and each was stimulated in part by the directionalist studies of Blackett's and Runcorn's groups. Each generally regarded palaeomagnetic studies as powerful evidence in favor of Drift, with or without polar wandering. Each had a different composition and institutional setting. It must be remembered in comparing the attitudes evinced at these symposia that directionalists extended and refined their assumptions, data and interpretations from 1956 through the early 1960s; i.e., the participants in the three symposia were not responding to exactly the same body of theory and information. Nonetheless, the developments in palaeomagnetism were not so major as to render comparisons invalid.

Carey arranged a symposium on Drift for 1956 at the University of Tasmania. Many of the papers trotted out biogeographical arguments which were agreed by the participants to be inconclusive. In contrast, directionalist studies were seen to offer a way out of the impasse over Drift. Recognition of their potential was not unmixed with skepticism in respect of the interpretations offerred by Runcorn and Blackett (e.g., Voisey, 1958). Irving (1958: 53) concluded that although a case for Drift had not yet been firmly established, 'there are many palaeomagnetic data which are strikingly consistent with certain aspects of the drift hypothesis'. In the opening address the venerable Chester Longwell (1958a) had suggested that the palaeomagnetic results could thus far be explained in terms of polar wandering without Drift and expressed a critical attitude toward Drift on other grounds. His view had altered slightly by the end of the symposium. He showed little enthusiasm for most of the papers but singled out palaeomagnetism as significant: 'the divergence between ancient polar locations . . . seems too consistent to be accidental. Readers will see much merit in Irving's analysis and will regard rock magnetism as a highly promising source of information on ancient geographies.' He judged that the palaeomagnetic data did not constitute 'firm proof' of Drift: more data must be gathered and the assumptions underpinning the techniques and conclusions scrutinized.[12]

The tenor of the symposium in respect of palaeomagnetism could be described as guarded: it was promising but more work was required before the claims of its practitioners could be recognized. This caution was evident even in the remarks of Jaeger who iterated some of the technical problems associated with palaeomagnetic studies.[13] The volume of *Proceedings* was, however, dominated by Carey's case for 'Expansionism' and this may have overshadowed the discussions of palaeomagnetism. Carey's case (1958b), which in published form occupied half of the *Proceedings*, advanced tectonic arguments for an expanding earth. Most of the traditional sorts of evidence for Drift; e.g., the 'jigsaw fit', geological pattern-matching between opposed coasts, and palaeobiogeography and biogeography, apparently could be reconciled with his model. Some of his tectonic concepts (e.g., 'oroclines') and his fitting together of the continents generated interest.[14]

Alberta, Canada, poles away from Hobart, Australia, provided the setting for a 1958 symposium on polar wandering and Drift. Geologists and geophysicists from all over the world had been drawn to Alberta by the

Canadian oil boom. Many joined the Alberta Society of Petroleum Geologists and it was this professional industrial association which sponsored the symposium. Few of the participants mentioned the economic aspects of polar wandering or Drift but all were aware of their possible relevance to petroleum geology. Both polar wandering and Drift bore directly on palaeoclimatology. A knowledge of ancient climatic zones would be of enormous utility in oil exploration. According to the 'reef theory' oil deposits had been laid down in association with equatorial belts of coral reefs. If formers positions of the earth's equator could be determined by palaeoclimatic reconstructions using Drift or polar wandering, petroleum geologists could pinpoint sites for exploration (Gussow, 1958: 47; Raasch, 1958b: 3). The symposium, unlike its Hobart counterpart, was not an academic exercise.[15] None, excepting perhaps E.R. Deutsch, were specialists in palaeomagnetism yet it was the focus of most of the papers. This symposium, like the subsequent one of 1960, provides an instructive example of how the empirical work and theoretical convictions of a local network or sub-specialty diffuse among other specialists. The point of contact was not palaeomagnetism *per se* but the interest of both groups in the related sub-field of palaeoclimatology. If palaeomagnetism provided new and useful information, it had a claim on the attention of petroleum geologists. Since its results had been interpreted in terms of polar wandering and Drift, which also bore on palaeogeography, those theories were also of interest.

The results of directionalist studies were greeted with enthusiasm; palaeomagnetism 'presents a new kind of evidence, capable of rendering a decision' on polar wandering and Drift (Deutsch, 1958a). Several spoke of palaeomagnetism as confirming the reality of Drift (Patterson, 1958) or as rendering it much more plausible (Scheidegger, 1958b). The aim of the 1958 symposium was to gain familiarity with the work of the directionalists and to evaluate it with respect to the problems and issues of concern to the members of the Alberta Society, not to break new ground in either palaeomagnetism or global theorizing. Deutsch, who was versed in the techniques, did little more than provide a summary (Deutsch, 1958a, b). D.W.R. Wilson (1958) drew on palaeomagnetism to analyze the former positions and rotations of landmasses with reference to Carey's orocline concept. Scheidegger (1958b) adopted a more favorable stance toward Drift than he had in his monograph on geophysics published earlier that year, and attempted to explain palaeomagnetic results with convection currents (Scheidegger, 1958a: 179–83). D.J.C. Laming (1958), perhaps alerted by the directionalists' suggestion that palaeoclimatic investigations might reinforce their interpretations, claimed that his study of aeolian sandstones (fossilized sand dunes) revealed a pattern of ancient trade winds conformable to and serving as independent 'confirmation' of the palaeomagnetic reconstructions.[16]

Polar wandering and Drift were the subjects of a symposium in Atlantic City, New Jersey, held as part of the 1960 meeting of the Society of Economic Palaeontologists and Mineralogists.[17] For most participants, Drift was 'given'. Carey was the lead speaker. He had been giving courses at Yale at Longwell's invitation and had spoken in favor of continental displacement (through earth expansion) at North American universities. He recalls that at the symposium the overflow audience had engaged in discussion of Drift long after midnight (Carey, MS). Interest centered on palaeomagnetism and its interpretation in terms of Drift and polar wandering. Munyan (1963b: 3), the organizer of the session, captured this well: 'Probably no other single line of investigation has done so much as [palaeomagnetism] . . . to advance the cause of continental shifting and polar wandering.' Runcorn briefly recapped the assumptions behind and difficulties inherent in palaeomagnetism, including a fairly technical discussion of the application of statistics (in the form of 'circles of confidence' for polar locations). He also discussed the fitting together of polar-wandering curves from different continents which had led him to adopt Drift (Runcorn, 1963). No other specialists in palaeomagnetism took part in the program. With the exception of Deutsch (1963a) who provided a lengthy history and summary, most of the participants seem to have been less interested in palaeomagnetism *per se* than in how their own repertoires of techniques and lines of evidence reinforced – and were reinforced by – the conclusions of Blackett and Runcorn. These ranged from traditional palaeobotanical approaches (Fulford, 1963) to palaeoclimatology (Nordeng, 1963) and glaciology (Ewing & Donn, 1963) to the plotting of deep-focus earthquakes (Harrington, 1963).[18]

Assessments of palaeomagnetism and Drift in the early 1960s

By 1960 what had begun with Runcorn and Blackett as a technique to test physical theories had yielded results relevant to Drift which attracted the attention of workers in other problem-fields and specialties. The diffusion of the work of the 'directionalists' into other specialties not only stimulated research on topics traditionally associated with Drift, it also promoted the development of other techniques, methods, and research projects. Drift enjoyed a favorable press in the later 1950s and early 1960s and for this, as Munyan suggested, palaeomagnetism was largely responsible. Nonetheless, there were still reservations not only about Drift with or without polar wandering, but also about the techniques and interpretations of the directionalists. Most geologists, unless they were themselves specialists in palaeomagnetism, were unlikely to have read the relevant papers or to have first-hand familiarity with the new techniques, instrumentation, and so on. There was no direct unequivocal demonstration that rock magnetism, as measured in the laboratory, provided a trustworthy stable record of the earth's magnetic field. Blackett, Runcorn and most other workers in the area were not impressed by such objections. Naturally they had little sympathy with hints that their techniques might be faulty. One can conceive of complex mechanisms which might account for divergent magnetic orientations of a few rocks but to admit that these were widespread would be to undermine their own investment in palaeomagnetic studies which were predicated on an 'imprinting' of the earth's magnetic field as the cause of

Figure 7.4. Polar-wandering curves from different continents. Proponents of the directionalist approach argued that these curves could be brought into agreement if Drift were invoked. These curves displayed in graphic, easily assimilable form the results of numerous separate measurements and computations but without indicating the many underlying assumptions (after Runcorn, 1962b: 24).

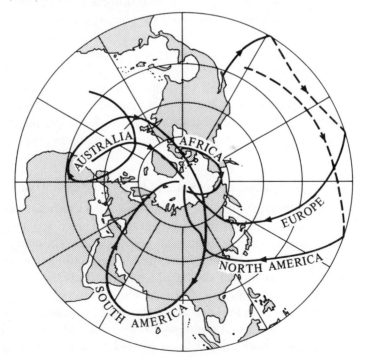

rock magnetism.[19] As for a demonstration of the validity of the palaeomagnetic data and their interpretation of it, that was not possible in isolation, but various lines of corroborating evidence from other specialties could be adduced – though these too were equivocal or inconclusive. This was certainly the case for palaeoclimatology. There was a sizable body of palaeoclimatic data which could be used to support large-scale drift. However, in the 1940s and early 1950s Chaney and Durham had argued forcefully that the climatic zones of the Tertiary coincided with those of the present distribution of the continents and oceans. This constituted strong *prima facie* evidence against substantial recent drifts or polar wandering, as Chaney (1940: 498) and Durham (1950: 1244; 1952: 321) were both aware, and their claims received respectful attention for some years (e.g., Cox & Doell 1960: 756).

We can reconstruct choices for geologists: if one's confidence in palaeomagnetic techniques were sufficiently strong– or if one were predisposed to Drift on other grounds, then one could follow the argument leading from certain assumptions and data and conclude in favor of Drift despite well-known objections. If one had little confidence in directionalist studies or if one were predisposed against Drift or both, then it might be preferable to adopt complex mechanisms of rock magnetism or to challenge other assumptions or techniques of palaeomagnetism and maintain the relative stability of continents. Or, one could adopt a 'wait-and-see' attitude. The last option was probably the most popular. Directionalist studies did not cause a sudden massive reversal of opinion about Drift. They had, however, contributed significantly to breaking the impasse which had existed in the Drift debate during the 1940s and early 1950s.

Allan Cox and Richard Doell's (1960) review article reflects enthusiasm for the potential use of palaeomagnetism in settling the question of Drift, discusses reservations about the data and interpretations offered thus far, and stresses the desirability of more – and more reliable – data. They first systematically rehearsed the theory that lay behind rock magnetism and its measurements, dealt with the objections raised by Néel and others, discussed stability tests, and summarized the mathematical techniques used to estimate the reliability of the data and to calculate pole positions and their reliability. They then collated and evaluated rock magnetic data previously gathered and presented them in tabular form. After these preliminaries, which added to the authoritativeness of their presentation, they drew their conclusions. Cox and Doell were not directionalists. Both had studied geophysics at Berkeley and were working under the auspices of the US Geological Survey on a project to establish a time-scale of geomagnetic reversals (*cf.* Glen, 1982). This may have contributed to the influence of their review: they were experts in palaeomagnetism but could be considered 'hostile witnesses' on the question of the support given to Drift by the directionalist approach. Although some have suggested that Cox and Doell showed scant sympathy with studies relating palaeomagnetism to Drift and polar wandering (e.g., Glen, 1982: 94n; Wood, 1985: 116), this was not the case. Their remarks must be seen in the context of a general review of the whole field and the global body of data collected. Their concern was probably not to establish the merits of global theories but to give a sober dispassionate defense of the techniques of palaeomagnetism – upon which they themselves relied. They did not accept the case for Drift in the Mesozoic and early Tertiary nor for Drift or polar wandering in the late Tertiary and after (Cox & Doell, 1960: 762–3). However, in their opinion the body of data cited by Runcorn and his group from the Carboniferous and Permian in Europe and North America was internally quite consistent and 'indicates a magnetic field configuration vastly different from the present. . . . [And] these results constitute a strong case for polar wandering' (Cox & Doell, 1960: 760). The Australian data for these same periods were even more impressive: they 'constitute evidence for a relative displacement of Australia with respect to North America and Europe which cannot be ignored' (Cox & Doell, 1960: 762).

The verdict on Drift was in the balance but it could be decided by further investigations. Palaeomagnetism was a reliable tool for the reconstruction of past geographies; misgivings on that score should be laid to rest. What was needed 'to test properly the hypotheses of large-scale continental drift' in times more recent than the Permian was to eliminate ambiguities in the existing data by the collection of new samples from well-dated strata in a broader range of geographical locations. Further, they endorsed biogeographical studies as an auxiliary line of research (Cox & Doell, 1960: 763–4). Only a year later, they expressed themselves even more positively toward the growing body of palaeomagnetic evidence purportedly showing that drift had occurred: not only was there now 'little doubt that the magnetic pole has wandered', if one made 'reasonable assumptions about the earth's field in the past, it is difficult to explain all the presently available palaeomagnetic data without invoking continental drift' (Doell & Cox, 1961: 302). Directionalists who assumed Drift as reality or as a working hypotheses would have been unlikely to take umbrage at Cox and Doell's remarks: they were not an enthusiastic endorsement of Drift, but they sanctioned the program on which Blackett and Runcorn had embarked and gave grudging concurrence to some of their conclusions.

Directionalists continued to pursue their agenda – with some success – in the early 1960s. A month before Cox and Doell's review appeared, Blackett gave his own survey of the field. He focussed upon alternative interpretations of the data. He ruled out as improbable processes other than the earth's magnetic field as the source of rock magnetism: the data from his and Runcorn's program were consistent yet taken from different types of rocks, different locations and different epochs. The more data amassed, the more improbable alternatives became. He then considered whether the earth's crust might have shifted with respect to the magnetic poles of vice versa. These versions could not be distinguished palaeomagnetically. However, he argued that polar wandering was insufficient since polar-wandering curves did not coincide unless Drift was also invoked. Next, he took up the suggestion that the earth's magnetic field had not always been an axial dipole field (that is, a simple N–S field more or less coincident with the earth's axis of rotation). Against this he offered both a theoretical argument; namely, that none of the extant theories of the earth's magnetism are consistent with a non-dipole field, and a phenomenological argument: to embrace a non-dipole field is to accept that the present alignment of the magnetic and rotational poles is purely coincidental and atypical. By a process of exhaustion, he leaves Drift as the most plausible theory for explicating rock magnetic data. He urged, however, that restraint be exercised in drawing up polar-wandering curves. Palaeomagnetic data alone could not yield the relative longitudes of two landmasses; all that could safely be compared were their apparent latitudes and orientation (Blackett *et al.*, 1960).

Blackett, Cox & Doell were not in complete agreement about the interpretation of data but the disagreement was one of degree. Palaeomagnetism by the early 1960s had become a 'black box'. There was no longer substantial overt concern with the instrumentation or the theory and techniques of its design and operation, with sampling techniques (except at a very general level), or with the mathematical reduction of data. This in itself was a significant accomplishment for the agenda initiated a decade earlier. Bullard later remarked (1975a: 14; *cf.* Menard, 1986: 87–9): 'the clarity which was finally achieved in the interpretation of palaeomagnetism should not obscure the complexity and difficulty of the route by which it was attained . . . the surprising thing is that by 1960 the case was substantially complete'. In the next issue of the journal in which Cox and Doell's review had appeared, Collinson and Runcorn (1960) presented results of further studies on North American samples. These, they claimed, showed that North America and Europe had drifted apart by nearly 30° since the Mesozoic.

Over the next few years Runcorn's group at Newcastle-upon-Tyne followed several lines of research to further their agenda. Alan Nairn, for example, who had earlier conducted palaeomagnetic research in connection with Drift (e.g., Nairn, 1956; 1957; 1960), sought to enlist palaeoclimatology. He not only helped disseminate knowledge of the directionalists' work among more orthodox palaeoclimatologists (Nairn, 1961), he also pointed out that conclusions about ancient geography reached by a novel palaeoclimatic technique of palaeotemperature measurements were in accord with those reached by palaeomagentism. Both the new technique and palaeomagnetism could benefit from this linkage. Additional importance and utility accrued to the novel technique in palaeoclimatology when it was linked via Drift to palaeomagnetism; similarly, confidence in palaeomagnetism would be enhanced by this development of an independent method which yielded supporting results. And, of course, both provided ammunition for Drift (Nairn & Thorley, 1961). Runcorn plunged into geophysics, especially convection-current theory. He had been guided by convection-current theories of the origin of the earth's magnetic field in his initial investigation of palaeomagnetism. Convection currents could now serve dual and complementary explanatory roles: they might not only account for the earth's magnetic field but also provide a mechanism for the Drift indicated by magnetic studies. He sketched out these ideas early in 1962 (Runcorn, 1962a). This was a major theme in a collection of papers later that same year which spanned many topics in geophysics but focussed on convection currents as related to palaeomagnetism and Drift. The volume was boldly titled *Continental Drift* (Runcorn, 1962b). Through it Runcorn again moved from one specific problem-field or subspecialty to address and seek support from a broader segment of the geological community. He was not isolated from the data, the problems and the preferred theories and programs of other parts of geology. The apparent conflict between Drift and geophysics had to be removed or at least softened before his own work in palaeomagnetism would be acceptable to those outside his problem-field.

It is a moot question whether or not directionalist studies alone or even in concert with palaeoclimatology and palaeobiography would have been sufficient to lead to widespread acceptance of Drift. The line of research pioneered and sustained by Blackett, Runcorn and their associates played a pivotal role among land-based geologists in bringing Drift out of the doldrums in which it existed in the 1940s and 1950s. It sparked renewed interest in Drift which transcended the specialty in which they worked. It gave the impression that Drift was progressing empirically and theoretically.[20] Palaeobiogeographers, for example, had reached a near-stalemate over Drift by the 1950s. The data were too unreliable, and too fragmentary to conclude on that basis alone whether or not drift had occurred (Frankel, 1981). In the late 1950s and early 1960s specialists in this field, including some opponents of Drift, noted that directionalist studies had at least reopened the question (Frankel, 1985: 170–1; 178–80). However, it was not the directionalist studies alone, but a confluence in the mid-1960s with other theoretical and empirical developments which produced a fundamental reorientation and a new 'global geology' which incorporated Drift.

Voice-Over

Directionalist palaeomagnetic studies evolved in just over a decade from a minor problem-field on the fringes of geophysics and physics to an emergent specialty of recognized importance for the evaluation of rival global geological theories. We can draw distinctions between 'social' and 'technical' or 'cognitive' aspects of this evolution but this does not imply a tension or opposition between them; they seem to have operated in concert. Moreover, these categories of 'social' and 'cognitive' likely were indistinguishable to the participants. I impose them only to facilitate our discussion of the nature of scientific change and of rationalist and social models of science.

A cognitive account of the evolution of palaeomagnetic studies would emphasize shifting aims, problem-solving ability and relationships to other disciplines and to rival global theories. Palaeomagnetism seems to have been directionless until Johnson, Graham, Blackett and Runcorn applied it to the evaluation of competing theories of geomagnetism. The directionalist approach first crystallized around this aim and the new instruments and techniques which provided the means to achieve it. The role of technical interests is well illustrated by Blackett's development of the astatic magnetometer to directly test his theory of geomagnetism, followed by his use of it to test his theory indirectly and finally his application of it to Drift. These developments were – at least in part – shaped by the interaction between theory and empirical results. Blackett's and Runcorn's redefinition of their agenda to link it with Drift occurred for good reasons. Both had prior acquaintance with Drift, and Drift alone of the competing programs seemed compatible with their data.

The relationship between directionalist studies and Drift was a symbiotic one. Drift offered a rationale. If Drift were rejected, then the patterns – the divergent polar-wandering paths – elicited from the scatter of apparent ancient pole-positions were meaningless; perhaps researchers would have turned to other phenomena or perhaps this field of investigation would have reverted to a marginal problem-field of little consequence. We note here the persuasive power of such representations as polar-wandering paths which summarized in graphic form large amounts of esoteric data and at the same time placed a theoretical construction on that data. Drift also served a heuristic role; i.e., it suggested ways in which the directionalists could develop their agenda. At a theoretical level, Drift afforded an interpretation of diverging curves in terms of relative movement of landmasses and motivated Runcorn's work on convection currents. At an empirical level, Drift directed attention to the employment of palaeoclimatic data as a cross-check on and reinforcement of palaeomagnetic data. At a 'technique' level, Drift offered guidelines as to appropriate locations and approximate dates of deposits which might be studied. Conversely, directionalist research provided fresh arguments for Drift.

Partisans of Lakatosian, Laudanian or Kuhnian views of scientific change can fit some aspects of the revival of Drift occasioned by palaeomagnetic studies to those models. Both Kuhn (1977: 322) and Lakatos (1978: 49, 52, 69, 112) suggest that scientists' ability to solve in a research program new, unanticipated problems – problems it was not invented to solve – is a powerful reason to prefer that program. For Blackett, Runcorn and others Drift solved the problem of reconciling divergent polar-wandering paths from different continents, a problem which did not arise until four decades after Drift had first been proposed. L. Laudan's model also applies here. For Drift, directionalist studies created a whole new class of 'solved problems'; for its rivals, a new class of empirical 'anomalies'. Additionally, the linking of palaeomagnetics with other specialties may have given greater weight – or at least new visibility – to some of the older 'anomalies' facing the rivals; e.g., the great southern glaciation. Laudan (1977: 143) would also suggest that what is especially relevant to the appraisal of global theories in this case is that there was no dependence of the 'problem' on one or other of the rival programs: the existence of rock magnetism, its detection and measurement, the plotting of apparent ancient poles and the construction of polar-wandering curves could be performed without using assumptions drawn from the rival programs. One could stretch a point and claim that the directionalist program yielded a 'successful prediction' which, for Lakatos (1978: 16–7) and Laudan (1977: 43), serves to confirm all the assumptions used in deriving the prediction. Directionalists themselves seem to have taken something of this view: for those who pursued Drift, the construction of polar-wandering curves which could be fitted together through Drift was regarded as a confirmation of Drift and, inferentially, of their collateral assumptions about the stability of magnetism, the unproblematic characters of reversals, the axial dipole view of the earth's magnetic field and so on.

This view was not dismissed out-of-hand even by those less sympathetic to Drift; e.g., Cox and Doell.

There are complications: aside from other difficulties with the notion of 'prediction' as applied to historical geology, there remained grounds for reservations about some of the collateral assumptions of directionalists. Their data and the steps in their argument were theory-laden. This was, however, acknowledged by them and by their opponents. Both they and their opponents agreed that these reservations might be met and the theory-ladenness mitigated by further empirical procedures. This is fully consonant with the models of Lakatos and Laudan, though not with strong forms of Kuhnian incommensurability nor with Feyerabend's descriptions of scientific argumentation as purely rhetorical. As we can see, a plausible reconstruction of the evolution of the directionalist agenda and its connection with Drift can be given which is incompatible with – at least at a crude level – Laudan's model and to a lesser extent those of Lakatos and Kuhn.

A simplistic social analyst would also derive some satisfaction from our study. We may suppose that the credibility of the directionalist approach among geologists and within the larger scientific community was enhanced by the prestige of some of the individuals associated with it. Bullard was Director of the National Physical Laboratory from 1950 to 1955 and then head of the Department of Geodesy and Geophysics at Cambridge. Blackett was a Nobel laureate in physics and held several government appointments. In the early 1950s R.A. Fisher, a distinguished geneticist and statistician at Cambridge, provided directionalists with statistical methods for handling results, which may have conferred additional authority (as well as meeting some technical objections). The institutions with which many of the researchers were affiliated had prestige and presumably some would 'rub off' to give their work greater authority. Conversely, declining support for Graham's work within his institution may have effectively silenced one potential voice of opposition. Though these kinds of 'social' factors should not be slighted, a more fine-grained analysis which invokes both intellectual and social elements – and which gives due attention to the possible technical and social 'interests' of the participants – appears warranted.

Both cognitive and social stakes were high. 'Palaeomagicians' might accomplish what other groups of researchers had been unable to accomplish: the breaking of the stalemate among Drift and its rivals. If this were done, then palaeomagnetism would presumably have become a central geological specialty and its practitioners – since they were the ones who had pioneered the approach and who possessed the technical skills – recognized authorities in a reshaped and redirected discipline. It is conceivable that the data and interpretations of such other specialties as palaeontology and palaeoclimatology would have been judged on the basis of their consistency with the data and interpretations of directionalists and Drift. This would constitute a *coup d'état* by physicists and geophysicists. In Britain, palaeomagicians were initially on the margin of the geological community and congregated around such institutions as the Royal Society and the Royal Astronomical Society rather than the Geological Society; in North America, the American Geophysical Union rather than the Geological Society of America. Their backgrounds were in physics or perhaps geophysics rather than in geology and they seldom occupied academic positions in geology departments. Blackett's 1956 conference, for example, was held under the auspices of the International Union of Pure and Applied Physics. Their publications, until the Cox and Doell (1960) review, appeared in *Advances in Physics*, the *Proceedings* or the *Philosophical Transactions* of the Royal Society, the *Journal of Geophysical Research* and only rarely in the mainstream geology journals which were dominated by more orthodox approaches and interests. Their methods and techniques were more akin to those of physicists than to those of most geologists. In the 1950s many geology texts, for example, which made reference to methodological matters still flew the banner of empiricism and naive inductivism (e.g., Garrels, 1951: 33–4; Kirkaldy, 1954: 11, 13) rather than that of the method of hypothesis urged by physicists and geophysicists (e.g., Scheidegger, 1958a; Howell, 1959). These circumstances may have made the task for directionalists of gaining credibility for their work all the more daunting – but all the more worth attempting.

Within the specialty, directionalists were arguing in effect for a redefinition of its aims and the adoption of new techniques. Their agenda as it developed took as unproblematic the problems which were central to the agenda of other palaeomagnetists, e.g., reversals. At a 'social' level, this could be seen as a struggle for authority and recognition within the field; at a cognitive level, in the publications of the participants, this resulted in conflicts over what counted as problems, acceptable solutions, appropriate methods, proper techniques and desirable aims. These issues were not debated upon nor do they seem to have been resolved on the basis of the prestige of the protagonists. Theories, data and the interaction among them both constrained and featured in disagreement and agreement. According to Kuhn (1970: 53–8; 1977: 174–5), scientists will not readily accept empirical results unless the are 'anticipated and sought in advance' and require 'no adjustment, adaptation, and assimilation from the profession'. If the results be truly novel, they will be accepted – if at all – only after critical scrutiny.[21] This might describe the reaction of those working within a Permanentist 'paradigm' and it was the case that the choice of global theories was not based simply on the 'facts of the matter'. The 'facts' put forward by directionalists were blurry, open to challenge and subjected to critical scrutiny. Kuhn's model, however, does not take into account the theoretical pluralism which existed in geology. For the directionalist–Drifter, many of the results were anticipated and sought in advance and readily assimilated. Moreover, scientists with different theoretical preferences were able to reach some agreement on empirical matters. Even if there were a multitude of objections to directionalists' data, techniques, methods, instruments and the representations of results as polar-wandering curves, and the interpretation of these curves as evidence for Drift, they had not been conjured up out of thin air, nor were they so treated. Runcorn had opposed Blackett on several theoretical issues. Nonetheless, Runcorn, reluctantly and in a piecemeal fashion, came to agree with Blackett's endorsement of Drift in response to the data that Runcorn himself had collected and reanalyzed and in response to the progress that Blackett had made by using Drift to make sense of his data. Similarly, Doell had by the mid-1950s abandoned his pursuit of the directionalist program because of what he thought to be severe empirical and conceptual problems with it (cf. Glen, 1982: 147–9). However, in his review with Cox he was sufficiently impressed by some of the data and its interpretation to concede that there was a strong case for polar wandering and for the drift of Australia. We might conjecture that within the specialty, disputes over the interpretation of results in terms of Drift were of less import than disputes over the means by which those results had been obtained.

Local and specialist interests

In our account we have emphasized the development of the directionalist agenda and its relation to a variety of technical and specialist interests. Such interests were involved in the decisions of Blackett and Runcorn to weld their palaeomagnetic studies to Drift. Both were 'outsiders' in respect of geological specialties and much of geophysics. They were not completely ignorant of the theories, data, methods and problems central to other problem-fields and specialties, nor were they unaware of widespread opposition to Drift. Nonetheless, these matters seemed to have weighed far less heavily for them than the apparent value of Drift in solving their problems within their problem-field. Although they may have opted for Drift on 'localist' grounds, they did not ignore the prevailing views in and central concerns of other specialties and disciplines. They sought to publicize their work to a broader spectrum of geologists; to demonstrate that their techniques, results, and interpretations met some notional minimal standards in the problem-field, the discipline, and other disciplines; to show that there was compatibility between the data and theories of their problem-field and others and to show that palaeomagnetism and Drift could be of use in addressing central problems of other specialties. This may represent a role for local and specialist interests complementary to that noted previously. Some scientists narrowed the grounds for evaluating global theories to the issue of their relative usefulness in their own problem-fields. Blackett, Runcorn and their colleagues broadened their arguments beyond their problem-field to assemble a more general case for Drift. Had they failed to do so, their approach may have stagnated or been ignored.

There were, of course, other agendas in palaeomagnetics. These too may be connected with local or specialist interests. Hospers's concentration on reversals followed on from his investigations of reversed lavas. Néel's work on mechanisms of palaeomagnetism and reversals formed part of his more general investigations of magnetism. Cox and Doell's agenda of developing a geological time-scale owed much to Cox's concern with geochronology and their familiarity with the young-rock dating research at Berkeley (Glen, 1982: 173–4). For each of these agendas we could construct an account of the technical and social interests which underlay them.

Informal networks

The evolution of the directionalist agenda was facilitated by an informal network of shared interests in which the links were partly social and partly cognitive.[22] Runcorn, for example, had studied under Bullard at Cambridge, had lectured in Blackett's department at Manchester and had returned to Cambridge where he and two of his students, Irving and Creer, liased with Blackett in their palaeomagnetic survey work. Bullard was influenced in his choice of research topics by Richard Field (who as we shall later see helped shape the interests in geophysics and marine geophysics not only of Bullard but also of Maurice Ewing, Harry Hess and J. Tuzo Wilson). Field invited Bullard to the US, fired him with enthusiasm for marine geophysics and introduced him to Maurice Ewing (Wood, 1985: 130–1). Bullard met Wilson while at Toronto in 1948–9 and Bullard was an associate of both Runcorn and Blackett. Indeed, Bullard had discussed with Blackett during their war service a dynamo model of the earth's magnetic field but had set it aside. Subsequently, after Blackett had proposed his 'fundamental' theory, Bullard had taken up the dynamo theory again as a possible explanation of variations in, not of the origin of, geomagnetism. Later, after learning of

Elsasser's work and determining that Elsasser's equations were not 'quite right', he reverted to the dynamo theory to explain the origin of the earth's magnetism. The manifold social and cognitive ways in which these informal networks can function fit well with the 'interest' and 'struggle' social analyses but is missed both by number-crunching sociometric-cum-socio-logical models of scientific change[23] and thus far – by most of the rationalistic models. It may be that the latter can incorporate informal networks but thus far those models have for the most part not been applied to the 'coalface' of daily scientific practice.

Constructing black boxes

The evolution of palaeomagnetic studies was marked by the invention of new instrumentation and disagreements over techniques. By the early 1960s directionalist palaeomagnetics had emerged as a black box: a routinized, unproblematic set of instruments and techniques into one end of which samples were fed and from the other end pole positions were spewed out. This came about through a complex process of negotiation and the resolution of disagreements over a series of theoretical, empirical and methodological issues which ranged from geology into physics, chemistry and mathematics (see Latour, 1987). Many of the instruments and techniques now routinely used by technicians may be the material results of disputes over and eventual consensus on their empirical and theoretical underpinnings. Lakatos (1978: 14–5, 44–5) acknowledges the theory-ladenness of instruments and Latour and Woolgar (1979: 63–6; also, Latour, 1987) describe instruments as 'reified theory'.

Palaeomagnetism is a rich field for investigating these matters. In respect of the palaeomagnetic orientation of sedimentary samples there were a series of issues to be resolved as we saw with respect to the tests proposed by Graham. There is the seemingly trivial question of whether the spinner and astatic magnetometers were measuring something that pre-existed or were 'creating' the phenomena they were supposedly detecting.[24] The question is not an absurd one. A Kuhnian (1970: 24, 192–4) might claim that palaeomagnetism of sedimentary rocks might not have been investigated and the necessary instruments invented but for the existence of theories in physics, geophysics and geology which directed attention to them. At a more subtle level one might claim that without these instruments and associated techniques, ancient polar positions and polar-wandering curves would not exist for scientists: 'the phenomena are thoroughly constituted by the material setting of the laboratory' (Latour & Woolgar, 1979: 64; Latour, 1987), they are not given to us by nature. For researchers who did not have access to the necessary equipment, the phenomena in an important sense did not exist – whether those researchers opposed or supported the agenda and interpretations. Those scientists who did have access may have thereby gained expertise, authority and even immunity from some types of criticism but they faced a difficult task in persuading others of the well-foundedness of their interpretations especially if they went against firmly-held beliefs.

Consider such a minor technique as the 'cleaning' of samples. By the late 1950s palaeomagicians as a matter of routine exposed their samples to either heat or an alternating-current magnetic field. The purpose was to remove 'extraneous' magnetization which might have been acquired since the formation of the rock and which overlaid or disguised its original permanent magnetization. Practitioners might have every confidence that this technique did not alter the 'original' magnetism; outsiders who did not have first-hand familiarity with the technique might well be dubious. Every step from sample collection to plotting poles could be analyzed in similar terms; every step involved a welter of empirical and theoretical assumptions. These were potential sources of divergent interpretation and disagreement.

Palaeomagnetists, especially directionalists, eventually gained acceptance for their instruments and techniques. Perhaps, in part, this was aided by practitioners' prestige and credibility inside and outside the field. However, what formed the substance of the emerging consensus were the internal consistency of their results, consistency with results obtained by other methods in other fields, the provision of a coherent theoretical account of the instrumentation and techniques and their relation to the phenomena they supposedly measured; the provision of theories of geomagnetism and how these related to palaeomagnetism and, of course, the explication of their results in terms of a global theory. The construction of the palaeomagnetics black box was complete when this consensus had emerged.

Return of the Pretender

Drift in the 1950s had not been suddenly rediscovered or reinvented. It had not disappeared from the textbooks (e.g., Holmes, 1944) to which Blackett and others turned and it had active proponents in Britain. It was a recognized global research program, even if an unpopular one, and could be used to interpret new data, to construct solutions to new problems and to serve as a rally-point for opposition to orthodoxy. It played these roles for Blackett, Runcorn and other directionalists. At the same time, palaeomagnetics was cited in a number of textbooks in the late 1950s and early 1960s as having given Drift new momentum; little notice was given of possible flaws in its techniques. Further, those who took up the discipline in the 1940s, '50s, and '60s had not had first-hand experience of the earlier debates over Drift and the reasons for its rejection, nor had they the same personal or professional stakes in ruling it out as those who had been victorious in routing it.

Notes

1 A useful introduction to technical aspects of theories of the earth's magnetism and of palaeomagnetism is Takeuchi, Uyeda & Kanamori (1967: 94–189).
2 A compass needle gives the N–S direction and, if freely suspended, the inclination which can be used to determine the apparent latitude. The needle will point straight up at the south magnetic pole, straight down at the north magnetic pole and will be level at the equator.
3 The Department of Terrestrial Magnetism had concentrated on the collection and collation of data on fluctuations in the present geomagnetic field. The impetus for palaeomagnetic studies seems to have come from Johnson and, later, Graham. Soon after Johnson's departure in 1948, the Carnegie research stagnated, partly because of a lack of support for Graham from his superior.
4 Glen (1982) gives a wealth of information on the development and application of geophysical and especially palaeomagnetic techniques in the United States to polar reversals, rock dating and the establishment of geological time scales.
5 Major topics at the 1954 annual international meeting of palaeomagnetists were the stability and reliability of rock magnetism as indicators of past magnetic pole positions and the vexed question of whether pole reversals were due to reversals of the geomagnetic field or to special characteristics of or processes in the rocks

Figure 7.5. Determination of latitude by magnetic inclination. The arrows represent the 'dip' at different latitudes of a freely-suspended, north-seeking compass needle.

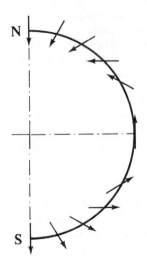

themselves. *Cf.* the papers from this conference published in *Journal of Geomagnetism and Geoelectricity*, 6, no. 4. *Cf.* for a parallel case Edge (1976).

6 Jan Hospers, a Dutch geologist, took issue with polar-wandering data and interpretations. His Ph.D. work at Cambridge, in which Runcorn assisted, involved a study of Icelandic lavas containing bands of normally- and reversely-magnetized rock (Runcorn *et al.*, 1975: 156). His problem-field was reversals in igneous rocks. He cautioned against the use of sedimentary rock samples and claimed that his measurements of samples from western Europe and Iceland gave a maximum polar wander of 5–10° – far less than that found by Runcorn in his studies. He closed with the admonition that future studies evaluating Drift be confined to igneous rocks but that even for these, there was no assurance of long-term stability (Hospers, 1955: 72). In other words, even if further work on the type of rocks he specified yielded results conformable to Drift, the data would still be open to question. His work did, however, support field reversals as the cause of reversed rock magnetism and was the first to use Fisher's statistical methods.

7 See Glen (1982: 119–29) for further dimensions of the controversy.

8 It is not clear to what extent Blackett's group was directly inspired by Irving's work to carry out studies on Indian rocks or to support palaeomagnetic results with palaeoclimatic evidence. This may have been coincidental; however, the first publication of the group on Indian rocks cited Irving's thesis on Torridonian sandstones, noted that he was the first to work on Indian samples and gave a vague summary of his findings (Clegg, *et al.*, 1956).

9 Munk (1956); T. Gold (1955) had recently reached a similar conclusion. See also Wood (1985: 113–5) for a brief discussion of polar-wandering theories up through 1955.

10 Runcorn in this paper cited Irving (1956) and it may be conjectured that his re-examination of his earlier data and conclusions was in partial response to Irving's work.

11 Verhoogen (1985: 403) noted that he and others found especially compelling the case put forward by Irving and Creer for the motion of Australia.

12 Wood (1985: 117–8) emphasizes Longwell's criticisms and passes over his comments in respect of palaeomagnetism.

13 Jaeger's comments are reported in 'Summary of Continental Drift symposium, Hobart, March 1956 by M.A. Condon and A.A. Opik', Opik Papers, Basser Library, Australian Academy of Sciences.

14 The paper which Carey actually delivered at his symposium seems to have centered on the use of the 'orocline concept' for palaeogeographical reconstructions in the context of Drift ('Summary', Opik Papers). Oroclines were patterns of folding produced by rotations of crustal blocks; for example, the rotation of Spain away from France to form the Bay of Biscay produced compression and folding of the Pyrenees at the 'hinge' of the rotation. More generally, an orocline is an organic or mountain belt with a horizontal curvature or sharp bend attributed to the action of horizontal forces operating within the crust. However, subsequent to the symposium, Carey had discarded Drift in favor of the expanding earth as his preferred global theory. As editor of the *Proceedings* he was able to incorporate a very long paper setting forth his new Expansionism in lieu of the paper he had read at the symposium.

15 Runcorn was the only exception, but it was not his paper, but a report of and commentary on it – by a petroleum geologist – which was included in the published record (Patterson, 1958).

16 The subsequent paper of Opdyke and Runcorn (1959) on aeolian sandstones does not cite Laming though presumably Runcorn would have seen the relevant number of the Society's *Journal*.

17 The session was organized in 1958, the same year as the Alberta symposium and the publication of the Hobart proceedings; the papers from Atlantic City were not published until 1963.

18 Ewing and Donn (1963) drew on palaeomagnetism to underpin their palaeoclimatic argument for polar wandering but made no reference to Drift; Nordeng (1963) advanced conclusions consonant with Drift but did not explicitly mention it.

19 The rationale for the program pursued by Cox and Doell bore precisely on this point: to collect enough samples from different locations and different geological eras to establish whether or not reversals were a general feature of the earth's magnetic field or of the rocks themselves. See Glen (1982).

20 Among those who commented in this vein in textbooks and monographs were: Goguel (1952t), Scheidegger (1958a), Dury (1959), Jacobs, Russell & Wilson, (1959), Kummel (1961), L.C. King (1962), Kirkaldy (1962), Longwell & Flint (1962), Read & Watson (1962), Scheidegger (1963), Zumberge (1963), Holmes (1965), Ramsay & Burckley (1965), Foster (1966), Read & Watson (1966).

21 Presumably empirical results which manage to surmount this hurdle will carry considerable weight in the evaluation of competing theories or programs.

22 Mulkay, Gilbert & Woolgar (1975: 187) sketch a three-stage development of a 'research area' as follows: an exploratory stage marked by a lack of effective communication among participants and by the pursuit of imprecisely defined problems; a stage of rapid growth and consensus; and a final stage of decline and disbandment of the network. There are correspondences between their model and the development of directionalist studies; one apparent discrepancy is that even in the early stages of the directionalist agenda there was much informal communication and a considerable measure of consensus on methods and aims even as those aims shifted.

23 I share Edge's (1979) reservations concering this approach particularly claims of its proponents that the methods are objective, scientific and therefore to be preferred to others. The importance of informal networks and informal communication are further reasons to question the value of such techniques as citation analysis other than as one of many tools at the disposal of investigators of scientific change. Edge himself (1979: 116–17) gives Bullard's work as an example of how sociometric methods could be systematically misleading.

24 This has some parallels with the old chestnut of whether or not a tree falling in the forest makes a noise if there is no one to hear it.

8

Patterns and puzzles from the sea

The early campaign for Drift was waged on land; success was to come at sea. Before World War II three-quarters of the surface of the globe was marginal to geological theories. The ocean floors were invisible to geologists; they were the province of speculation rather than extensive empirical investigation. Lyell (1830: 81) had realized the limitations this place on land-bound geologists:

> if the great ocean were our domain, instead of the narrow limits of land, our difficulties would be considerably lessened; while on the other hand, there can be little doubt, . . . that an amphibious being, who should possess our faculties, would still more easily arrive at sound theoretical opinions in geology. . . .

In the US close connections between national defense and oceanography developed during the war. These gave new directions and techniques to marine research. After the war there was explosive growth in marine geology and geophysics, fueled by government funding and the further development of remote-sensing devices. American geologists, with the assistance of the US Office of Naval Research (ONR), took an early lead. The ONR was the major source of research funding through the 1950s. This was supplemented by other government agencies such as the National Science Foundation (NSF).

The Navy was not in the business of advancing empirical and theoretical geology *per se*. The oceans and ocean floors were potential theatres of war, particularly submarine warfare. One effect of the Navy's interest was the financing of the ever higher costs of marine geology and the development of sophisticated techniques for the exploration of the ocean floors – techniques which often related to the detection of submarines. A second effect was to give prominence to the few established marine geologists and geophysicists and to attract younger geologists to studies of the ocean floors. There was an influx of recruits into marine geology and geophysics. It is an obvious point but not a trivial one that the population of the geological community as a whole and the relative popularity and importance of its specialties had changed markedly by the mid-1960s from that of the 1920s and 1930s (Menard, 1971: 168–71). The new recruits, who had not been seasoned in the earlier battles over Drift, were attracted to these expanding specialties partly by the lavish funding and opportunities for employment offered through the ONR and the NSF. Moreover, these fields compared favorably in glamor and excitement with aeronautics, space exploration and molecular biology.

Until the early 1960s most of the researchers, with a few such exceptions as Vening-Meinesz, seem to have devoted their efforts to gathering data – to making the oceans more transparent to geologists. Our emphasis in this chapter will be to summarize this harvest of data from the sea. The 'catch' was so varied and bountiful in seismology, sonar studies of topography, gravimetric and magnetic surveys, heat-flow studies, sedimentology and other lines of research based on specific techniques, regions and structures that geologists were hard-pressed to keep up with data in their own bailiwicks. Few were immediately concerned with how this information might relate to problems on land or to rival global research programs. Harry Hess, one of the leaders in the field, complained (Hess, 1954: 341) that 'the energy and ingenuity of the scientists concerned has gone largely into development of new techniques and the acquisition of data, with comparatively little time spent on meditation on the broader aspects of the meaning of the results.' It was not until the late 1950s and early 1960s that some geologists realized that the puzzles and problems which emerged from the sea drew together and cut across highly specialized lines of research and did bear directly and importantly on the competing global models. The oceans of data, in the absence of a global theory to organize and direct research and to make sense of the information being gathered, were not of obvious relevance during the 1950s to the rivalry between Drift, Contractionism, Permanentism and Expansionism. The version of Drift which eventually imposed order on chaos, co-ordinated patterns and solved puzzles did not emerge from a simple inspection of that data.

Wartime explosions

Before World War I, marine research in Europe was conducted largely under the aegis of governmental fisheries and navigation agencies. In the US, sponsorship by private individuals and institutions was more common but funding more meagre. Research was not orientated toward marine geology or the mapping of the ocean floors. There were severe technical constraints on such projects. The *Challenger* expedition which began in 1872 and which extended over three years found that the 'Dolphin Rise' was a mid-Atlantic ridge or mountain range extending for thousands of kilometers. It was a remarkable achievement considering that in the whole cruise only a few hundred deep-ocean soundings were made: the problems of plumbing the depths from an unstable platform using vast lengths of weighted line were formidable indeed!

World War I and unrestricted submarine warfare provoked co-operation between marine scientists and the military in the United States, France, Britain and elsewhere. The first significant remote-sensing technology, echo-ranging, which had been developed just before the war for iceberg detection and underwater message transmission, was applied to submarine detection. After the war relationships between scientists and the various navies were for the most part dissolved.[1] In the US, government funds were replaced by those of philanthropic foundations. In 1922 the Scripps Institution of Oceanography was founded; in 1930, the Woods Hole Oceanographic Institution. Echo-sounding devices were quickly adapted to mapping the ocean floors; these could be used while a ship was in motion and in all sea conditions to generate a continuous profile of the floor under a passing ship. The German *Meteor* expedition in the mid-1920s was, in a time-span shorter than the *Challenger* expedition, able to make over 30 000 soundings. On the eve of the Second World War, much progress had been made in topographical mapping. British oceanographers, for example, maintained a special interest in the Indian Ocean and Red Sea. Seymour Sewell and John Wiseman, who had headed up an expedition to these waters in the 1930s found a ridge system similar to the Mid Atlantic Ridge. They summarized some of the more interesting findings and argued that 'each of the major oceans, especially in the southern hemisphere, is traversed by tremendous mountain-chains' (Sewell & Wiseman, 1938). They were uncertain how these structures should be interpreted; they concluded that the data were too sparse to reach any judgement on the various hypotheses. They proposed further research: topographical surveys of a small section of a ridge coupled with bottom samples, sediment corings, and geophysical measurements. This agenda was shelved when the oceans became battleground of navies rather than theories.

Geophysical mapping yielded intriguing patterns. Maurice Ewing in the 1930s pioneered the techniques of marine seismic reflection and refraction. Ewing had been trained in physics but had had experience with the basic techniques which were used in oil exploration: an explosive charge was detonated on the surface and the resulting pattern of shockwaves, reflected off subsurface features and recorded by a seismograph, indicated the underlying structure. Richard Field, a Princeton geologist and a mainstay of the infant American Geophysical Union, was one of the first geologists to press for geophysical studies of the ocean basins. He convinced Ewing in 1934 to adapt his techniques to ocean-floor mapping.[2] The ocean floors were in some regions bare rock; in others, covered by layers of sediment up to five kilometers thick. Seismic reflection and refraction enabled Ewing to chart the sediments, the basaltic crust and the structure of that crust. Bullard applied Ewing's techniques to British waters. At the outbreak of World War II this research was yielding a crude picture of what had been inaccessible terrain.

The war cemented ties between the US government and marine geology and geophysics. In the post-war years, these ties sustained a rapid expansion of research. During the war, most American and British oceanographers served in the armed forces or engaged in defense-related research. Concomitant with the challenges posed by magnetic mines, submarine detection, and amphibious warfare was an exponential growth of oceanography. The US Naval Research Laboratory increased its staff from 400 to 2000. Woods Hole grew from a summer complement of 60 to a year-round staff of 335. Scripps in 1941 employed 26 staff, owned one small ship, and had a budget of $91 345; by 1948 it had grown to a staff of 250, had four ships, and had a budget of nearly $1 000 000. Moreover, the nature of the war and the problems it posed for oceanographers drew attention to some areas which had been neglected and relegated others to the background. Physical studies; i.e., marine geology and geophysics, received more emphasis; marine biology, less. One result was an immigration of physicists into research on sonar, marine acoustics, and wave-motion prediction. Another was that the government accorded a more elevated status to physical studies in marine geology compared with the position that such studies occupied in land geology. The higher status was maintained after the war.[3]

Maintaining the boom

The huge research establishments built during the war could not be maintained by private funding. Marine geology and geophysics, were they not to wither, required massive government funding. The end of the war saw the outbreak of the 'Cold War' and there was concern that the US maintain the world position it had so recently acquired, a position due in part to the atomic bomb and other scientific research. In 1946 the ONR was established to fund and oversee long-term research projects. This was followed in 1950 by the creation of the NSF, partially in response to the worries of some scientists that their specialties were becoming too dependent upon military funding. In the 1950s and 1960s marine geology and geophysics were the geological equivalents of 'big-machine physics'. They were exploratory, glamorous, expensive and kept afloat by government funding. The abortive 'Project Mohole', the object of which was to drill through the ocean floor to the mantle,[4] and the more successful American effort in the International Geophysical Year (1957–8)[5] were two sides of the same government coin.

The nature of marine geology rendered inappropriate the pecking order of specialties developed on land. Gregory's dictum that geology was based on 'the interpretation of the rocks' sank like a stone in the face of practical difficulties. Geologists could not apply traditional field methods to the sea floors. The could not wander about with sketchbooks and hammers doing stratigraphical surveys, breaking off samples, establishing correlations, examining road and rail cuttings. We speculate that this was one reason why few geologists before the war gave much attention to marine geology or to devising new methods of investigation. There was also an unstated presumption that there were no problems or data unique to the ocean floors. Any general theory which could cope with the continents could surely be extended sight unseen to the sea floor. Why do underwater what could be done more easily and more cheaply on land?

The development of remote-sensing techniques helped circumvent the limitations of more traditional approaches. They also subtly altered the traditional hierarchy of geological specialties and the empirical base used by marine scientists to construct their own theories or to evaluate those of landlocked geologists. This could be regarded as a form of localism dictated by the environment itself. Remote sensing was dominated by geophysicists and physicists. Physical and geophysical data were of primary importance to the empirical and theoretical research agendas of marine geology. Geophysics, buttressed by generous funding, came to occupy the chief place in the pantheon of specialties of seafaring geologists. The emphasis on physical studies was underlined by the formation at Scripps in the immediate postwar years of the Marine Physical Laboratory and, in concert with UCLA, of the Institute of Geophysics.[6] On the east coast, Ewing in 1949 set up the Lamont-Doherty Geological Observatory as part of Columbia University. Over the next two decades the triumvirate of Scripps, Woods Hole, and Lamont played important roles not only in the resumption and further sophistication of sonar, gravimetric, and seismic reflection studies but also in pioneering measurements of the heat-flow and magnetic properties of the ocean floors.[7] Ewing conceived of Lamont as the data bank against which theories and programs would be evaluated.

A more direct effect of war-time co-operation was the development of remote-sensing devices. The techniques of and interest in marine magnetism originated from and were sustained by military interests. In the late 1930s the Gulf Oil Company began to carry out research on magnetic fuses for mines. Victor Vacquier (later at Scripps) was employed on this and realized that the approach could also have applications for geological surveying and could be adapted to moving platforms; e.g., ships and airplanes. In 1942 the US National Defense Research Council pushed for the further refinement of this technology for detecting submarines. The resulting instrument could measure the total magnetic field of an area and could be carried in an airplane: MAD (Magnetic Airborne Detector). After the war, the techniques were further refined so that the instrument could be towed behind a ship to give a continuous plot of the magnetic patterns on the sea floor. In 1955 the US Navy was persuaded by scientists associated with Scripps to include a magnetic survey using a new type of instrument as part of a detailed topographical survey off the west coast of the US. We might conjecture that little persuasion was needed: since magnetic detection systems could be used against submarines, maps of the 'natural magnetism' of the sea floor might be of use in concealing friendly submarines or detecting hostile ones. Oceanographers probably had no particular theoretical expectations of what such magnetic surveys might reveal. Nonetheless, any new, systematic physical measurements would add to the meagre stock of information about this gigantic geological province. Of course, this would also further reinforce the primacy of geophysics in that domain (*cf.* Speis, 1980; Wood, 1985: 143–4).

Earthquake seismological techniques gave a data set which could be exploited by marine and land-based geologists alike. This area too benefited from military patronage.[8] There had been several attempts to form a worldwide network of seismic stations prior to the First World War; the only one to survive into the 1950s had been organized by the Jesuit order. In 1954 this network seismically detected the hydrogen bomb test at Bikini Atoll: seismology now had practical applications. In 1958 negotiations between the USSR and the US on a nuclear test-ban treaty were begun. A sticking-point was verification: how could underground tests be detected? The answer pursued in the United States was the expansion and sophistication of seismological techniques and the establishment of seismic stations. Beginning in 1961 a network of some 120 stations, the World-Wide Standardized Seismograph Network (WWSSN) was constructed. The data gathered on earthquakes by this network were collected, collated, analyzed and plotted on a global scale. Unlike many other geological specialties, seismology dealt not with the history of the earth but with present-day processes and their correlation with known geological structures.

The harvest from the sea

Prior to the 1950s, the continents were familiar; the ocean floors, mysterious. By the mid-1960s, when new versions of Drift were being put forward by marine scientists, the ocean floors were better-mapped geophysically than the continents. Bill Menard, a leading marine geologist at Scripps, described the enormous advances in empirical knowledge as follows (Menard, 1986: 41):

> By 1956 Lamont, using Navy submarines, had tripled the number of gravity observations at sea. . . . There were no heat-flow observations in the oceans in 1946, but by 1963 Scripps had taken about 300. . . . There were perhaps 100 cores of sediment from deep-ocean basins in 1948. By 1956 Lamont had taken 1195 . . . by 1962 Scripps had about 1000. There had been not seismic stations in the deep sea, and by 1965 there were hundreds. . . . The number of deep-sea soundings had increased by about 10^8 and the number of plotted soundings by 10^5. Nothing comparable to shipboard magnetic profiles had ever been known, and Lamont and Scripps had towed magnetometers for hundreds of thousands of kilometers. . . . Even in 1964 I was only half jesting when I wrote of a 'digression from the familiar ocean basins to the mysterious continents. . . .'

The data which flowed so abundantly to the surface revealed that the ocean floors were far simpler than the continents but rather more complex than anticipated. Some major relief features; e.g., trenches, the East Pacific Rise and the Mid Atlantic Ridge, had been partially charted before the explosion of marine science. Most geologists in the mid-1940s expected the floors to be otherwise rather featureless. They were presumed to be very old (even Drifters presumed the Pacific to be quite ancient), to be smooth and gently sloping except for the occasional volcanic island, and to be buried under five or more kilometers of sediment (the product of millions or even billions of years of continental erosion and of the detritus of marine life). The actual oceanic crust was thought to be tens of kilometers thick. All of these expectations were to prove incorrect. None the less, the floors were to prove geophysically and geologically simpler than the continents. There were striking regularities, patterns, correlations between topographical features and geophysical data which contrasted with the chaos of the continents.

All of the ocean floors seemed to be similar to each other and to differ in thickness and composition from the continents.[9] By the mid-1950s Ewing had reaffirmed, 'contrary to the prior assumptions of almost all geologists and geophysicists', that all the ocean basins were similar: they were composed of basalt and were only about six kilometers thick (Wertenbaker, 1974: 98). The continents were formed mostly of granite and were over 40 kilometers thick. The abyssal plains were flatter and larger than any corresponding part of the continents. Yet, cut by deep trenches, thrust up into mountain chains and peaks, carved by canyons and valleys, the sea floor was a very uneven floor indeed. The floors appeared, however, to display unusual order and regularity. The results of remote sensing, as well as the data garnered from dredging and shallow cores, revealed unexpected patterns associated with this topography. Earthquake distributions, sediments and thermal, magnetic, and gravimetric anomalies were neither uniform nor randomly scattered. Surprisingly, the ocean floors by the mid-1950s seemed to be quite young compared with the continents: except for a few odd samples, the sediments, cores and scattered rocks yielded going back at most to the Cretaceous (about 140 MYA). There was disagreement about the interpretation of data; nonetheless, the apparent regularities and symmetries stood in striking contrast to the continents.

The most spectacular topographical features were the mid-ocean rises and ridges ascending thousands of meters from the floor. The Mid Atlantic Ridge was not unique. By the later 1950s a globe-girdling network of ridges extending over 60 000 kilometers had been mapped. Their origin was unclear: were they the result of lateral compressional forces generated by a contracting, collapsing earth squeezing up the crust; tensional features where the crust was being pulled apart; the product of volcanic outpourings; or the scars of continental fragmentation and drift? Many of the ridge segments possessed a symmetry of form: they were bisected along their

Figure 8.1. The world-wide system of oceanic ridges including the East African Rift Valley and the San Andreas Fault. Compare with the plots of shallow-focus earthquakes in Figure 8.4. (after Runcorn, 1962b: 38).

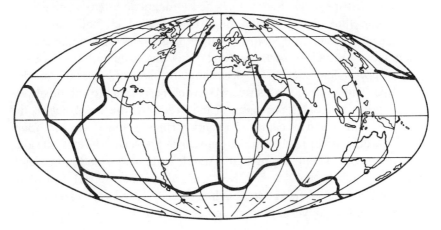

length by what appeared to be a valley not unlike the great Rift Valley in Africa. Indeed, the ridge system in the Indian Ocean appeared to extend through the Gulf of Aden and to connect up with the East African Rift (Ewing & Heezen, 1956a). By the later 1950s and early 1960s the ridges were most often interpreted as tensional in character.

The ridges were associated with geophysical regularities. The floors in general emitted roughly the same amount of heat per unit area as the continents. This in itself was surprising since from the early work of Holmes in the 1920s it was thought that radioactive materials – the presumed source of most of this heat – were concentrated in the cooled scum, the continents, rather than in the ocean basins. The highest flow of heat was on the ridges; their flanks were relatively cool. This gave some support to the view that they were volcanic in origin. This was reinforced by the slim evidence obtained from dredging and coring: samples from ridges were almost invariably young volcanic rocks; layers of sediment were almost entirely absent. On the other hand, true oceanic sediments tended to be thicker and older near the continents: quite the reverse of what would be expected on Wegener's version of Drift. The ridges were among the youngest features of the youthful ocean floors.

Earthquake plots and magnetic anomalies revealed other patterns. The crests of the ridges and the median valleys were associated with shallow-focus earthquakes (originating in the upper 70 kilometers of the earth). So much was this the case that Ewing had been led originally to link up the heretofore separate ridges into a network and to predict the locations of ridges in the Antarctic by seismic plots (Ewing & Heezen, 1956b). Deep-focus earthquakes (more than 300 kilometers deep) were almost entirely absent from ridges. Magnetic surveys mapped out regions of magnetic anomalies: alternating strips of sea floor which had a magnetic intensity greater or lesser than that of the earth's mean field and which had patterns of magnetization different from those of their neighbours. The source of these anomalies was unclear; perhaps it had something to do with the composition of the sea floor such as the occasional inclusion of bodies of iron-rich minerals. These alternating strips were roughly parallel to the ridges.

The other striking topographical features were the great deeps, the trenches, which had so captured Vening-Meinesz's and Hess's attention. Trenches did not form a continuous network but seemed to be located along the west coasts of Central and South America and near islands and island-arcs such as Japan, the Aleutians, New Zealand, the East Indies, New Guinea and the Marianas. Vening-Meinesz had already noted the correlation of trenches and negative gravity anomalies and this, together with some

Figure 8.2. Map of magnetic anomalies in the vicinity of the Juan de Fuca Ridge. The 'stripes' represent zones of stronger (dark) or weaker (light) magnetism relative to the earth's field. This ridge segment and the magnetic anomaly patterns associated with it were later used as key evidence for a new version of Drift. (after Raff & Mason, 1961: 1268, by permission of the Geological Society of America).

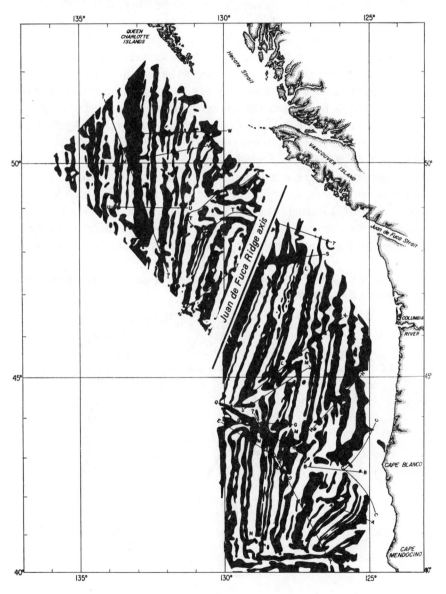

limited seismic data, had led him to identify the trenches as areas of active crustal deformation and down-buckling. Other regularities and correlations were soon established. Trenches seemed to be associated with active volcanos, especially those arranged in the circum-Pacific 'Ring of Fire'. Gutenberg and Richter's elegant plots of earthquake foci (Gutenberg & Richter, 1949; 1954) confirmed and extended Vening-Meinesz's observation that seismic activity was associated with trenches in the Dutch East Indies and the Caribbean. They found that intermediate and deep-focus earthquakes were – with few exceptions – all located near trenches.

There are far more consensus on the nature of the trenches than on ridges. Trenches, most agreed, did represent substantial down-buckling or subsidence of the oceanic crust. This measure of agreement may have been in part because this view was assimilable to two of the major theories of orogeny. Advocates of the geosynclinal theory of mountain building, usually part of a 'Fixist' global program, pointed to trenches as examples of how local subsidence of the ocean floor could produce a deep pit which could be filled with sediment and then squeezed to form a coastal mountain range. 'Mobilists' could interpret the trenches in the manner of Vening-Meinesz and enter them as further evidence for the existence and geological importance of convection currents. There was, of course, considerable disagreement as to the mechanism responsible for trenches; proponents of

Figure 8.3. Location of the major oceanic trenches or 'deeps'. Compare with the plots of deep-focus earthquakes in Figure 8.4.

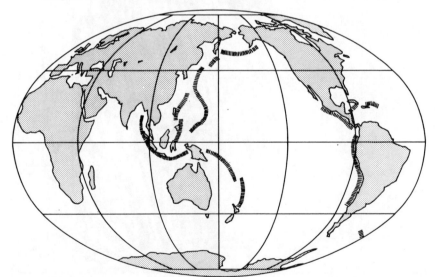

geosynclines usually argued in favor of 'block-faulting' (i.e., the sinking of slabs of crust); those who favored convection currents, of the 'sucking down' of crust through the action of currents beneath that crust. Many geologists regarded such currents as hypothetical in the extreme, but support was much stronger among geophysicists.

Cleaning the catch

The data gathered had yielded unexpected patterns and regularities in the ocean floors. Marine geologists and geophysicists did not regard the regularities associated differentially with trenches and ridges to be purely coincidental. They were taken to be causally related. The question which provoked disagreement was what was (or were) the cause(s)? Although we have passed over the issue in this chapter, the data was of course neither gathered nor arranged in a theory-free environment. 'Fixists' and 'mobilists' alike delved into the oceans and displayed their catch according to their theoretical preferences. Even so, it was not clear how most of this data or the

Figure 8.4. Plots of shallow- and deep-focus earthquakes. The centers of earthquakes do not seem to be scattered randomly; most are clustered in fairly well-defined patterns. Many more, and more precise, plots were subsequently made with the aid of the World-Wide Standardized Seismographic Network in the 1960s. (after Gutenberg & Richter, 1954: 14–15).

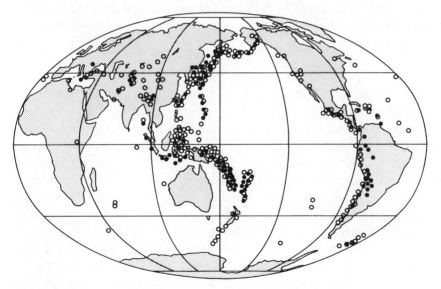

o Large shallow-focus earthquakes
• Large deep-focus earthquakes

patterns could be fitted to continent geology or the theories evolved on land. Ewing described the mood of delegates to the first International Oceanographic Congress, held in 1959, as follows:

> Most believed that the Mid-Ocean Ridges must be the source and cause of some sort of wholesale motion of the earth's crust, like continental drift. What form of motion was quite unknown – except that it was not Wegener's – and the subject of the liveliest curiosity. Too tentative to get more than passing mention in the papers delivered, the question was discussed in the halls (Quoted from Wertenbaker, 1974: 176).

Should the topography of the ridges be explained by one theory, the magnetic anomalies by another, the trenches by another, and the gravity anomalies by still another, or was there one theory which could integrate all in a coherent manner? Revelle in a 1959 speech listed a series of questions about ridges, trenches, sediments and so forth which he hoped further research would answer (Revelle Papers, SIO, Box 29, Folder 29: 11–13). There is no clear suggestion that these questions might be interlinked or answerable by one or even a small number of theories. These data and questions posed formidable challenges to the rival research programs in geology, programs which had heretofore been relatively landlocked. This also created fresh problems in a number of specialties. By the late 1950s and early 1960s geologists were attempting to harmonize the new findings with the old theories or to develop new versions of those theories. The revival of interest in Drift, stirred by palaeomagnetic studies, swelled into a second controversy over global theories, a controversy in which the ocean floors were central.

Voice-Over

The rapid growth of submarine geology and geophysics after World War II was dependent upon and spurred by government funding. This was given in the belief that the research purchased was of value in the military sphere. It may well be true that much of the research conducted had no overt bearing on defense matters and that still other projects invoked 'national defense' or 'national interest' simply as magic words to open the door to government treasuries. The question is nonetheless raised: to what extent did 'external' factors such as sources of funding affect the development of marine geology as a theoretical and empirical field in the 1950s and early 1960s? This is, of course, a more specific form of the question of the extent to which science is influenced by or is a product of the society to which it is practised. On this question there is a broad spectrum of convictions ranging from the view that science is completely autonomous to the view that scientific knowledge is, at bottom, a social construct so that just as we can talk in terms of a culture's music or theology or art we can also talk about that culture's science; i.e., its network of beliefs about the 'natural world'. The extremes of complete autonomy and cultural construction do not seem applicable to the development of marine geology and geophysics and their connection with the theoretical conflicts of the 1960s.

To the defender of autonomy we can make the obvious reply: the rate of growth of oceanographic research was directly dependent upon massive funding which was tied to particular historical and social circumstances (submarine warfare, inter-service rivalries, foreign and domestic politics, personal contacts, lobbying efforts, and so forth). A similar argument could be made with regard to the differential recruitment of scientists into the various specialties comprising geology and the development of increasingly sophisticated equipment. Moreover, it was not only the rate of growth that was affected but also the directions of research pursued; the importance assigned geophysical work, for instance. These changes in the rate and direction of growth, and indirectly, the circumstances which brought them about, had more subtle effects and implications. They may, for example, have had a delicate influence on the course and outcome of the Drift debates in the 1960s.

In the matter of timing, the wealth of new data began to filter into the broader geological community about the time that palaeomagnetic directionalist studies were stirring interest in Drift. We can only speculate how the revival of Drift would have proceeded – if it had proceeded further – had these waves of data struck five or ten or more years later. Alternatively, we might ask if Drift interpretations would have been offered for ridges, trenches, and associated phenomena had not Drift been thrust into prominence by Blackett, Runcorn, and others. Some Permanentists, Contractionist and Expansionist accounts of the marine data were offered and others might have followed and gained wide currency. It could be argued that both the palaeomagnetic and the marine data would have been gathered 'sooner or later' and that, therefore, the 'modern revolution' was 'inevitable'. Amongst the many problems of such an argument, it ignores what may have been a crucial juxtaposition within a short span of time of both sets of data. The apparently rapid progress of Drift – compared with its rivals – in coping with both sets was likely an important factor in the rapidity with which it came to occupy a position of dominance in the discipline.

In the matter of the direction of growth, changes in the perceived hierarchies of geological specialties and problem-fields were probably relevant to the acceptance of Drift. If Drift seemed to be an effective problem-solver within certain specialties or with respect to certain sets of problems and if those specialties and problem-fields grew in prestige, importance, and size at the expense of others in which Drift seemed not so useful, then the chances of Drift being accorded a more sympathetic hearing are surely increased. In this manner such an 'external' factor as funding could at least indirectly affect the evaluation of competing theories and research programs. This, as we shall see in the ensuing chapters, is not a purely hypothetical argument.

The organization and institutions of a discipline or field of research may also affect the rate and direction of development. Much of the funding for US marine research was channeled into Scripps, Woods Hole, and Lamont. These institutions were the major foci for data gathering and many of the pre-eminent marine geologists and geophysicists were affiliated with them. These centers had by the early 1960s won authority and prestige in empirical oceanography. They not only received substantial government funding themselves but, through the peer-review process, would have exercised some control on the directions and rate of research by individuals not affiliated with them. Bullard, for example, wrote to Scripps in 1948 that he had been interested for a long time in measuring heat-flows on the floors but that he had not bothered to design the needed equipment since there was no ship available in England for the purpose.[10] The subtle hint was picked up and Scripps supported his studies.

Maurice Ewing, the director of Lamont, was quite blunt in expressing his preference for the geophysical approach as opposed to the methods and attitudes of more traditional geologists; the latter he characterized as: 'Annoying fellows . . . [who] spend their time poking around trying to explain this or that little detail. I keep wanting to say, "Why don't you try to see what's making it all happen?"' (Quoted from Wood, 1985: 138). At the level of techniques and the relative importance of specialties, Ewing exercised a dominant influence on his institution. At the level of aims and strategies he was equally dominant. He preferred Permanentism but took the view that only the collection of data rather than the propagation of or debates about theories could resolve the issue. Despite his own convictions, he did not turn Lamont into a citadel of fixism. As we shall see, Lamont in the late 1950s and early 1960s harbored considerable theoretical diversity. Members of its staff entertained, employed, pursued and accepted a range of programs and theories. Its greatest importance lay in the vast treasures of data meticulously stored up during Ewing's tenure. When Lamont fixists turned to that data in the mid- and late 1960s to demolish Drift, they found themselves instead demolishing fixism. This pattern was roughly the same at Scripps and Woods Hole. The aims, activities, and emphases of the leading formal institutions of oceanography reinforced and were reinforced by those of their sources of funding. This is hardly suprising but does not constitute further evidence in favor of the manifold effects of the 'extra-scientific' context in which science is practised. With respect to the role of the institutions mentioned above, it seems that those effects fell short of a 'social construction' of the content of science except in the sense that preference was given to data gathering over theorizing and to some form of data gathering over others.

All this does not warrant a strong 'social constructionist' interpretation. We may fairly conclude that descriptive models of scientific change and development should not ignore the social context in which science is practised but it does not necessarily follow that those models must take science to be purely an artefact of social production. Trenches, ridges, magnetic anomalies, and so forth were not social artefacts manufactured by the military. As we shall see, the empirical harvest of the 1950s and 1960s provided challenges to and constraints on the theories and programs defended, advanced, and abandoned. And, those theories and programs were overtly evaluated less on the basis of nationality, institutional affiliations, sources of funding, ideology, and so forth than on the basis of how well they worked in specific problem-fields and specialties and how well they could make sense of the global patterns emergent on land and sea. Finally, the substantial external funding for marine studies was not directly theory-specific. What was prized and rewarded was the acquisition of more data, not speculation on how that data might relate to global theories or even how those theories might solve problems of national defense. However, if marine geologists and geophysicists received preferential funding and a consequent increase in status within the discipline and if many of them preferred Drift, then indirectly there may have been a connection between funding and theoretical issues.

Thanks for the memories

We have noted earlier (pp. 32–33) some of the problems of 'practitioner' histories. Similar strictures apply to 'participant' histories and to the use of scientist's recollections set out in autobiographical sketches, printed reminiscences and interviews. Such 'memories' can correct and supplement formal papers, correspondence, conference proceeedings and other documentary sources. 'Memories' may disclose (stated) motivations, frank judgments of the work and credibility of other participants, informal networks and other social and cognitive aspects of science which are not preserved except in memories. 'Memories' are a resource – but not a specially privileged one – on which I and others (e.g., Glen, 1982) have drawn in constructing histories of recent changes in geology. Since many of the participants are still active, we may expect that more 'memories' will be forthcoming and that these will add detail, color and new insights to the events of the past three or four decades.

These 'memories' cannot be taken at face value. Participant accounts are not a record of 'what really happened' or the 'true story'. We should beware, for example, of any historian or sociologist or philosopher who, after interviewing a scientist, makes a special claim for accuracy or authenticity on the basis of 'getting it straight from the horse's (scientist's) mouth'. There are obvious pitfalls: the reconstruction of the past in light of subsequent developments, the selectivity and vagaries of memory, the familiar problem of 'eyewitness' testimony, the interpretative glosses given consciously or subconsciously by the participant (or the interviewer) and so on. More fundamentally, no historical account of science (with or without the use of 'memories') – or even science itself – should be conceived of as a neutral recital of the facts. 'Facts' related by participants are no less problematic than 'facts' set down in scientific publications or other sorts of 'facts' used to investigate the nature of scientific change – but they are no less welcome.[11]

Notes

1 One notable exception was the co-operation which Vening-Meinesz secured from the Dutch navy; another was the joint Princeton–US Navy expedition to the Caribbean in which Hess and Vening-Meinesz participated.

2 Field exhorted physicists and geophysicists to turn their attention to the geophysical exploration of the oceans. Three proponents of Drift in the early 1960s – Hess, Bullard, and J. Tuzo Wilson – were inspired by Field's fervent preaching. See Wood (1985: 125–32) for further details.

3 Schlee (1973) gives an overview of these developments; at a more detailed level, a brief sketch of Roger Revelle's career (Day, 1985) illustrates the impact of the war and of government funding on the career of an influential oceanographer and marine geologist.

4 Amid charges of influence-peddling, a 2500% increase in estimated cost and acrimonious disputes among scientists, the project was terminated by an Act of Congress (the only basic research project thus far to win that dubious distinction). Cf. Greenberg (1967).

5 R. Laudan (1986a) makes the point that a thorough 'testing' of Drift in the 1920s and 1930s would have required an enormous commitment of resources and international co-operation for which at the time there was insufficient justification. The IGY some three decades later represented such a commitment and such co-operation but was aimed not at testing theories but at data gathering – in accordance with the expressed priorities of most geologists.

6 The Scripps archives testify to the importance of the links forged during the war between science and government: Day (1985); SIO Office of the Director Records (1943–55) (Folder 11) states that for a two-year period ending 1 July 1952 government contracts amounted to nearly three-quarters of the research budget of the Institute.

7 J.Tuzo Wilson (1957: 73) commented that geology and geophysics would not have become estranged had geological methods of observing the visible part of the earth not developed so much earlier than the physical methods required to study the rest of it.

8 Wood (1985: 155–9) gives a useful overview.

9 This had been assumed in Wegener's distinction between sial (granite) and sima (basalt). This distinction had, however, been challenged. Gutenberg had claimed that seismic evidence indicated that the Pacific floor was basaltic but that the Atlantic and Indian floors were granitic and therefore similar to the continents (Gutenberg, 1936). This was in harmony with his theory of 'continental spreading' as opposed to 'drifting'. His claim received credence, perhaps all the more so because it made more plausible sunken bridges or continents despite the absence of large gravity anomalies over the bulk of the floors which granitic strata would have produced.

10 Bullard to Eckart, SIO Office of the Director (82–106), Records 1943–55, Folder 7.

11 See Thompson (1978: esp. 91–137) for a sympathetic discussion of oral history; Murphy (1986), for more critical remarks.

9

Shifting theories

The patterns and puzzles of the ocean floors opened a new theatre for competition among the programs and theories of continent geology. Geologists interested in *terra firma* remained aloof into the early 1960s. Marine geologists and geophysicists took the lead making the old views amphibious. By the early 1960s marine scientists had advanced global theories aimed at accounting for the sea floors. The geology of the continents was appended almost as an afterthought, a stunning reversal of the traditional order of priority observed by land-based specialists.

George Lees, President of the Geological Society of London, auditioned as King Canute at a 1953 symposium on the ocean basins. He was convinced that the continents had been shaped by contraction. This, he contended, fit the 'seen' geological evidence on land, implying that the 'unseen' evidence from the oceans was of minor importance. In any case, he claimed that there was 'presumptive evidence' that his model fitted the ocean basins as well. He rejected the permanence of the ocean basins and the existence of any fundamental differences in composition between continents and ocean floors. As for the tide of data from those floors and the interpretations of it offered by geophysicists which conflicted with his own land-locked theory, let them be rolled back: 'deductions drawn from geophysical observations – seismic, gravity, and magnetic – have unduly influenced geological opinion' (Lees, 1953: 217). The backbone of geology was stratigraphy and correlation of formation; geophysics was a distracting frippery. He expressed these same sentiments at a Royal Society discussion 'On the Floor of the Atlantic Ocean' (Lees, 1954).

Lees's, aspersions were challenged by sea-faring geologists. At the Royal Society, Hess and Bullard defended marine studies. Hess flatly contradicted Lees on a key point: the floors *are* different in kind and structure from the continents. Further, Hess outlined several explanations of features of that floor which made no reference to the hypotheses of continent geologists (Hess, 1954). Bullard affirmed the centrality of geophysical data for marine scientists: 'The true picture of the oceans must fit the considerable body of geophysical facts about the ocean floor' as well as other evidence. Only then did he add, inverting Lees's order of importance, that the true picture 'must fit satisfactorily with our much more detailed knowledge of the continents' (Bullard, 1954). For Bullard as for Hess, the oceanic crust was different from continents: it was not simply drowned land.

Continent geologists had their own research agenda in which the oceans played a very minor part. Those land-based geologists who did evince interest in the harvest from the sea mostly sought to pick out bits of the 'catch' which would fit into their land-based theories and programs. Ridges were simply underwater equivalents of the Alps or the Appalachians; trenches and submarine canyons, equivalents of geosynclines and valleys. Kenneth Landes in his 1951 Vice-Presidential Address to the geology section of the American Association for the Advancement of Science welcomed the new findings. He accepted Contractionism though he admitted it might be shrinking in popularity (Landes, 1952: 225). Instead of commanding the tide to roll back, he sought to channel it to support his preferred program. There were differences in the composition of continents and ocean crust; gravity had caused the latter to subside. Trenches were likened to rift valleys and, like them, were due to differential collapse of the crust. Deep-focus earthquakes near the trenches were produced by the pressure of collapsing seafloors against neighboring continental masses. Ridges and island arcs were likened to mountain ranges: all were products of lateral forces generated by contraction and subsidence. His aim was less to account for the features of the ocean basins than to apply Contractionism to glaciation, palaeobiogeography, epicontinental seas and other landlocked processes. This harmonized with a version of Contractionism which incorporated a geosynclinal account of orogeny and the notion of continental accretion. By the mid-1950s many North American geologists, both Permanentists and Contractionists, had adopted this view (cf. Dott, 1974, 1978) in which the present continents were the result of gradual additions to the margins of permanent, stable 'cratons' identified with the ancient 'shield zones'. The Canadian Shield, for example, was an ancient and stable formation which had undergone little deformation since the Pre-Cambrian (590 MYA). On the margins, geosynclines formed by subsidence, filled with sediments, and then squeezed by lateral forces produced new mountains which were added to the Shield. This tallied well with the results of continental rock-dating which indicated that the shields were generally much older than the surrounding landmasses and mountains.[1]

J. Tuzo Wilson, a Canadian geophysicist, in the 1950s followed an approach similar to Landes. Wilson as an undergraduate had switched from physics to geology; his professors were 'appalled'. Ernest Rutherford, the famous physicist who had worked at McGill University in Canada, had compared geology to 'postage-stamp collecting' in that 'it consists of making maps by identifying and locating rocks and fossils'. It lacked general theories, 'which were scorned as "armchair geology"' (quoted in J.T. Wilson, 1985: 143). Wilson was in the 1960s to help change this image but his early career gave little hint of that. He received an M.A. at Cambridge where he imbibed Jeffreys's Contractionism. He then studied for a Ph.D. at Princeton with Field and Hess. He returned to Canada where he was employed with the Geological Survey until the war. His Contractionism was solidified by his extensive fieldwork on the Canadian Shield. As he put it, 'nothing could appear more rigid or unchanging' than the region in which he had such a strong interest and expertise (J.T. Wilson, 1966a). Despite his field exper-

Figure 9.1. In the geosynclinal version of continental evolution, the Canadian Shield was the ancient core or 'craton' of North America to which had been added more continental crust through the filling and uplift of geosynclines on its margins. (Reproduced from Holmes, 1944: 388, by permission of Van Nostrand Reinhold (UK).)

ience he was not a philatelic geologist. Field methods would not reveal 'the processes that have given the earth a history', for that, 'a consideration of physical laws is necessary' (J.T. Wilson, 1951: 85). After the war he was appointed Professor of Geophysics at the University of Toronto.

His skills at formulating grand generalizations embracing geophysics and traditional geology soon came to the fore. In 1950 he drew upon seismic and structural parallels between islands arcs and young mountain ranges to argue for a cooling earth as the cause of orogeny (J.T. Wilson, 1950). He praised the new geophysical instruments and techniques and foresaw their corrosive effect on older geological theories. In 1957, for example, he summarized the new data garnered from the ocean floors and claimed that the 'new discoveries in geophysics are demanding a reconsideration of much geological dogma handed down from the last century . . .' (J.T. Wilson, 1957: 73). His reconsideration led him to a new version of Contractionism. Still under the spell of the Canadian Shield (Jacobs *et al.*, 1959: 348–9), he argued for accretion by mountain building through geosynclines around stable continental nuclei. However, he rejected a simple cooling of the earth as the mechanism. The new knowledge of the ocean basins required a new version of his preferred global theory (Jacobs *et al.*, 1959: 230): 'Until very recent times, theories about the structure of the crust depended upon observations made upon the . . . accessible parts of the earth: the continents and islands. The theories were correspondingly deficient and biased.' He aimed to provide an account of the four principal types of structures of the earth's surface: ocean floors, ridges, active island arcs and mountains, and the continents themselves.

Wilson's global model divides the earth's surface – continents and ocean basins alike – into segments, the edges of which are defined by 'fracture systems' marked by volcanic, seismic, and orogenic activity. The mid-oceanic 'fracture system' is marked by the ridges and was an 'old, fundamental, and immobile feature of the earth's surface upon which the mid-ocean ridges had grown by the slow extrusion of basalt from the uppermost layers of the mantle' (Jacobs *et al.*, 1959: 347). The ridge system segments the ocean floors. The continental fracture system is made up of two active belts, one of which extends around the Pacific Ocean from Indonesia up and over through the Aleutians and down the west coasts of North and South America; the other from Morocco through the Alps and the Himalayas and south-west Asia till it meets the other belt in the Celebes (Jacobs *et al.*, 1959: 290). The primary mountains are formed from volcanos bringing up magma from the lower layers of the mantle; secondary mountains, by erosion, the deposition of eroded material in geosynclines, and the compression of those geosynclines as the earth contracts (Jacobs *et al.*, 1959: 345). Continents are formed from deeper mantle material and grow through the addition of mountains on the edges of shields. The continents differ in composition from the ocean floors. The collapse or contraction of the earth occurs through an expulsion of the mantle material onto the 'original' surface of the earth which is marked by the Mohorovicic Discontinuity.[2]

Wilson gave a tour of the earth's surface in which the major features of each segment were explained in terms of the theory. His theory was global in scope and scale; it is comparable with those of Suess, Carey and Wegener. Moreover, he attempted to incorporate the latest findings from both marine and continent geology. He addressed himself not only to the data, but also to other versions of Contractionism and also to such rivals as Drift (which he distinguished from convection-current theories of crustal mobility). He rehearsed the usual physical arguments against Drift and claimed that all the evidence for it and all its problem-solutions could be incorporated into his theory. His views were unlikely to be dismissed lightly. He had acquired a substantial reputation as a geophysicist, and in 1957, the start of the International Geophysical Year (IGY), he had been elected President of the International Union of Geodesy and Geology, a key organization in the IGY.

Lester King looked to the new data for evidence for Drift. Wegener and Daly had pointed to the Mid Atlantic Ridge as a scar left by the split of the New World from the Old. King claimed that the global network of ridges were scars of continental separation. The ridges and associated patterns were for him of little interest; the ridges were merely templates to be used to fit the continents back together: an 'independent' means of reconstructing Gondwana which could add weight to arguments based on stratigraphic correlation and palaeoclimatology (L.C. King, 1958a,b,c; 1961).

Throughout the 1950s most marine geologists and geophysicists were too submerged in the successive waves of empirical information to fashion or refashion theoretical structures to absorb them. There were often long delays in the publication of data. Marine scientists gathered reams of measurements on a voyage, returned to port, and then set sail again before publishing or even digesting previous results which would often be super-seded by the next batch. Much of the information as a consequence was not readily available; it circulated almost by word-of-mouth among kindred specialists (Menard, 1986: 49). The theories which were used were often borrowed from programs contending for supremacy on land. By the late 1950s and early 1960s, however, some had begun to adumbrate solutions for marine patterns and problems. Most of these solutions were modelled on those developed within landlocked programs. The ridges, perhaps because of their magnitude and apparent likeness to mountains, were the focus of special attention.

At Lamont there was no unanimity on how oceanic structures and regularities were to be explained. Ewing was above all concerned with gathering data. His attitude seems to have been that it was fruitless to worry about theory until all possible data had been gathered. He seldom theorized after the mid-1950s; when he did, he usually opposed Drift. Even when in 1960 he proposed a speculative 'polar control' theory of glaciation, according to which the poles of the earth had shifted, he explained this as a shift of the crust as a whole relative to the poles: the relative positions of the continents were unchanged (Ewing & Donn, 1963). His attention was fixed on the oceanic crust and ridges. His own work had convinced him that the crust was thin but strong and rigid: continents could not plough through it. As for the ridges, he seems not to have considered them unusual: they were like the mountains on land. He had drained the water from the oceans and found nothing unfamiliar. Others at Lamont entertained views on such as Expansionism and Drift which Ewing thought wildly radical but tolerated.

A swelling earth

Bruce Heezen at Lamont reached a very different conclusion from Ewing's. He acknowledged in a 1958 symposium on oceanography that palaeomagnetism and other studies had reawakened interest in Drift but contended that the trenches, ridges, and the phenomena associated with them could be better explained by a program other than Drift. He favored Expansionism. The trenches were tensional features stretchings and thinnings of the crust as the earth grew larger. The ridges themselves had median rift valleys and, like rift valleys on land, were extensional structures (Heezen, 1959b). Heezen, like Carey, thought on a scale appropriate to global theories: rather than restricting himself to a single mountain range or canyon, his theory encompassed at least three-quarters of the earth's surface. Unlike Carey, who had reached his conclusion by more traditional methods such as spherical projection and map-fitting, Heezen had reached his by grappling with the problems posed by the seafloors. He campaigned for a growing earth, despite Ewing's preference for Fixism. In 1959 he took up polar-wandering curves alleged to be evidence for Drift. From the viewpoint of marine geology they were not persuasive: there were no 'wakes' of moving continents left on the ocean floors. However, Heezen hinted that an expanding earth could accommodate palaeomagnetic data as well as marine data though he cautioned 'in view of the meagre evidence now available, this may seem too drastic and may itself have other more serious objections of astronomical nature' (Heezen, 1959a). By 1960 he seems to have been more confident of Expansionism. In a semi-popular exposition he criticized Drift and sought to integrate many of the diverse, newly-discovered features of the seafloor into Expansionism (Heezen, 1960). His discussions were, however, never more than sketches. He neither supported his speculations with a wealth of geological material as Carey did nor do more than pay lip-service to the physics which might underpin them.

Heezen was one of several advocates of Expansionism in the late 1950s and early 1960s. While Expansionism had been advocated by a few European geologists and geophysicists in the 1920s and 1930s, Carey and Heezen seem to have arrived at it in ignorance of this European tradition. Heezen, apparently independently of Carey, had proposed it on the basis of marine geology. Carey had come to it through reflecting on large-scale features of the continents. Expansionism offered solutions to what he regarded as problems of continent geology: marine features were of secondary importance. Carey in 1958 published a volume of papers from his Tasmanian symposium on Drift which included his lengthy case for Expansionism as an alternative to Drift. He not only developed the geological arguments for an expanding earth in much greater detail than Heezen, he endeavored (and endeavors) to answer objections arising from apparent conflicts with physics. He also proselytized for the program in his publications (Carey, 1958b, 1963, 1976, 1981a) and in person. Both Carey and Heezen advocated a 'rapid expansion' of several centimeters per year. Laszlo Egyed, a Hungarian physicist and geophysicist at the Eötvös University in Budapest, immediately prior to Heezen and Carey had presented a formal geophysical case for expansion (Egyed, 1956, 1957). He based it upon a novel theory of the inner structure of the earth and supported it with some geophysical evidence. Egyed proposed a 'slow expansion' of the earth's radius of several millimeters per year. A few years later he suggested a palaeomagnetic test (Egyed, 1960). *If* the palaeomagnetic co-ordinates of samples of deposits from two widely-separated points on a continent were measured precisely and *if* these deposits were formed at the same time such that they acquired their magnetism simultaneously and *if* there had been no change in the distance (no deformation of the crust) between these points since the deposits were formed, *then* the radius of the earth at the time of their formation could be calculated and compared with the present radius. Several sets of tests were performed but were inconclusive. They seemed to rule out rapid expansion but not Egyed's slow expansion (Verhoogen, 1985: 405). As we might expect, opponents of Expansionism stress the negative results of the tests made to date; proponents, including proponents of rapid expansion, point out that it is unlikely that all of Egyed's conditions were or can be met; e.g., that the crust between sample sites has undergone no deformation or that the samples are exactly contemporaneous.

At roughly the same time that Egyed, Carey and Heezen each for different reasons proposed Expansionism on geological grounds despite apparent conflicts with physical theory, some physicists were led by novel physical theories to look to geology for evidence that the earth had expanded. Pascual Jordan, a physicist at Hamburg, and R.H. Dicke, a physicist at Princeton, independently put forward Expansionism on cosmological grounds. Both had taken up a suggestion made in the 1930s by the English Nobel laureate and theoretical physicist P.A.M. Dirac that the universal gravitational constant (G) might not be constant after all. If so, gravity might decrease with time. Jordan and Dicke pursued the implications for gravitational theory and concluded that one effect – which might be measurable – was that the earth should have undergone a slow expansion (Dicke, 1957; cf. Jordan, 1966t). Dicke suggested several geophysical phenomena which might be associated with an expanding earth but was diffident about the role of expansion in explaining the gradual separation of continents, the mid-ocean ridges or the generation of new seafloor (Dicke, 1957, 1962).[3] Jordan was more enthusiastic: data from the oceanfloors and the traditional empirical arguments for Drift could be explained by an expanding earth. He disclaimed competence as a geologist; the earth was of interest only as a testing-ground for his gravitational theory (Jordan, 1966t: x–xii).

This confluence of geology and physics produced a flurry of interest in Expansionism in the late 1950s and early to mid-1960s and for a time it appears to have been a serious rival to Drift. Most of the evidence cited in favor of Drift could be accommodated by Expansionism as could some of the new evidence from the ocean floors; e.g., the network of ridges with their central rift valleys were likened to 'expansion cracks'. J. Tuzo Wilson, a *quondam* Contractionist, adopted Expansionism around 1959 and defended it primarily on the basis of ridges and other oceanic phenomena (J.T. Wilson, 1960; cf. Creer, 1965; R. Laudan, 1980b;). By the mid-1960s enthusiasm subsided, perhaps because of the emergence of newer versions of Drift or due to gradual waning of interest by physicists in novel gravitational theories.[4] The program did not disappear: it was pursued by a few geologists as an alternative to Drift and its other rivals.

Convection currents

Hess and Vening-Meinesz in the 1930s had offered convection-current explanations for trenches and the associated phenomena of island arcs, seismic and volcanic activity, and gravity deficiencies. They pursued this approach in the 1940s and 1950s. Hess concentrated more on articulation of the theory to incorporate new data and patterns, and on strengthening its empirical base; Vening-Meinesz, on mathematical modelling of the currents themselves.[5] One of Vening-Meinesz's aims in his gravity expeditions in the 1930s had been to examine Holmes's proposal of convection currents acting as 'conveyor belts' in the mantle to carry the continents through the ocean floor. He, like Holmes, continued to regard currents as a possible mechanism for Drift. Hess was silent on this matter in his publications. He generally characterized the oceanic crust as rigid and this would presumably be incompatible with Drift. None the less in the 1940s and 1950s currents were often linked with Drift and were seen as a promising explanation for trenches and orogeny. Except for the trenches, scant application had been made to the ocean basins.

Hess extended the theory to the oceanic ridges.[6] In 1954 he advanced three hypotheses to explain their formation and composition. In all three his specialist interests as a petrologist in the formation and composition of the rock making up these structures were clearly evident. One theory involved a simple outpouring of basalt which piles up as the seafloor sagged beneath it. Two others invoked convection currents, in one (the Mid Atlantic Ridge) a rising limb; in the other (the Walvis Ridge) a descending limb. Different explanations might apply to different ridges. The ridges in different parts of the oceans were similar but were not yet linked into a network. Not all segments need be formed in the same way. No more than the basins themselves were the ridges a system to be explained by a single theory.

By the 1959 International Oceanographic Congress, Hess had thought further about the ridges and floor. He now proposed that the oceanic crust was cooled and hydrated mantle; the ridges were belts extruded over the upward limbs of convection cells (Hess, 1959). The ridges and the floors were, to the eye of a petrologist, one and the same. Unlike mountains on land, ridges were not the product of folding and compression, but were a direct expression of mantle processes. The oceans had generated an oceanic theory which owed little to land-based relatives. Hess admitted that he was puzzled by the thinness of sediments on the crust. Did this mean that the oceans had only recently come into being, that sediments were somehow scraped away or that sedimentation had not taken place in past times?

The field of marine geophysics was in ferment at the time of this first Congress. The seafloors seemed increasingly different from anything comparable on land. Detailed mapping of the East Pacific Rise revealed, for example, that it was crossed at right angles by fracture zones. The magnetic anomalies which ran more or less parallel to the ridge seemed to be displaced along these zones. Vacquier and Menard at Scripps speculated that these fracture zones represented a sideways shift of portions of the floor: in one part, lateral displacements of up to 150 kilometers; at the Mendocino Fault, upwards of 1200 kilometers! Menard tied these to convection currents. The theories discussed at the Congress ranged from convection currents to Permanentism to Expansionism to Drift. There was little consensus except that a simple adaptation of continent theories seemed inappropriate.

Seafloor spreading

A year after the Congress Hess concocted a theory which encompassed the whole surface of the globe as an integrated system. It incorporated ridges, trenches, oceanic crust and convection currents into a global theory unmistakably arrived at by a scientist with feet planted firmly on the seafloor. The theory later received the sobriquet of 'seafloor spreading' but for Hess it was 'geopoetry'. The view of the earth that he presented was creative, orderly but highly speculative (Hess, 1962): 'The birth of the oceans is a matter of conjecture, the subsequent history is obscure, and the present structure is just beginning to be understood.' His frankness about the uncertainties in his scheme and his labeling it 'geopoetry' may have been rhetorical devices to ward off criticism from more orthodox geologists. If so, the tactic worked, for his heterodox model generated little immediate opposition.

The nub of his theory was that new seafloor was generated at ridges by the upwelling of mantle material. The old seafloor gradually moved from the ridges and was eventually dragged down at the trenches and reconverted into mantle. The cycle was driven by convection currents rising under ridges and descending under trenches. The continents were passive passengers seated on dynamic ocean floors. Hess's model provided a solution to the conceptual problem with which Drifters had wrestled for fifty years: how could the continents drift through the ocean crust? Hess's answer was that they moved with the crust, not through it. The poetry scanned well with Hess's methodological criterion of a good theory as one which coordinated 'many geologic facts previously merely a collection of observations with no apparent relationship to one another' (Hess, 1938: 75). It also fitted his aim of generating theories which were effective problem-solvers (cf. Frankel, 1980: 361–2). Ridges and trenches received a unified explanation. He now answered his earlier question about the thinness of oceanic sediments: the ocean floors were young, they were being constantly created and destroyed. Gravity deficiencies associated with trenches were due not merely to the downbuckling of the crust but to 'subduction'; i.e., the swallowing of the lighter crustal layer into the mantle. Patterns of heat flow in trenches and ridges were due to convection cells. The 'rift valleys' down the centers of ridges were precisely that: rifts which were constantly forming and being filled by transformed mantle. Ridges became the floors so of course they were of identical composition. The odd regularity of the Mid-Atlantic Ridge – that it ran down midway between the Old World and the New – was not odd. A limb of a convection cell rose under the ridge and caused the floor (and the continents embedded on it) on each side of the ridge to move from it at the rate of a centimeter or two each year. Hess's theory thus offered a unified view of the ocean basins as a dynamic system in which solutions were proposed to many of the patterns and puzzles from the sea. He observed that Expansionism could explain much incorporated in his version of Drift but he refused to 'accept this easy way out' because of the implausibility of a swelling earth and because – true to his marine interests – it would require the creation of huge amounts of water to maintain the depths of the oceans (Hess, 1962: 610).

Hess's model is similar to Holmes's but the differences are significant. Hess explicitly claimed that not only had he arrived at his model from entirely different considerations but also that similarities between the two were superficial (Hess, 1968). Holmes's model was a dry land model; the ocean basins played a minor role and their nature was almost wholly unaddressed. It reflected the abysmal ignorance at that time about the abyssal seas. Hess's model was directed primarily at the oceans and secondarily at the other quarter of the earth's surface. It incorporated much of the vast mass of empirical data about the seafloors gathered since World War II. A crucial

Figure 9.2. Hess's seafloor-spreading version of Drift. New seafloor spreads from the central rift valley; old seafloor is consumed in the trench. Both processes give rise to volcanic and seismic activity. The continent is carried along with the seafloor in which it is embedded from the ridge toward the trench. Compare with Holmes's model in Figure 6.4. (after Sullivan, 1974: 69, by permission of McGraw–Hill Book Co.)

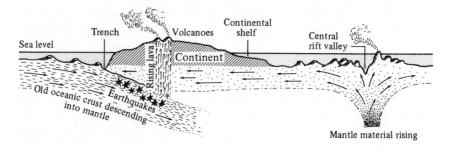

difference is that Holmes's continents moved through a rigid seafloor; Hess's, with the seafloor. Holmes's textbook (1944: 506) is much more definite on this point than some of his earlier writings. In 1933, for example, he suggested that new ocean floor may be created over upwelling currents but this occurred through a passive infilling of the stretched and thinned seafloor left behind the moving continents (Holmes, 1933). Holmes's view was by 1960 inconsistent with the new knowledge of the seafloors: there was no evidence of 'wakes' left behind moving continents.

Hess's poetry did have a few verses on continent geology. The continents were less dense than the oceanic crust and did not descend with it into trenches. However, in 'riding over' trenches volcanic activity and orogeny could occur in a manner not dissimilar to that envisaged by Holmes. Propelled along with the floors, the landmasses could collide with one another, be fused together, thrust up mountains, and be broken apart. Although Hess's model was basically an oceanic one, it could draw upon the continent evidence used in defending older versions of Drift. Hess was not insensitive to this. In his 'poem' he referred, for example, to the pro-Drift arguments of directionalists.

Hess did not immediately publish his hypothesis. It finally appeared in print in 1962 but differed little in substance from the views that he had been propounding since 1960 (Hess, 1962). He had disseminated these in preprint form, as an ONR report, and in lectures, seminars, and personal discussions. Fred Vine, for example, who soon was to use spreading to explain magnetic anomaly patterns, heard Hess lecture on his model at Cambridge in 1962 (Vine *et al.*, 1977). Hess also discussed it with Robert Dietz in 1960 (Hess, 1968). The first 'public' announcement of the model came from Dietz, then at the US Navy Electronics Laboratory, a facility linked with Scripps. In 1961 he outlined in *Nature* a version he dubbed the 'spreading sea-floor theory', soon to become 'seafloor spreading' (Dietz, 1961).

Dietz's path to seafloor spreading was different from Hess's. Dietz was a naval hydrographer and had focussed on bathymetric data rather than on petrology or the structure and composition of the ridges and trenches. The bathymetric patterns must have fallen into place only when Hess proposed his model to him, for Dietz gave Hess 'full credit' for spreading on the grounds of 'priority and for fully and elegantly laying down the basic premises'. His own contribution was the coining of the succinct and graphic term 'sea-floor spreading' (Dietz, 1968). Dietz's model differed little from Hess's. Dietz spelled out the role of the trenches, a point on which Hess (1962) was coy until 1963 (Fisher & Hess, 1963). Dietz's version was put forward explicitly as a 'less radical' solution to apparent seafloor rifting than Carey's and Heezen's Expansionism (Dietz, 1961). Like Hess, he emphasized applications to the patterns and puzzles of the sea; the implications for continents geology were relegated to the background. Dietz added magnetic anomaly patterns to the list of phenomena discussed by Hess (Dietz, 1961). Menard had suggested that breaks in these patterns might be due to lateral displacements along faults intersecting the ridges. Dietz took up this suggestion and interpreted magnetic stripes as stress patterns caused by convection currents under the ridges. His remarks on this, as well as Hess's reference to directionalist studies, mark the beginning of the union of land and sea Drifters which was to prove highly potent in the near future.

Spreading initially was received coolly, if not ignored, by mainland geologists. It co-ordinated an array of observations and phenomena and suggested solutions and explanations for many of the problems and patterns of marine geology and geophysics. However, it wedded two concepts each of which was widely thought to be speculative, suspect, and highly problematic: continental drift and a stable global system of mantle convection currents. Few skeptics would accept that two 'wrongs' could be added together to make a 'right' theory. Few continent geologists were likely to be well-disposed to this marriage. It arguably removed or at least by-passed the problem of a mechanism for Drift but it seemed to offer no particular advantages over previous versions of Drift with respect to solving problems of concern to them.

For marine geologists and geophysicists, spreading offered promise. Some opponents of Drift realized that this new version had to be taken seriously. Maurice Ewing's initial reaction was that spreading was probably right. He directed a substantial portion of Lamont's efforts over the next few years to testing the theory (Menard, 1986: 177–8). If spreading occurred, he reasoned, then the farther from the ridges, the older the ocean floors and, consequently, the farther from the ridges, the thicker should be the layer of oceanic sediment accumulated on the floors. He proposed to test this new version of a global program through one specific technique: the measurement of sediment thickness by seismic reflection. He found that the ridges themselves were virtually bare of sediment (as one might expect if they were a recent feature), but otherwise there seemed to be no obvious relationship between thickness of sediment and distance from a ridge. He concluded that there were so many variables 'that the validity of such a test is open to question'. Even so, he believed that the results told against spreading. The masses of data accumulated by Lamont could be explained in terms of a permanent ocean floor swept by a complex system of marine currents;

furthermore, such patterns of sediment distribution that did emerge did not seem to fit with spreading (Menard, 1986: 202–4).[7]

Dietz's and Hess's comments on Expansionism seem to have caught the attention of J. Tuzo Wilson. Wilson (1961), in response to Dietz's, *Nature* article and to Hess's version which he had read in preprint form, gave an extended gloss of the implications of seafloor spreading for continent problems in which he had long been interested. He noted, for instance, that the processes outlined could be adapted to explain growth of continents by accretion at the margins, a view which he had found useful in interpreting Canadian structures. He was reluctant to accept the theory *in toto* and offered a compromise which transferred spreading seafloors onto an expanding earth. In retrospect, spreading and Expansionism were not mutually exclusive. If old seafloor were not devoured in trenches, then the production of new seafloor at the ridges required a relatively rapid expansion of the earth. Expansionists could claim 'proofs' of spreading at ridges as 'proofs' of their theory. Or, as a compromise, one might admit that some subduction occurred but that a slow expansion nevertheless took place.

Accelerating Drift

The momentum of Drift rapidly increased. Runcorn (1962b) orchestrated in *Continental Drift* a combined operation of land and sea geologists and geophysicists. The Drift navy was a motley crew with varying theoretical views: Dietz spoke on seafloor spreading; Heezen argued for spreading and drifting on the surface of an expanding earth; Vening-Meinesz pressed for convection currents; and Vacquier restated his claim that offsets of magnetic anomaly patterns plainly showed large-scale lateral displacements of the seafloor. There was agreement among these marine specialists that drift had happened but each proposed or employed different explanations. The land-based contingent, which included Runcorn and Opdyke, stressed a Drift interpretation of recent palaeomagnetic and palaeoclimatic work. They paid scant attention to the sea. No grand synthesis was effected: they Drifted together but were divided by their specialist and local interests and their adherence to different versions of the mobilist program. Nonetheless, the solutions which Drift could offer to a range of problems on sea and land, the continuing dissemination of Drift theories, and the influx of additional observations and phenomena which were amenable to Drift interpretations not only confirmed many workers in their allegiance to Drift but also began to attract new adherents.

J. Tuzo Wilson, for example, in 1963 set aside Expansionism in favor of Drift. Following a hint from Hess (Hess, 1962: 611–2) he hit upon a novel test: *if* the crust spread from the ridges and *if* islands were formed near those ridges and *if* those islands moved with the crust away from those ridges, then the farther an island was from a ridge, the older it should be. A scan of the literature yielded results which generally squared with this deduction. Wilson had proposed a new pattern discernible in the seas, had found it, and had explained it in terms of the latest version of Drift. He could have reconciled this with Expansionism but discussed it in relation to only two programs: Permanentism, which he rejected, and Drift powered by convection currents (J. T. Wilson, 1963a). This was soon followed by a piece blandly titled 'Hypothesis of Earth's Behaviour' in which he summarized a mix of evidence and arguments mostly concerning the ridges and associated phenomena. He also mentioned that such major continental faults as the San Andreas Fault showed crustal mobility of a magnitude comparable to that inferred from the offset of seafloor magnetic anomaly patterns (J.T. Wilson, 1963b). He next gave a popularized exposition of the 'Hypothesis', now more frankly titled 'Continental Drift' in the pages of *Scientific American* (J.T. Wilson, 1963c). That Hess's 'geopoetry' could find its way so quickly into a mass-circulation periodical which dealt mostly with empirical science gives a sign of the gathering momentum of Drift.

Bullard, in 1963, commended Drift as a hypothesis which had come of age to the Geological Society of London, a bastion of traditional geology.[8] He explicitly set out his criteria for a hypothesis worthy of pursuit (Bullard, 1964: 24):

> It is sufficient if it is not clearly absurd either geologically or physically and if it suggests relations and other hypotheses that can be tested. The idea of continental drift is exceptionally fruitful in this way; it suggests large programs of investigation on land, at sea, and in the laboratory.

In his view in the 1920s and 1930s the geological evidence for Drift had been inconclusive because of 'ambiguities and gaps' and 'the lack of direct evidence for movement'. Until recently so little was known of the oceans that any theory involving the globe as a whole was inescapably arbitrary in its extension to three-quarters of the surface. These difficulties had now, he proclaimed, been overcome. For 'direct evidence for movement' on continents, he cited recent work on the San Andreas Fault which indicated that the land on the Pacific side of the Fault was moving northward about 5 centimeters per year. At sea, magnetic anomaly patterns on either side of faults cutting across the ridge off the Californian coast could be matched up if it was accepted that the crust had been displaced nearly 1200 kilometers. Bullard insisted that the displacements of this magnitude made Drift plausible despite any arguments that such displacements were physically impossible. He discussed palaeomagnetic directionalist studies extensively. He admitted that there were uncertainties in technique and interpretation but none the less these also constituted unambiguous evidence for large-scale relative motions of continents. Further, palaeomagnetic data gave precision to and reinforcement of the older palaeoclimatic and palaeogeographical arguments for Drift. He re-examined the accuracy of the 'fit' between Africa and South America. Wegener's reconstruction was open to charges of 'fudging'. Citing Carey's 'fit' and fits generated by mathematical techniques, he claimed that the 'fit' was remarkably good; indeed, the fit of the present continents along their continental shelves is much better even than Wegener had thought (Bullard, 1964: 19).

After this careful preparation, Bullard turned to issues of theory and mechanism. Echoing Rastall's argument of 35 years earlier, he pointed out that many phenomena such as ice ages and the geomagnetic field 'undoubt-edly occur but which, if they had not been observed, would probably not have been predicted by physical theory' (Bullard, 1964: 19). Drift, he implied, was now in this category. That there was no accepted mechanism was no reason to reject the occurrence of drift. Thermal convection currents provided a plausible mechanism though there might be disagreement on matters of detail. He concluded by indicating how convection currents, as employed in the seafloor-spreading version, could co-ordinate and account for diverse phenomena of the ocean floor and could yield testable predic-tions. Bullard had publicly cast down the gauntlet to Drift's opponents and competitors. His address was an insightful summary of the case for Drift just prior to the sea-change in opinion which occurred in the mid- to late-1960s.

The discussion at the Society after Bullard's address reveals that en-trenched opposition to Drift was unshaken in the face of the new theories, arguments and data. Much of this opposition was rooted in specialist concerns. W.D. Gill dismissed geophysics as an arena of conflicting theories unconstrained by facts and suggested that real geologists should fasten their faith to the bedrock of structural geology and avoid 'accepting theories that did not even explain the facts we had so long and diligently established as a basis of our knowledge of the history of the continents' (Bullard, 1964: 29). Gill likely spoke for many field geologists who had devoted their careers to the mapping and correlation of regional structures. Gill was, however, in the minority among the discussants. Most were at least receptive to Bullard's arguments. Nonetheless, for traditional geologists each line of evidence for Drift was individually open to rebuttal. The linking of diverse phenomena through seafloor spreading might be impressive to those who thought on a global scale, perhaps, but such patterns, correlations, and their incorporation in a unifying master theory constituted no 'proof' of the existence of convection currents, seafloor spreading or Drift.

Blackett, Bullard, and Runcorn the following year organized a two-day symposium on Drift for the Royal Society. None of the three convenors had received a traditional geological education or worked in a mainline geological specialty. Oceanography and geophysics were by now the 'cutting edges' in the Drift program and this reflected in the papers and discussion at the symposium. Only one paper dealt with the traditional pattern-matching case for Drift. The rest covered palaeomagnetism, seafloors, and convection currents. Perhaps the most striking single piece of evidence put forward for Drift was the 'Bullard Fit'. Bullard had been impressed by Carey's reconstruction of Pangaea (Carey, 1958b) and was annoyed – as was Carey (Carey, 1958b: 218–24; 1976: 6–8) – by Jeffreys's insistence that the jigsaw fit was quite poor (Blackett, Bullard & Runcorn, 1965: 41; Bullard, 1975a: 21). Bullard believed that the question of whether or not the continents could be convincingly reassembled was important to the Drift case. He sought to resolve it once and for all by using 'objective' mathematical methods (Blackett *et al.*, 1965: 41). He and two colleagues applied Euler's theorem (according to which motion on a sphere could be expressed as a rotation about some pole of that sphere) to determine the 'best fit' of the Atlantic continental shelves. Though adjustments had to be made to the data, there was an aura of objectivity about the enterprise: instead of manually fitting together outlines of the shelves, the data were fed into a suitably programmed computer. The result 'exceeded our expectations and fully confirms the work of Carey' (Blackett *et al.*, 1965: 42).

The 'Bullard Fit' did not escape criticism. M.G. Rutten, from the University of Utrecht, commented that 'Bullard, to obtain a nice fit of continents, does away with Iceland, "because it is young", but keeps the Rockall bank, "because it fits nicely"' (Blackett *et al.*, 1965: 321). Bullard had massaged his data and, though the term 'garbage in, garbage out' had not been coined, the charge could be made against his use of a computer. The diagrams of the 'Bullard Fit' were none the less widely cited and reproduced.

Most who gave papers or supplied comments to the symposium favored convection-current versions of Drift. Despite the variety of approaches explored and the types of evidence put forward, the tenor was that although Drift was now the preferred program of many British physicists and geophysicists, there was no compelling demonstration that Drift had occurred or that convection currents could propel continents or even that convection currents existed. Ruttens's summation, though perhaps somewhat idiosyn-

Figure 9.3. The 'Bullard Fit', made using a computer, did not require significant distortion of the present continents to achieve an impressive jigsaw match of continental shelves. It was graphic evidence in support of Drift. Compare with Figures 2.1. and 3.1. (Reproduced from Bullard, Everett & Smith, 1965: 48–9, by permission of the Royal Society of London).

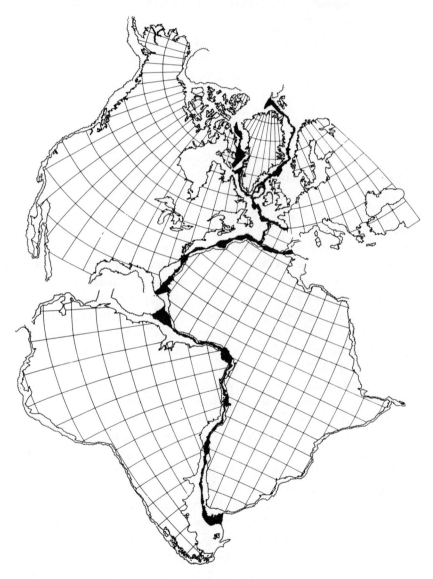

cratic, was probably not far off the mark and would have been assented to by many geologists (Blackett *et al.*, 1965: 321):

> It still depends on which part of the geological data one finds most strongly heuristic, if one is a 'drifter' or a 'fixist'. It is only the measurements of palaeomagnetism which have introduced a really new set of values. The only way to remain fixist now, is to disbelieve palaeomagnetism, a position which becomes more and more awkward as its methods tend to become better substantiated.

Bullard (1964: 24) had conjectured that a sustained program of research would in 10 or 20 years probably yield a verdict for Drift. The resolution came much earlier than that. It came from a union, not a simple juxtaposition, of palaeomagnetic research and seafloor-spreading via convection currents. The 'confirmation' of the 'Vine–Matthews Hypothesis' heralded a rapid shift of allegiance.

Stripes, spreading, and reversals

Fred Vine and Drummond Matthews, both of the Department of Geodesy and Geophysics at Cambridge, in 1963 tied together seafloor-spreading and palaeomagnetism in a speculative proposal aimed at explaining the patterns of magnetic anomalies found parallel to ridges. They proposed as an articulation of Hess's spreading that magnetic stripes were associated with the generation of ocean floor at the ridges and with geomagnetic field reversals.[9] Vine and Matthews were strongly inclined toward Drift at the outset of their work; Matthews, as a result of first-hand geological mapping in Antarctica and the Falklands (two pieces of Gondwana); Vine, by his own account, from a depiction of the fitting of the continents and the influence of one of his Cambridge tutors (Frankel, 1982: 12–3). Both were Cambridge products and had a solid grounding in physics and geophysics.[10] They were familiar with palaeomagnetic research in which Cambridge was a leading center. Cambridge acted as a magnet for geophysicists, drawing them from elsewhere; Vine and Matthews kept abreast of the latest research and brainstorms of Hess, Wilson, and others. Vine, for example, was much struck as an undergraduate by Hess's lecture on seafloor spreading and himself gave a talk to the student geology club on 'HypotHESSes' (Vine *et al.*, 1977).

Vine, as a graduate student, was set the task by Matthews, his supervisor, of reducing the magnetic survey data collected in late 1962 over the Carlsberg Ridge, an oceanic ridge in the Indian Ocean. He began with a belief in Drift, seafloor spreading and reversals of the geomagnetic field and he 'was particularly looking for some record of drift and spreading with the ocean basins'. His mind was filled with preconceptions as he approached the data. If Hess's model were correct, the ocean floor should be of uniform composition: stripes could not be strips of materials with different magnetic susceptibilities.[11] He concluded that the patterns of magnetization in the vicinity of the ridge were due to alternations of reversed and normal magnetization of the rock forming the ridge and floor. These reversals were produced by reversals of the earth's field, not local circumstances. He coupled this with seafloor spreading; as the emergent seafloor cooled at the ridges, it acquired a magnetization in accord with the geomagnetic field prevailing at the time. Later, as this proportion of the seafloor was displaced by newer material, if the earth's field reversed, the newer material as it cooled would be magnetized in the opposite direction to the adjacent, older seafloor. The alternation of normally reversely magnetized strips of ocean floor was a 'fossil' record of geomagnetic field reversals. Vine tried the idea on Matthews who arranged for publication of a short paper in *Nature* (Vine & Matthews, 1963) even though there was little hard evidence in favor of the proposal (Vine, 1966).

Vine and Matthews's conjecture stirred little interest initially. Each element of their synthesis was open to challenge. Seafloor spreading was not generally accepted even by proponents of Drift, much less by other geologists. Fixists of course rejected spreading; any theory which assumed spreading was equally unacceptable. There were still doubts that reversals of the geomagnetic field were the principal cause of reversed polarity in rocks. Even if it were agreed that geomagnetic field reversals had taken place, it was not evident that this was how the magnetic anomaly patterns were generated. The prevailing view was that the stripes represented rocks of different

Figure 9.4. Vine and Matthews claimed that the ocean floor recorded geomagnetic field reversals. As new ocean floor is formed at the ridge, it cools, 'locking in' the direction of the geomagnetic field at that time (here, dark for 'normal' magnetization). The alternating dark and light (reversed from 'normal') bands represent old sections of the ocean floor moving symmetrically from the ridge and magnetized in accord with previous orientations of the geomagnetic field.

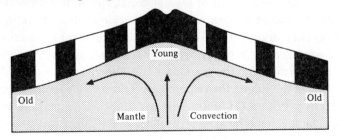

composition. This could not be decided without extensive sampling. Researchers at Lamont, for instance, worked diligently in the mid-1960s on the Reykjanes Ridge off Iceland and on the East Pacific Rise to solve the striping problem within a Fixist program. They also argued that the actual magnetic profiles did not fit the Vine–Matthews model (Heirtzler, Le Pichon & Baron, 1966; cf. Frankel, 1982: 23–6). Setting aside these difficulties, it was not clear what further evidence could be brought to bear on the Vine–Matthews model. They themselves had not set out a research agenda. At the 1964 Royal Society symposium little interest was expressed in it despite intense interest in Drift and seafloor spreading. Only Vacquier mentioned it: he commented that it failed to account for the Pacific patterns with which he was familiar (Blackett *et al.*, 1965, 77–82).

Gordon Macdonald, a geophysicist at UCLA, launched a general attack on spreading. He based it on a mathematical model of the deep structure of the continents. He had an impressive reputation and by the age of 32 was a full professor and Fellow of the American Academy of Arts and Sciences and of the National Academy of Sciences. He drew upon a wealth of geophysical data to argue that the structure of the continents extended 500 kilometers below the surface of the earth. If this were the case, 'it is no longer possible to imagine thin continental blocks sailing over a fluid mantle' (Macdonald, 1964: 928–9). Seafloor spreading was geophysically impossible. His argument was taken seriously, at least by those not disposed to Drift (Menard, 1986: 226–9).

Spreading the theory

Vine and Matthews persisted with their model despite its lacklustre appeal to geologists and marine scientists. In early 1965 Vine and J. Tuzo Wilson made major revisions and extensions. Both Hess and Wilson visited Cambridge where they collaborated with Vine. Hess was enthusiastic about the 1963 Vine–Matthews paper; Wilson had yet to read it. The first fruit of this collaboration in early 1965 was Wilson's development of the notion of 'transform faults' upon which he constructed a new version of Drift. The second was the spelling out of several corollaries of the Vine–Matthews model which enabled it to draw upon 'hard' evidence from the ocean floors.

Wilson concentrated on the fracture zones which cut across the ridges. Here too a pattern had recently been found: seismic activity was connected with the ridges and with the portions of the faults which connected the offset portions of the ridges; the remaining portions of the faults were seismically inactive. Wilson (1965a) christened these active parts of the faults 'tranform faults'; here the active spreading at the ridges was 'transformed' into lateral displacement of the ridge itself and then 'transformed' again into spreading. He used the notion of a transform fault to change his 1959 conception of world-wise fracture systems. Mountains, trenches and island arcs lay along 'belts' of mobility and seismic activity. Each may be transformed at its termination into one of the other types of mobile belts. These belts thus form a continuous network which divides up the entire surface of the earth and defines the edges of several large, rigid 'plates'. He proposed that the San Andreas Fault was not a transcurrent fault, a simple slippage between two pieces of the crust, but a 'transform fault' forming part of a mobile belt and connecting a suspected ridge off Vancouver and the 'end' of the East Pacific Rise near the Gulf of California.

Wilson's conjectural transform faults and plates were far more than a simple extension of spreading; spreading became but one expression – albeit a crucial one – of a unified global tectonics. He also formalized and grounded in theory a division between continent and marine geology. With the exception of parts of continents in which mobile belts were transformed onto dry land – e.g., the San Andreas Fault or the East African Rift Valley – the oceans were the site of all major tectonic processes. One learned about how the 'earth is behaving' by studying the ocean basins. The continents by and large were, as Hess had thought, passive passengers. They recorded the effects of such processes. One learned about how the 'earth had behaved' in the more distant past by studying the continents. Marine geology was 'presentist'; continent geology, 'historical'.

Figure 9.5. A section of ridge (vertical double lines) offset by Wilson's 'transform faults' (horizontal dashed lines). Seismic activity occurs in the shaded areas. Magnetic anomaly patterns on either side of the ridge segments could be matched up by shifting the offset bits of ridge back together (after J.T. Wilson, 1965a: 345, by permission of Macmillan Journals Ltd.).

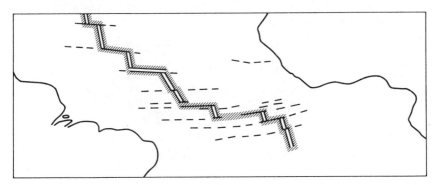

Vine, encouraged by Hess and Wilson, refined his 'hypothesis'. Vine and Matthews had not initially seen the implications for their model of an assumption of a uniform rate of spreading. Further, they had not drawn the inference that if the median rift valley marked the center of an upwelling convection current which divided under the ridge, then the magnetic anomalies should ideally be distributed symmetrically on either side of the ridge (Vine, 1968: 82). The first of these opened the possibility of applying dating techniques to the stripes themselves; the second, a direct comparison of the measured magnetic profiles of the two sides of a ridge: they should be mirror images of one another. The development of these two corollaries of the original model in 1965 and 1966 enabled Vine to support his model with the 'concrete evidence' lacking in 1963.

Vine and Wilson late in 1965 applied their new ideas to the structure and magnetic patterns of the newly-described Juan de Fuca Ridge off Vancouver Island. Wilson (1965b) argued that it was a short piece of ridge bounded on each end by transform faults. He drew attention to the complex patterns of magnetic anomalies associated with this system: bilateral symmetry of the anomalies in the immediate vicinity of the ridge seemed to fit Vine's model. Vine and Wilson collaborated on an accompanying paper (Vine & Wilson, 1965) which treated the patterns in more detail. They spelled out the symmetry corollary: the anomaly pattern on one side of the ridge should mirror that of the other side. Graphical representations of some Juan de Fuca patterns appeared to show a high degree of bilateral symmetry about the ridge. They admitted that other traverses across the ridge had given patterns

Figure 9.6. J. Tuzo Wilson claimed that the San Andreas Fault was a transform fault linking a section of the East Pacific Rise (ridge) off the Baja Peninsula with the Juan de Fuca Ridge off Washington. This formed one small section of a globe-girdling system of zones of mobility and rigid plates. Displacement of several centimeters per year occurred along the Fault – as much as some versions of Drift required (after Wilson, 1965a: 346, by permission of Macmillan Journals Ltd.).

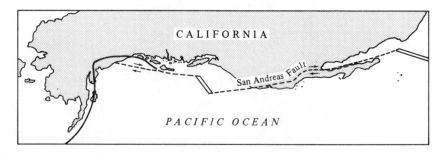

which were less symmetrical. The second major argument of the paper provided a second way to the tie Vine–Matthews to 'concrete evidence', in so far as geophysical data could be construed to be 'concrete'. If spreading occur at a uniform rate, then there should be a correlation between the widths of the normally- and reversely-magnetized stripes and their ages. If the floor forms at 4 centimeters per year (2 centimeters on each side of the ridge), then a strip of floor 200 meters wide represents a record of 10 000 years of the earth's magnetic history. These ages should match time-scales based on continent rocks for reversals of the geomagnetic field. With these assumptions and a recently published time-scale of reversals (Cox, Doell &

Figure 9.7 The bilateral symmetry of magnetic anomaly patterns across the Juan de Fuca Ridge is suggested by a comparison of an actual profile (a) with its mirror image (b). The horizontal line represents '0' magnetic anomaly; above the line, positive anomaly; below the line, negative anomaly. The ridge axis is indicated by the centerpoint of the horizontal scale. Vine and Wilson associated positive anomalies with 'normal' magnetization; negative anomalies with 'reversed' magnetization. Compare with the map of anomalies given in Figure 8.2. (Reproduced from Vine & J.T. Wilson, 1965: 488, by arrangement with *Science* copyright 1965 by the AAAS).

(a) Profile 'b' reversed

(b) Observed profile 'b'

(c) Model 2

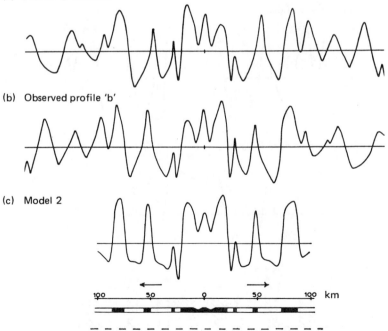

Dalrymple, 1964), Vine and Wilson constructed a model of ridge and its anomaly patterns. This closely matched the Juan de Fuca patterns. The symmetry corollary and the incorporation of a time-scale based on polarity reversals provoked interest from specialists on whose work seafloor spreading had until then impinged marginally if at all. The Vine–Wilson paper

Figure 9.8. Vine and J. Tuzo Wilson compared profiles for (a) East Pacific Rise and (b) the Juan de Fuca with (c) an ideal profile generated from their model of seafloor spreading. The model (bottom) was based on a uniform spreading rate and the mechanism depicted in Figure 9.4. Note that although bilateral symmetry is not as obvious in (b) as in (a), there are correspondences between (a) and (b). (Reproduced from Vine & J.T. Wilson, 1965: 487, by arrangement with *Science* copyright 1965 by the AAAS).

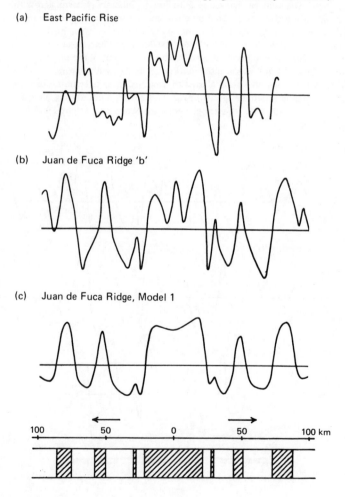

(a)   East Pacific Rise

(b)   Juan de Fuca Ridge 'b'

(c)   Juan de Fuca Ridge, Model 1

100   50   0   50   100 km

drew upon and gave additional importance to research agendas in young-rock dating and polarity reversals which had hitherto been pursued independently of Hess's and Vine and Matthews's theories.

Magnetic reversals and geomagnetic time-scales

Cox and Doell in 1959 had set a research agenda which drew together rock-dating and reversal studies.[12] Both had worked as students with John Verhoogen at Berkeley. Verhoogen had lectured on Drift and convection currents from the 1940s. Both Cox and Doell were skeptical of the interpretations of and assumptions embedded in directionalists' research. Directionalists assumed reversed rocks were the result of global field reversals and that in plotting polar-wandering curves the 'reversed' polar positions could be simply rotated 180°. How could such an assumption be justified? Cox and Doell joined two lines of research to solve this problem.

One line was dating of progressively younger rocks by the Potassium–Argon (K–Ar) method. In the 1910s Holmes had developed a method based on the decay of uranium to lead. His method was limited, its margin of error large, and it could not be readily applied to 'young' rocks. In the 1950s similar methods using other isotopes, particularly the decay of K^{40} to Ar^{40}, extended the absolute scale to younger rocks.[13] Berkeley became a world center for co-operative effort among physicists, geophysicists, and geologists. The techniques were exported to Canada, Australia and elsewhere. At Berkeley the focus gradually shifted from older to younger rocks. In part, this was conceived as a critical test of the techniques themselves but in part it came as a response to local interests in regional geology. The Berkeley department was oriented to field studies of California geology. As one member of the department expressed it, geology 'stops when you get to the Nevada border' (Glen, 1982: 57). Many of the regional structures were in geological terms quite young. A further aspect of this 'localism' was that some of the local rocks, particularly basalts, were difficult to date even with the K–Ar methods then used; much effort was expended to develop 'whole-rock' dating so that the methods could be applied to these as well.[14] The second line of research was the study of reversely-magnetized rocks. Cox preferred a general explanation in terms of reversals of the geomagnetic field to special explanations in terms of self-reversals. If palaeomagnetic reversals were due to a global cause, then all the rocks formed at a time when the earth's field was reversed should show reversed magnetism.

Cox and Doell fused rock-dating with reversals into a new research agenda. The agenda was straightforward: rocks of the same age of formation should show the same polarity. If a large number of rocks from different regions and different epochs were dated accurately and their polarity measured, then the question of reversals could be resolved. Moreover, the result would a be a time-scale of reversals which could provide a key to the solution of stratigraphic and correlational problems in continent geology. Brent Dalrymple, a K–Ar specialist trained at Berkeley, provided expertise in dating. The first chronology of reversals appeared in June 1963 (Cox, Doell & Dalrymple, 1963). Their efforts were spurred by the Australian National University (ANU) group headed by Ian McDougall and Don Tarling. The first ANU scale appeared only a few months later (McDougall & Tarling, 1963). These scales were followed by more complex and detailed ones as more 'events' (very brief reversals) were found to occur within more sustained 'epochs' of normal- or reversed-magnetization.

Corroboration and conversions

Vine in November 1965 discussed the application of his theory to the Juan de Fuca Ridge at the American Geological Society meeting. Dalrymple was also there and gave a paper on a revised reversal time-scale. He told Vine that the group had found a new 'event' about 0.9 MYA (later named the Jaramillo Event). Vine was electrified by the news (Glen, 1982: 310).

> I realized immediately that with that new time scale, the Juan de Fuca Ridge could be interpreted in terms of a constant spreading rate. And that was fantastic, because we realized that the record was more clearly written than we had anticipated. Now we had evidence of constant spreading, that was very important.

The Jaramillo Event enabled Vine to get a better match between selected profiles of magnetic anomalies and the model that he and Wilson had constructed using the 1963 reversal scale.

The Lamont group had put up a nearly-united front against the original Vine–Matthews hypothesis. Several papers on the Reykjanes Ridge pointed out that although the anomaly patterns near the center of the ridge were fairly well marked, those on the flanks were less clear and showed what they regarded as significant differences. Their account of both the patterns and the structure rested on Permanentist assumptions. To explain striping they invoked self-reversals or, alternatively, differences in magnetic susceptibility due to compositional differences. Neil Opdyke dissented. He had trained under Runcorn, had carried out directionalist studies, and in the late 1950s had accepted a Drift interpretation of polar-wandering curves. He had come to Lamont in 1964 to conduct palaeomagnetic studies on sediment cores under the supervision of James Heirtzler, a Permanentist, but Opdyke maintained his Drift convictions. He was conversant with many of the recent arguments for Drift and in particular with the papers of Hess, Dietz, Vine and Matthews, and was sensitive to data favoring Drift.

In late 1965 Opdyke was working on the Lamont databank of Pacific cores. He identified an 'event' about 0.9 MYA. At the same time a colleague, Walter Pitman, was processing the magnetic anomaly data gathered by the research ship *Eltanin* over the East Pacific Rise. Pitman noticed the similarities between the data gathered from three traverses of the ridge, *Eltanin*-19, -20, and -21, and the profile just published by Vine and J. Tuzo Wilson for the Juan de Fuca Ridge. When Opdyke saw Pitman's profiles, he later recalled, 'suddenly the whole thing fell together'. With respect to his own project of sediment core palaeomagnetism, 'I could see that the magnetic stratigraphy of the cores could be used to verify the pattern that Pitman was getting in the magnetic anomaly profiles. He was sitting in the next office to me, and what he knew, I knew within a day' (Glen, 1982: 331). Pitman recounted the ensuing developments as follows (Glen, 1982: 335):

> The bilateral symmetry of *Eltanin* 19 was the absolute crucial thing. Once Opdyke saw that he said, 'That's it – you've got it!' Opdyke was a very important catalyst at that point. He immediately saw what it meant to become our advocate. . . . Heirtzler took a while to come around; at first he tried to explain it away by electrical currents.

The bilateral symmetry across the ridge was so nearly perfect that the two sides could be folded on top of one another across the ridge axis.

The symmetry was so striking that some at Lamont initially dismissed it as 'too perfect'. This was Heirtzler's first reaction. He then showed the profile to a colleague who had a similar response. Pitman recalled, '(Joe) Worzel looked at it for a while and finally said, "Well, that knocks the seafloor spreading nonsense into a cocked hat." I said, "What do you mean, Joe?" He said, "It's too perfect" and walked out of the room' (Glen, 1982: 336). Vine visited Lamont and was shown the newly-constructed *Eltanin*-19 profile, the most symmetrical. Vine too immediately recognized its significance for his theory and the possible impact on the large community: 'It was all over but the shouting' (Glen, 1982: 336). His confidence was justified: in February 1966 most US geologists, with the exception of some marine geologists and geophysicists, were still opposed to or undecided about Drift. Beginning in April 1966 opposition to Drift began to crumble and the undecided began to pursue or accept it.

The American Geophysical Union met in April 1966 in Washington. Cox chaired the session in which Heirtzler unveiled *Eltanin*-19. Cox had had a sneak preview and recalled that he had immediately seen the implications

Figure 9.9. Magnetic anomaly profiles collected by the research vessel *Eltanin* became persuasive evidence for seafloor spreading. Walter Pitman at Lamont was especially struck by the bilateral symmetry of traverse number 19. *Eltanin*-19 was singled out for attention; the other profiles were set aside. (Reproduced from Pitman & Heirtzler, 1966: 1165, by arrangement with *Science* copyright 1966 by the AAAS).

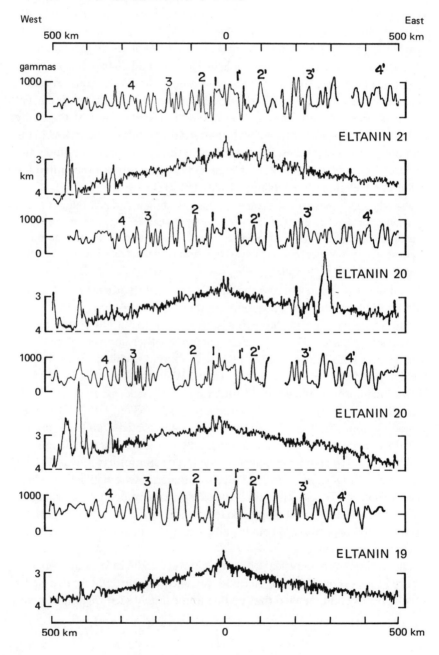

for his own work on reversal time-scales: the profile not only exhibited the reversals which he had previously documented but also some which he only suspected and others of which he had no inkling and no evidence. He later recalled, 'that was the most exciting year of my life, because in 1966, there was just no question any more that the seafloor-spreading idea was right'. Doell's reaction on seeing *Eltanin*-19 was similar: 'It's so good it can't possibly be true, but it is' (Glen, 1982: 337–9). After protracted negotiations between Vine and Lamont over the magic profiles, Pitman and Heirtzler published them with an interpretation based on Vine's revised theory. Their conclusion was unambiguous: 'these results strongly support the essential features of the Vine and Matthews hypothesis and of ocean-floor spreading as postulated by Dietz and Hess' (Pitman & Heirtzler, 1966). The walls of Lamont had been breached by the sea. Vine two weeks later cited four widely-separated segments of ridge to which his and J. Tuzo Wilson's version applied: Juan de Fuca, East Pacific Rise, Carlsberg, and Reykjanes. He also applied the theory to the Red Sea rift. Further, he answered the Lamont

Figure 9.10. The symmetry of *Eltanin*-19 can be seen by comparing the actual profile (center) with its mirror image (upper). Pitman and Heirtzler also calculated an ideal profile (lower) based upon the latest reversal age-scale and uniform spreading. The match between the observed profile and the theoretical one impressed many marine scientists. *Eltanin*-19 soon became the standard illustration for the Vine–Matthews and Vine–Wilson versions of Drift. (Reproduced from Pitman & Heirtzler, 1966: 1166, by arrangement with *Science* copyright 1966 by the AAAS).

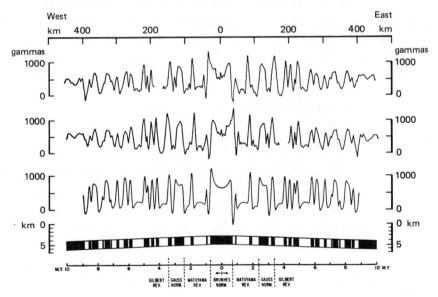

objection of 1965 that the different patterns on a ridge and its flanks were *prima facie* evidence against his theory. He explained that this was due simply to a different frequency of reversals in the more distant past as compared with the more recent past. Finally, he pointed out that the rates of spreading deduced from magnetic profiles and the reversal scale agreed with the rates of Drift earlier computed by continent geologists on the basis of traditional geological evidence (Vine, 1966).

The naval campaign for Drift was essentially complete. On land, there had been some skirmishing but probably most North American geologists – and many elsewhere – still accepted and worked within non-Drift programs. In the next phase, yet another version of Drift – plate tectonics – was put forward which incorporated spreading but which was directly applicable to some problems of continent geologists. By 1970 Drift in the form of plate tectonics was becoming orthodoxy.

Voice-Over

Our narrative until now has been relatively measured and stately. Even though we have been watching Drift, we have also kept an eye on the fortunes of its rivals. In this chapter and the next, our pace becomes more frenetic and our perspective narrower. Our attention is now fixed almost exclusively on Drift, rather than its competitors. We ignore the complex texture of theoretical and empirical developments in other areas of geology except as they bear on Drift. Lunar geology, economic geology, satellite mapping are set aside. In palaeontology, for example, we do not explore the profound implications of the punctuated equilibrium theory of evolution or the controversy which has swirled around the hypothesis of an extra-terrestrial cause for mass extinctions (see Raup, 1986). It is as though we were at the finish of an Olympic marathon; first one and then another competitor has gained the lead over a gruelling course, now, tens of kilometers later, as the leader enters the stadium, attention passes from the other runners to the lone figure sprinting the last few hundred meters to the finish line. In the 1960s Drift entered the home stretch – or so, at least, we depict this phase of our story. This change in pace and perspective to some extent reflects an acceleration of the Drift program: version building on version, each one taking in new evidence and new specialties without faltering or stumbling, winning applause from more and more of the audience. The other programs seemed to lag farther and farther behind. There is a danger that this is a pace and a perspective which we have imposed upon those years because we know that there was a 'finish line' and that Drift was approaching it. It is reassuring that some of the scientists at the time also judged a 'revolution' to be imminent, as we shall see in the next chapter.

In the 1960s neither the rival programs nor the database remained fixed. Geologists extended, modified, and refined theoretical expressions of the rival programs. With respect to the Drift program, for example, a fairly vague convection-current version was in vogue in the 1950s. In the period 1960–65 we can identify three further, successive expressions of that program: Hess's, Vine–Matthews's and Vine–Wilson's. The last was soon elaborated and refined into the plate tectonics version which became widely accepted among geologists from all specialties.

The development of each program was constrained by and directed toward the growing body of data and patterns gathered from the ocean floors. The database was constantly changing. Judgments of geologists on the programs – and the glimmering of a consensus – were not simple processes of determining 'the facts of the matter'. Geologists explicitly linked their stances toward the contending programs and theories with their relative ability to make sense of the data, to solve problems arising from conflicts between the new data and previously-held theories, and to solve or avoid conceptual problems. Scientists often reached different judgments – there was no unanimity. This is partly explicable in terms of specialist and localist interests. Spreading, for example, seems to have been more attractive than the continental accretion version of Contractionism to marine geophysicists; the former addressed problems of immediate concern to them. The reverse might well have been the case for 'hard-rock' geologists in Canada or Central Russia. In the mid-1960s some consensus on seafloor-spreading versions of Drift began to emerge among marine geologists and, later, among other specialists.

Facts, theories and consensus

Geologists are no more immune than other scientists to social and psychological considerations which may inform their judgment of theories and programs. There is, for example, a natural reluctance on the part of proponents of a theory, especially if they have attained positions of some authority in the field on the strength of achievements based on that theory, to admit that they were 'wrong' and to opt for a rival theory (cf. Bullard, 1975a). Moreover, there was certainly a 'struggle for authority' between those scientists with authority and technical interests in marine geology and geophysics and those with authority and technical interests in continent geology. This competition was influenced by funding. Educational background, institutional affiliations, employment, and similar elements of career trajectories may have been of greater or lesser importance. Even if these could be somehow 'factored out', the simple model of comparing theories with each other and with the data is too simplistic.

The simple model assumes that 'theories' and 'facts' are neatly packaged and unproblematic. One of the major themes of the history and the philosophy of science over the last quarter-century is the 'theory-ladenness' of facts or data (cf. Chalmers, 1982: 22–37). For example, one's theoretical views underpin decisions about which facts are relevant to a given problem or knowledge claim. The 'fact' of the jigsaw fit was highly relevant for many proponents of Drift. Opponents either dismissed the 'fit' as an irrelevant coincidence or argued that this 'fact' was not really a fact at all, i.e., that the 'fit' was quite poor. There was agreement among proponents and opponents of Drift that the Greenland data did not support Wegener's contention about Greenland's alleged drift; however, there was no consensus on how this might bear on the general question of Drift. Specialists in palaeomagnetism routinely 'cleaned' their samples and their data, whether the focus was directionalist or reversal studies or magnetic anomaly mapping. It was neither the 'raw' specimens nor the 'raw' data which found their way into the literature, but rather specimens and data adjusted according to various theoretical beliefs and expectations. Consider the famous *Eltanin-*19 profile. The data from which it was constructed were gathered using a complex network of theories and methodological presuppositions and equipment which itself was a concrete manifestation of theory. Marine scientists had no direct observations of magnetic stripes. The data were then processed and 'cleaned' in accordance with theory and so on. But, that to which I wish to draw attention is that the other profiles were tossed aside in favor of *Eltanin-*19: it demonstrated clearly the bilateral symmetry expected on the basis of Vine's refined model. In a sense, *Eltanin-*19 'made' Vine–Matthews and Vine–Matthews gave *Eltanin-*19 a significance it would not otherwise have possessed. It is an apparently clear-cut case of selecting the 'best' data where 'best' is defined as fitting most closely theoretical expectations and thereby 'confirming' them. The other profiles were judged to contain more 'noise'; i.e., the symmetry which they displayed was less striking, less convincing.

Given that theory is involved in the collection, selection, 'massaging' and interpretation of data, and that data gain significance only within some theoretical context, it is small wonder that a simple appeal to 'the facts of the matter' may often be insufficient to settle disputes or to single out one theory or program as the 'best' available. Further, proponents or different theories, as noted elsewhere, implicitly or explicitly may assign different weight to the 'same' data and may differ in their assessment of the importance of specific problems and the adequacy of proposed solutions (see L. Laudan, 1977: 31–40, 64–6). This is partly a function of their theoretical commitments, partly a determinant of those commitments; but also partly a function of their specialist and local interests which in turn also are determinants of their theory preferences. Kuhn (1970: 123–6), L. Laudan (1977: 15) and Lakatos (1978: 14) all acknowledge that empirical data are theory-laden and as such are not absolutely decisive in scientists' choices of programs.

How can some measure of consensus arise from this plethora of theories and programs, this welter of social considerations, and the multitude of specialist and local interests which seem to have characterized geology in the 1960s? A partial answer may be seen in the changing attitudes toward the succession of seafloor-spreading theories. This evolution within the Drift program incorporated more and more specialties and problem fields. With Wilson's version it impinged upon the central problems and concerns of a broad range of specialties. Those specialties might find not only that the newer versions solved outstanding problems but also that these versions gave significance to their research that it might not have otherwise possessed and, conjointly, additional authority and prestige within a redefined discipline. Even scientists who accepted other programs were aware of the rapid development of Drift. For example, the rigidity of the ocean floors had once been a major objection to Drift; now the origin and composition of those same floors were explicable through the latest versions of Drift. A counterexample had been turned into a solved problem (cf. L. Laudan, 1977: 31). In so far as these successive versions proved to be of value within these specialties and with respect to different problems, the successive versions were entertained, pursued or accepted by more and more workers from a growing array of specialties. Menard (1986: 238) expresses this succinctly: 'it will be a characteristic of this scientific revolution that most specialists were only convinced by observations related to their specialties'. The 'colonization' of specialties by Drift was probably enhanced by the rise to prominence of geophysics and marine geology in the 1950s and 1960s. Practitioners of those and related specialties gained a louder and perhaps more authoritative voice with respect to general questions of standards, aims, and characteristics of preferred theories and programs.

Underdetermination and consensus

Social interpretations of scientific change often invoke the 'underdetermination' of theories: for any given set of data, there is an infinite or at least indefinitely large set of possible theories compatible with it. Therefore, if scientists reach a high degree of consensus of any one theory or program, this cannot be explained in purely rational terms; e.g., that one theory is in best accord with the data. Recourse must therefore be had, it is claimed, to non-rational factors to explain consensus. Our historical narrative is not inconsistent with underdetermination. However, in the 'real world' of scientific practice that we are examining, underdetermination in its pure form does not seem applicable. Geologists did not have to hand all possible theories to account for ocean-floor data. They had a restricted set of three or four programs and a dozen or so versions of those programs from which to select or on which to build. The data base was not fixed but evolving; as it changed, so too did scientists' judgments of the relative merits and demerits of those versions and programs with respect to one another and to the data. It may be that social 'interests' outside the discipline played a role in theory choices but it is not clear that this need be invoked to account for what became a near-unanimous choice despite the principle of underdetermination. An examination of the social interests inside the discipline would likely be more rewarding.

This conclusion seems reasonable for Drift *vis-à-vis* Permanentism and

Figure 9.11. Successive versions of Drift in the 1960s drew in more and more specialities and problem-fields.

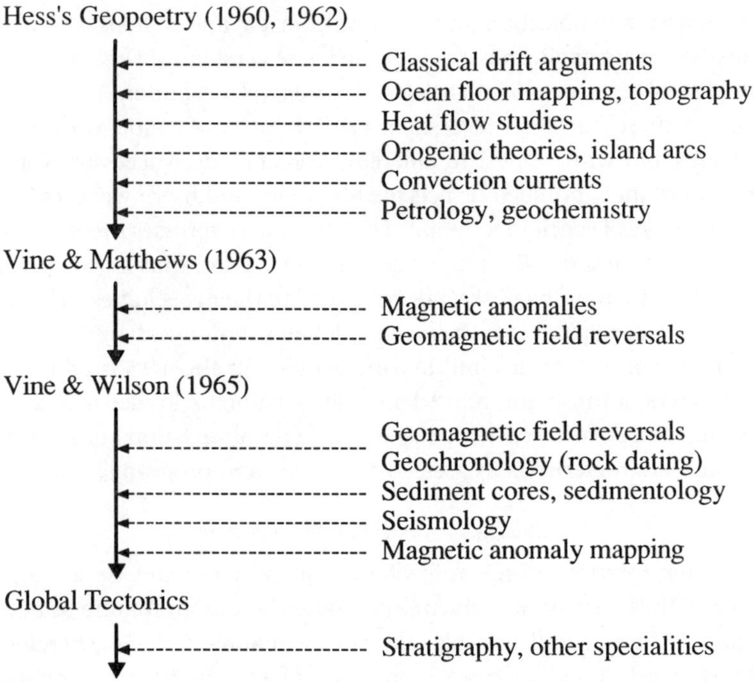

Hess's Geopoetry (1960, 1962)

------------------------ Classical drift arguments
------------------------ Ocean floor mapping, topography
------------------------ Heat flow studies
------------------------ Orogenic theories, island arcs
------------------------ Convection currents
------------------------ Petrology, geochemistry

Vine & Matthews (1963)

------------------------ Magnetic anomalies
------------------------ Geomagnetic field reversals

Vine & Wilson (1965)

------------------------ Geomagnetic field reversals
                         Geochronology (rock dating)
------------------------ Sediment cores, sedimentology
------------------------ Seismology
------------------------ Magnetic anomaly mapping

Global Tectonics

------------------------ Stratigraphy, other specialities

Contractionism but perhaps less so for Drift *vis-à-vis* Expansionism. 'Confirmations' of Vine–Matthews and of Wilson's transform faults were 'confirmations' of the formation of new oceanic crust through seafloor spreading. This could have been taken as evidence for Expansionism rather that Drift if one supposed that no subduction took place or that the rate of creation of new crust was greater than the rate of subduction of old crust.[15] Why was *Eltanin*-19 not immediately seized upon as evidence for Expansionism? Expansionism in the mid- to late 1960s seems to have slipped into the doldrums. At an empirical level, some geologists cited the null results of the test proposed by Egyed as a reason for rejecting the program. However, these tests were arguably inconclusive for rapid expansion and did not explicitly rule out slow expansion. It has been suggested (Nunan, 1986) that the diffidence of most geologists toward the idea of a swelling earth was due to its very novelty – at least for English-speaking geologists – and its apparent conflict with supposedly well-attested theories of physics and astronomy. The latter may have been reinforced in the late 1960s by a fading of interest among physicists in the gravitational theories of Dicke and others. Although these reasons for a marked preference of geologists for Drift over Expansion are not without merit, several less cognitive factors should not be ignored even though their effects are conjectural.

Drifters provided solutions to some of the puzzles and patterns from the sea before Expansionists. The public unveiling of *Eltanin*-19 was performed as a dramatic vindication of the seafloor-spreading version of Drift. Those who first presented and explained the profile did so in terms of the theories of Hess, Dietz, Vine, Matthews and Wilson.[16] Although *Eltanin*-19 was not the product of the Drift program, it first gained significance in the context of that program. Drifters therefore 'beat to the punch' Expansionists in solving this problem. In Laudan's scheme, Drifters simultaneously solved a problem in their program and created an anomaly for their rivals. A strict rationalist might argue that what should count in the long term is not which theory or program *first* solved a problem but which ones *could* and eventually *did* solve it. If Expansionists could make the case that the new data and seafloor spreading were compatible with their program – and Carey (1976; 1981b; MS) and other Expansionists; e.g., Jordan in 1971 (1966t) subsequently attempted to make that case – then it should not matter who first offered a solution (cf. Nunan, 1984: 278–9). The strict rationalist's argument may be sound but it ignores the psychological impact of being first. By the time that Expansionist solutions were forthcoming, sentiment had swung to the new versions of Drift. In a critical period Drift seemed to be the program making rapid progress and directing the attention of geologists to new and fruitful lines of empirical and theoretical work. As one possible consequence, Drifters may have been more likely to be successful than Expansionists in getting grants and ship-time to further develop their program, thus maintaining their momentum.

Proponents of the 'interests' or 'internal struggle' models could add that Expansionists were few in number, lacked significant authority and stature in the community and, perhaps most important, were inactive in the crucial years. Notable proponents of Expansionism in the English-speaking world – the scene of most of the major theoretical and empirical developments in the 1960s – were J. Tuzo Wilson, Carey, Heezen and Dicke.[17] Wilson had considerable authority but by 1963 had abandoned the program. Dicke disclaimed expertise in geology. His case for Expansion was restrained, he forebore detailed explanations of geological structures and did not contribute to the development of Expansionism on the geological front after 1962. It is difficult to estimate Carey's reputation and authority in the 1960s. He was geographically far from most of the Anglo-American centers of activity and turned out few disciples.[18] Although Carey proselytized for Expansionism, his publications in the 1960s would have done little to make the program competitive with the seafloor-spreading version of Drift. Carey was not inactive, but his papers for the most part did not appear in prestigious and widely circulated journals. In only a few (e.g., Carey, 1963) did he develop his 1958 version and in none of them did he deal with the rush of empirical and theoretical developments which were taking place in the mid-1960s. It was not until 1976 that Carey presented a formulation of Expansionism which discussed in detail such matters as palaeomagnetism and seafloor striping. Heezen was one of the few marine geologists to accept a swelling earth but he had not gained the status of a Hess or Maurice Ewing or Wilson. During the mid- to late 1960s Heezen allegedly had a falling-out with Ewing, which resulted in Heezen's being denied ship-time and also access to the flood of new data pouring into Lamont on ridges, sediment cores, striping and other key topics (Menard, 1986: 200–1). Whatever the reason, Heezen became a virtual spectator on the 'revolution' occurring around him and he did not further articulate the brief sketches of the Expansionist program he had already given (Heezen, 1959b; 1960; 1962). I do not imply that Drifters won out over Expansionists by default. These aspects of the emergence of a consensus in favor of Drift do, however, emphasize the point that research programs or traditions or paradigms are not self-perpetuating entities with life and dynamics of their own. They are sustained by and developed through the efforts of individual scientists engaged in the social system of science.

Consensus and priority

Concern about priority – about who was first in presenting new data or in proposing a new theory – arises when a knowledge claim has been accepted or seems likely to be accepted. The issue of who first proposed Drift became a 'live' one in the 1920s and 1930s when Drift was gaining some support. Those sympathetic to Drift sometimes put forward precursors of Wegener for chauvanistic reasons: 'Drift may be true, but it had already been suggested by an American (or a Frenchman or an Englishmen or . . .)'. Wegener's opponents sometimes put forward precursors as a means of undercutting his case: 'The idea of Drift is not new; it was put forward years ago by X and was ignored for good reasons; those reasons have not changed so we really ought to ignore Wegener'. When the prospects for Drift had dimmed, concern about priority waned. Taylor, for example, was quite happy to dissociate himself from what some Americans had termed the 'Taylor–Wegener Hypothesis'.[19] There are no great stakes, no credibility or prestige to be gained, in defending or claiming priority for rejected knowledge claims.[20]

There seems to have been no overt concern about priority for seafloor spreading relative to Holmes, Hess or Dietz in the early 1960s. Neither Hess (1962) nor Dietz (1961) cited Holmes's publications, nor did Hess or Dietz cite one another. Hess (1962: 607) did refer to Holmes as having 'suggested convection currents in the mantle to account for deformations of the Earth's crust' but not to engineer Drift! Arthur Meyerhoff advanced a priority claim on behalf of Holmes in 1968 when seafloor spreading was rapidly gaining acceptance (A.A. Meyerhoff, 1968). He may have done so as a means of undermining the theory, for he was a trenchant critic of spreading. Hess's (1968) and Dietz's (1968) comments about priority were in direct response to Meyerhoff. Hess drew a distinction between Holmes's seafloor-stretching and his own seafloor spreading. Dietz conferred priority on Hess. There the matter rested. It did not flare into controversy.[21] Perhaps it was because Holmes, Hess and Dietz were all working within the same general program – Drift; perhaps because this version of Drift was already obsolescent with respect to the Vine–Matthews, Vine–Wilson and plate tectonics versions. If a scientist's concern is choosing the theory or the program which is the most progressive or most promising or best problem-solver, the question of *who* first solved a problem within a given theory or proposed a theory within a given program is irrelevant. Within the social system of science, of course, such matters are highly relevant to the credibility, authority and prestige attributed to an individual scientist and this in turn may be relevant to the way future claims of that scientist are received. Similarly, there was no furore over priority between Morley and Vine and Matthews.[22] This may have been because they were working in the same program, because the hypothesis initially aroused little positive reaction, because it was Vine and not Morley who worked effectively to improve and to gain acceptance for the hypothesis and because by the time the hypothesis began to gain general acceptance it was being superseded by other versions of Drift. Vine, and to a lesser extent Matthews, have been awarded priority; the case for Morley has been put not by a geologist but by a philosopher (Frankel, 1982).

Notes

1 This approach, which represented something of a common ground between Contractionism and Permanentism, merits systematic investigation.

2 The 'Moho' marks the separation of crustal layer from the mantle as determined by the velocity of seismic waves. P – or compressional waves travel faster above the Moho than below; S – or shear waves travel faster below the Moho than above.

3 Dicke discussed his views with Hess (Dicke, 1957; 1962) and was apparently impressed with Hess's explanations of ridges and trenches. After 1960 Hess could give an account of these features and of continental displacement in terms of a growth of the seafloor which did not require expansion.

4 None of the calculations and tests conducted by Dicke and others gave results not easily explained without cosmological consequences by other physical theories.

5 Vening-Meinesz (1962) gives references to intervening papers.

6 Frankel (1980) presents a detailed exposition of Hess's successive views which culminated in 'seafloor spreading'.

7 It was only much later that the complexities of sediment distribution could be calculated on the basis of the more sophisticated plate tectonics version of Drift.

8 Bullard's remarks in a 1959 paper published (1961) seem to fall well short of a full endorsement of Drift. However, in his address to the Society, he specifically stated that he had 'announced' his 'conversion' in that paper. This may be a case of a memory reconstructed in the light of later events. Even in his 1963 address his tenor could be described as one of pursuit rather than acceptance. Perhaps this was appropriate for an audience which would hardly be described as friendly to Drift.

9 Frankel (1982) details the development and reception of the Vine–Matthews Hypothesis. Lawrence W. Morley, a Canadian geophysicist, combined seafloor-spreading and geomagnetic reversals in a note rejected by *Nature* several months prior to the acceptance by this journal of the similar seminal paper by Vine and Matthews. Morley also tried the *Journal of Geophysical Research* but to no avail. One referee commented 'This sort of thing you would talk about at a cocktail party, but you would not write a letter on it' (quoted from Frankel, 1982: 17). Morley's version of the 'Vine–Matthews' hypothesis did not appear until 1964.

10 Ronald Girdler, a Cambridge-trained geophysicist who concentrated on the Red Sea and the Gulf of Aden, arrived in the late 1950s and early 1960s at Drift interpretations. His attention was caught by a deep narrow trough running down the center of the Red Sea which seemed to be linked with the East African Rift Valley (Girdler, 1958). Girdler by 1962, influenced by Runcorn's (1962a) defense

of convection currents as the mechanism for Drift, argued that the Red Sea was an embryonic case of continental splitting caused by a newly-formed convection current which produced a ridge with a median rift valley (the trough) down the center (Girdler, 1962). He was puzzled by the magnetic anomalies associated with this structure and suggested that these were due to remanent and reversed magnetization of the basaltic intrusions and surrounding rocks, and idea similar to that of Vine and Matthews. He was apparently unaware of their hypothesis as late as early 1964. (Drake & Girdler, 1964).

11 Girdler had reached a similar conclusion but attributed it to self-reversal. Vine, on the other hand, may have been more thoroughly imbued with the opinion that spontaneous reversals were a relatively rare phenomenon, an opinion which underpinned the directionalist studies pursued by present and former members of his department.

12 Glen (1982) gives a very detailed narrative of this complex story. I present a simplified version.

13 Initially, researchers were especially keen to apply isotopic methods to Pre-Cambrian rocks. Pre-Cambrian formations contained very few fossils. There were horrendous problems in correlating Pre-Cambrian strata in different locales. There was also, inevitably, competition to find and date the oldest rocks in the world.

14 'Whole-rock' methods were not very accurate when applied to oceanic basalts. For marine rocks the normal procedure was to use the date of the lowest sediment layer resting on them.

15 See Nunan (1986) for an extended analysis of Expansionism versus Drift with reference to theory appraisal.

16 Recall too that the original version of seafloor spreading had been put forward as an explicit alternative to Expansionism.

17 Pascaul Jordan disclaimed expertise in geology. When in 1971 his monograph on earth expansion was translated into English, it was not only out-of-date in terms of its geology but it had been overtaken by the emergence of newer versions of Drift including plate tectonics.

18 Fakhruddin Ahmad's most widely available publications (1966; 1968) treated Drift favorably and did not press an Expansionist case.

19 In textbooks published since the mid-1970s Wegener – if anyone – is credited with being the first to put a real case for Drift. A few American texts give passing mention to Taylor, Baker and Snider-Pelligrini.

20 'Priority' is intimately associated with 'discovery'. Brannigan (1981: 71) suggests that 'discoveries' are attributions made by the scientific community on the basis of several criteria including the judgment that the knowledge claim is substantively true or valid. If this is the case, then a knowledge claim judged to be false or invalid has no status as a 'discovery' and no interest with respect to questions of priority.

21 Credibility, authority and prestige often seem to show the 'Matthew Effect': 'To him that hath shall be given. . . .' Hess received the lion's share of the credit for seafloor-spreading. This reflects not only his contributions to that theory but possibly also his standing in the early 1960s among marine geologists. A cursory examination of 20 texts published since 1980 bears this out. Those which include historical sketches credit Hess with spreading. The four which mention the similarity between Holmes's and Hess's version do not lodge a specific claim for

priority for Holmes. This may be because in most of these texts, there is a gap in the history between Wegener and the events of the 1950s and 1960s.

22 Scientists' standing in the community may affect the initial response to their knowledge claims: 'To him who hath shall be listened'. This may represent neither pure rationality nor the principle that knowledge claims should be evaluated without reference to their source, but it seems to be a common rule of thumb in the actual practice of science. Vine and Matthews were 'insiders': Vine was only a Ph.D. student but was in a prestigious Cambridge department; his co-author of the *Nature* paper, Matthews, was a well-known Cambridge geophysicist. Morley was an 'outsider': a Canadian with no reputation in geophysics when his note was rejected by *Nature*. Whatever the reason, texts often refer to Vine, Matthews and Wilson but Morley is not mentioned. More curiously, those directly involved in the development of the plate tectonics version of Drift are rarely singled out by name; the implication is perhaps that little originality was required or perhaps that it was a 'group discovery.'

10

The 'revolution' proclaimed

In 1965 most North American geologists and geophysicists accepted Permanentism or Contractionism. Beginning in 1966 sentiment swung rapidly to Drift. This seems to have been a ripple effect beginning among marine specialists and spreading to others. Recent converts are often the most enthusiastic; the American geological community was no exception. There were pockets of resistance but for the most part protesters were run over by the wheels of the rapidly-moving Drift juggernaut. Elsewhere the shift was less dramatic: large-scale mobilism of the earth's crust had been either widely accepted or recognized as an acceptable research program. By the early 1970s a new version of Drift, plate tectonics, had become the established research program. The 'revolution' was complete.

November 1966

The ministry to the unbelievers and the baptism of the converted began in November 1966: on the east coast of North America, at a special conference sponsored by the Goddard Space Institute and Columbia University and, on the west coast, at the annual meeting of the Geological Society of America (GSA). A corps of Drifters from marine geology and geophysics preached the new gospel. They were joined by several new converts, including Cox, Doell and Dalrymple and several continent geologists.

The Goddard symposium was dominated by Drift. Vine briefly reviewed his model and further developed it. If he were right in his argument that new ocean floor was created at ridges *and* that this process was recorded in the magnetic striping *and* that striping was produced by global magnetic reversals *and* that seafloor spreading occurred at a uniform rate on a given ridge *and* if the spreading rate for any given ridge were taken into account – a great many 'if's' – *then* the same pattern of magnetic striping should be found for every ridge. He compared the profiles from *Eltanin*-19 and for the Juan de Fuca Ridge with that prepared by Lamont for the Reykjanes Ridge and claimed that the patterns were the same (Vine, 1968). He displayed maps of the anomaly patterns to show their symmetry: those who had not seen these patterns were 'stunned' (Menard, 1986: 274). Bullard, Menard and Irving also took pro-Drift lines. Several geologists who specialized in continent topics presented modernized versions of older arguments for Drift. Marshall Kay, a stratigrapher at Columbia University, and his collaborator John Dewey, a Cambridge geologist, implied that Drift could explain detailed pattern-matching between the Canadian Appalachians and the British Caledonians (Dewey & Kay, 1968: 166). Patrick Hurley, a geochronologist at MIT, used modern isotopic dating methods and the Bullard Fit to construct a pattern-matching case for Drift. He stated that the Pre-Cambrian shield in Nigeria was made up of two major geological provinces: on one side of a sharp boundary, the basement rocks were about 2000 MYA; on the other, about 600 MYA. Using the Bullard Fit, he extrapolated this boundary into Brazil. He claimed that the dating of Brazilian rocks on either side of the projected boundary confirmed his extrapolation and that this was striking evidence for Drift (Hurley & Rand, 1968).

The pro-Drift atmosphere must have been palpable. Ewing sensed the tone of the meeting on the first day and (perhaps anxiously) asked Bullard, 'You don't believe all this rubbish, do you, Teddy?' (Bullard, 1975a: 20). Ewing and a Lamont colleague, Xavier Le Pichon, were the only two open opponents of Drift among the major speakers. Their effort to shore up Permanentism (Langseth, Le Pichon & Ewing, 1966) was ineffective. Opdyke described his work on Antarctic sediment cores. He did not defend Drift but stressed that two separate measures of the reversal age-scale derived respectively from sediment cores and continental rocks were mutually consistent (Opdyke, 1968: 61–2). This was indirect support for the revised Vine model of seafloor spreading which related the age-scale of continental rocks to magnetic anomaly patterns. Far more dramatic were the declarations in favor of Drift by two other Lamont researchers, Jim Heirtzler and Lynn Sykes. Heirtzler displayed a range of profiles across ridges and claimed that all were consistent with the most recent version of seafloor spreading (Heirtzler, 1968a: 94). The effect on those in the audience unacquainted with Heirtzler's recent change of heart concerning Drift would have been similar to that on the Geological Society of London had Lyell suddenly endorsed catastrophism. Sykes, a seismologist at Lamont, analyzed mechanisms for several earthquakes on the ocean ridges. His conclusion was unequivocal: 'The results support the hypothesis of ocean floor growth at the crest of the mid-oceanic ridges' (Sykes, 1968: 121)[1] Menard took seafloor spreading for granted and concentrated on residual problems; e.g., to explain the origin of transform faults he proposed that they might arise from the decay of a large convection cell into several smaller, more localized ones (Menard, 1968: 110).

The effect of these successive blows in favor of Drift must have been jolting. The 39 invited scientists included several newcomers to the field who were to develop further versions of Drift. Dan McKenzie, who had just completed his Ph.D. at Cambridge and who would be one of the architects of plate tectonics, described his reaction (quoted from Menard, 1986: 273–4):

> various papers clearly showed me for the first time that sea-floor spreading was a global phenomenon and had probably generated the ocean floor. I think the papers by Vine and by Sykes were the ones which really convinced me. I returned to England for about six weeks and immediately started to work on the subject.

Menard in a memorandum on this 'remarkable meeting' judged that 'Sea-floor creation and spreading centered on the mid-ocean ridge and rise system is now demonstrated' and that 'marine geology and geophysics are at a turning point. . . . It will be a very exciting time for participants but a sad time for onlookers' (Menard, 1986: 277–8). The mood was captured by Bullard. There were two summaries scheduled, one by Bullard in terms of Drift and one in terms of Fixism by Gordon MacDonald. When MacDonald cancelled, no one came forward to take his place. Bullard tried to speak to both points of view but he deemed his Fixist remarks 'unconvincing' and played these down in the published version (Bullard, 1975a: 20). Bullard singled out the agreement of field reversals on land, in sediment cores and in seafloor striping and their incorporation into seafloor spreading. More generally, the new version of Drift made sense of the diverse results of research on land and sea, from geophysics and from traditional geological specialties. Finally, Drift suggested 'things we might do', that it, a multitude of projects and problems. Its greatest drawback was that it said little about 'the great problems of the continental crust' (Bullard, 1968).

Drift did not get 'star billing' at the 1966 GSA meeting. Of some 500 abstracts submitted, most dealt with traditional topics in traditional ways: a single local structure or a small slice of the earth's history described in detail. Few took a broader global view. Only nine dealt explicitly with Drift or seafloor spreading or closely related topics. Cox, Doell and Dalrymple exhibited their latest reversal scale and suggested its application to sediment cores and to the unravelling of problems of glaciation in Iceland. Sykes, Vine, and Hurley covered the same ground as they had at the Goddard symposium.[2] It is difficult to estimate the impact of the few papers on Drift but two likely became more widely known through publication in *Science* (Vine, 1966;

Hurley *et al.*, 1967). The impact seemed neither immediate nor overwhelming. A year later at the 1967 GSA meeting of the more than 500 abstracts submitted again only nine dealt specifically with Drift or related matters and of these only four made even implicit reference to seafloor spreading.[3] Outside North America, interest and enthusiasm was building. The first Gondwana Symposium was held in 1967 in South America; several papers by stratigraphers and other mainline geologists were predicated on Drift.

The *Eltanin*-19 profile, the Goddard symposium, and the flurry of Drift-related publications met a more enthusiastic response from geophysicists. The program for the 1966 meeting of the American Geophysical Union (AGU) did not include a session on Drift or seafloor spreading. By 1967 the climate had altered. Wilson gave 'Transform Faults and Magnetic Anomalies in the Ocean Basins' as one of the four invited 'Frontiers' papers, a clear signal that his and Vine's version of Drift had 'arrived'. Just over 50 papers were scheduled on seafloor spreading and related topics. There were four sessions under the headings of island arcs, seafloor spreading, ridges and magnetic anomalies.[4] The new versions of Drift – at least in the US – still operated from a naval base.

Plating the earth

By mid-1967 the Vine–Wilson version had won support from many marine geologists and geophysicists. The new views were already percolating into classrooms. In one of the introductory courses in marine geology at Scripps the professor began by walking to the blackboard, saying, 'I've got to tell you about these magnetic anomalies' and drawing pictures (Menard, 1986: 281). As yet, however, the version had neither been shaped into a truly 'whole earth' theory nor did it have much to offer continent geologists. Over the next few years the race was on to exploit the theory at sea, to refine it into a 'new global tectonics' and to extend this – at last – to major continent structures.

The transformation from seafloor spreading to global tectonics was a transformation from two-dimensional representations and arguments based on maps and charts to three-dimensional arguments based on the globe. It was also a transformation from what some geophysicists would have regarded as a vague qualitative model of earth processes to a more rigorous, geometrical, quantitative one. Hess, Vine, Matthews, Wilson and others through the mid-1960s had advanced, refined and defended their views with the use of two-dimensional maps, graphs, magnetic profiles, earthquake plots, and similar evidence, often restricted to relatively small sections of the earth's crust. Drift was, however, a theory appropriate in scale to the 'big earth'.[5] Bullard's use of Euler's theorem to attack a problem of global (in the double sense of world-wide and spherical) geology established a significant precedent.

Bullard's approach, coupled with Wilson's transform faults and his division of the crust of the earth into rigid slabs bounded by zones of mobility, opened the way for a mathematical tectonics. This treated the entire surface of the globe, subsumed seafloor spreading and was validated by and gave validation to the seismological work of Sykes and others. Could the blocks of crust, mapped out by Gutenberg and Richter's seismological plots (see Figure 8.4) and confirmed and made more precise through the data gathered by the WWSSN, be plotted onto a globe such that their combined directions of motion (as determined by earthquake mechanism solutions) and their rates of motion (as determined on the basis of seafloor spreading and reversal chronology) formed a coherent whole embracing the ocean floors and the continents? The problem, the data and the mathematical tools were known: the prize for the answer would not be a Nobel Prize, but it would be substantial.[6]

The keenest competition to answer this question came from those associated with a familiar group of institutions: Cambridge, Princeton, Columbia University (Lamont–Doherty) and Scripps. These institutions had won pre-eminence in earlier phases of the 'new geology'; they continued in the forefront of the race. Dan McKenzie, who had in the pivotal year 1966 earned a Ph.D. from Cambridge with a thesis on the 'Shape of the Earth' was first across the finish line. At the end of 1967 he and Robert Parker at Scripps, who had devised a computer program for plotting data on map projections, published a 'plate tectonics' version of Drift (McKenzie & Parker, 1967). McKenzie restricted his comments primarily to the Pacific plate but his geometry portrayed the entire crust as an interlocking set of 'paving stones'. Sections of the crust and transform faults became geometrical entities. The Pacific slab or block or, as McKenzie termed it, 'plate' was ideally rigid and undeformable, like Wilson's rigid plates. Its topography, composition, indeed all of its geology was set aside. It became a platonic form functioning in the tidy world of geometry. Employing Euler, McKenzie fashioned a world in which transform faults became arcs of circles defined by the poles of rotation of the plate; ridges and trenches, 'lines along which crust is produced and destroyed' (McKenzie & Parker, 1967: 1276). For the Pacific plate, he found 'remarkable' agreement between the theory and such geophysical data as earthquake mechanism solutions; he confidently asserted that 'the paving stone theory is essentially correct' for that region.

Jason Morgan of Princeton followed a few months later with a model which divided the crust into 20 major and minor 'blocks', as he prosaically termed McKenzie's plates. Morgan had summarized his solution at the April 1967 meeting of the AGU. His submitted abstract dealt with trenches; ironically, McKenzie, who had attended most of the session, left before Morgan ignored his abstract and talked about 'blocks' (Menard, 1986: 284–6). The approach was the same as McKenzie's, with all the seeming arrogance of the physicist–geometer legislating the way the world must be: 'We have supposed that slow compressive systems are difficult to identify and have freely placed such boundaries at likely places'. He assumed that each block was completely rigid, despite contrary evidence laboriously gathered by field geologists over the previous century, for it was this assumption 'that gives this model mathematical rigor' (Morgan, 1968: 1960). He was a kindred spirit to Jeffreys even though Morgan was erecting 'a geometrical framework with which to describe present day continental drift' (Morgan, 1968: 1959) instead of attempting to prove its impossibility. He calculated the rate and direction of motion for each of his major blocks and found them to be mutually consistent. The motions of the giant interlocking pieces of lithosphere, perhaps 100 kilometers thick, could be harmonized on a globe. Le Pichon, a French geophysicist then working at Lamont, was next. He had with Ewing argued against seafloor spreading at the Goddard symposium. Now he adopted a 'simple earth model' of six large

Figure 10.1. Xavier Le Pichon divided the crust of the earth into an interlocking set of six blocks or plates. The continents were located on these and moved with them. This simplified picture facilitated quantitative studies of plate motions. (after Le Pichon, 1968: 3675).

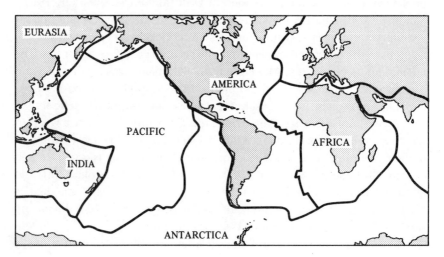

rigid 'blocks' which enabled him to give 'a mathematical solution' as a 'first approximation' to actual crustal displacements (Le Pichon, 1968). He shared the view then held at Lamont that seafloor spreading was episodic rather than continuous but was able to draw on magnetic anomaly patterns and reversal chronology to buttress Morgan's analysis and to reconstruct the relative positions of the continents over the past 120 MYA. In plate tectonics continents neither 'floated' on denser sima as Wegener had envisaged nor were they carried along on the backs of convection cells as Holmes had thought. They were embedded in huge, thick rigid chunks of lithosphere – the plates – which moved over a weaker aesthenosphere. The plates formed an interconnected global system. They were created at ridges, devoured in trenches, bumped uneasily past one another and sometimes collided. As Le Pichon emphasized, 'all movements are interconnected' and 'any major changes in the pattern of spreading must be global' (Le Pichon, 1968: 3693). The architecture of the globe as far as marine geologists and geophysicists and geometers were concerned was complete. The whole theory could be summarized in a cartoon.

Evidence that geologists would count as 'hard' evidence, geological evidence as opposed to geophysical evidence, for the key conception of seafloor spreading was produced in 1970 by the JOIDES (Joint Oceanographic Institutions for Deep Earth Sampling) project. Core samples of the oceanic crust down to the 'bottom' basaltic layer were obtained from nine sites. The

Figure 10.2. The globe according to the plate tectonics version of Drift. The plates were generated at ridges and consumed in trenches; their boundaries were marked by seismic activity. Compare with Figures 8.1, 8.3, and 8.4. Expansionists might visualize this 'exploded view' as corresponding to the 'real' earth: these pieces rest on an inflatable ball; as the earth expands, new material fills the 'expansion cracks'.

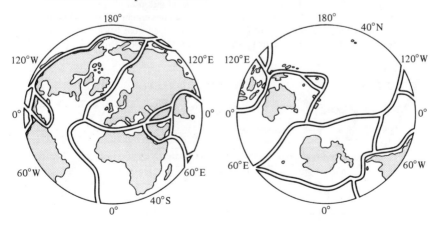

cores fitted the implication of seafloor spreading that the floors should increase in age with distance from ridges (Maxwell *et al.*, 1970). Moreover, they were in accord with the presumption of Vine, Wilson, and others that spreading was continuous and uniform rather than episodic as argued by Ewing and Le Pichon. What remained was the daunting task of invading the land: of conquering both the geology of the continents and land-bound geologists.

Climbing the mountains

By the late 1960s seafloor spreading and plate tectonics were probably orthodoxy for marine scientists. Although Wegener had put forward Drift to solve problems of continent geology, in the late 1960s few landlubbers had applied the theory to their specialist interests. Marshall Kay, who specialized in Appalachian geology, inclined to Drift as early as 1965 through the British palaeomagnetic studies and his own observations on stratigraphic correspondence between mountains in Newfoundland and in the British Isles. He had hesitated to accept it and had taken the cautious stance of calling for more evidence (Kay & Colbert, 1965: 452–69). At the Goddard symposium too he was cautious. His discussion with Dewey of pattern-matching between the Canadian Appalachians and the British Caledonians is reminiscent of E.B. Bailey's some forty years earlier. They stated only that

Figure 10.3. The plate tectonics version of Drift could be reduced to a cartoon, useful in disseminating and teaching the 'new global tectonics'. This oft-reproduced diagram depicts the processes occurring at the edges of plates – which J. Tuzo Wilson had earlier termed zones of mobility. The plates form the rigid lithosphere. New plate material is formed at a ridge (center) which is offset by two transform faults. A simple convergence of plates (right) results in the subduction and consumption of old plate material in a trench. A more complex pattern of convergence (left) results in two trenches joined by a transform fault. (Reproduced from Isacks, Oliver & Sykes, 1968: 5857, copyright by the American Geophysical Union).

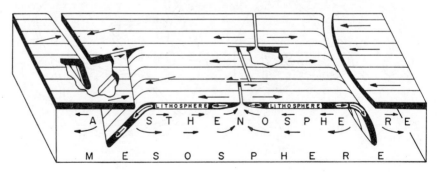

these mountain chains 'were once contiguous' (Dewey & Kay, 1968: 166). The following year Kay convened a symposium at Gander, Newfoundland, on the origin of the North Atlantic Ocean. Most of the participants, who came from North America, northern Europe and Britain, were either palaeontologists or stratigraphers. Most merely described in minute detail structures and strata on the opposing coasts and did not mention Drift. Reading the proceedings of the symposium gives a sense of *déja vu*: the arguments and conclusions, though showing a more thorough knowledge of regional geology, are similar to the discussions of the 1920s and 1930s. With few exceptions [7] even those who favored Drift did not distinguish between vague notions of continental separation and the new theories of seafloor spreading with their attendant geophysical arguments and lines of evidence. Kay modestly asserted in the beginning of his summary that the belts of opposing strata and structures 'are so similar that they must have been contiguous, or in the same trend' (Kay, 1969b; 965). He went on, however, to explain the orientation of the mountain chains on the opposing coasts in terms of seafloor spreading and to leave no doubt as to his own preference for this theory.

Dewey soon took full advantage of the 'new global tectonics' to construct a new general theory of orogenesis. A quondam collaborator with Kay, he also participated in the Gander symposium. There he linked his discussion of the British Caledonians to J. Tuzo Wilson's recent speculation (J.T. Wilson, 1966b) that an older 'Atlantic' had once divided America and Europe, had closed and had reopened along slightly different lines to form the present Atlantic (Dewey, 1969a).[8] After Gander, Dewey gave an account of the formation of the Appalachians and other mountains bordering the Atlantic in terms of the openings and closings of the Atlantic through plate movements: for example, a second collision of the North American plate with the African plate had produced mountains in Virginia and Pennsylvania (Dewey, 1969b). With John Bird, Dewey extended this analysis to other mountain ranges (Dewey & Bird, 1970). In Wegener's theory some mountain chains were due to collisions of continents but most were due to the crumplings of bows of sialic ships sailing through simatic seas. In the new global tectonics, the Alps, Andes and Rockies were also attributed to collisions. Moreover, Wegener's account of orogeny had been limited to mountains formed after the breakup of Pangaea. The new version could be applied to the whole history of the crust. Dewey's theory marked a successful invasion of the land by ocean-born versions of Drift. Drift carried off the glittering prize of continent geology, the prize for a global theory which could give a convincing explanation of orogensis.

Drifting into earth science

J. Tuzo Wilson (1964: 89) prophesied the imminence of a revolution in geology in language that smacked of Kuhn's (1962) account. In 1968 he proclaimed that the revolution had transpired, that its effects would be felt in all fields of geology and that this would require a drastic alteration in the teaching of the subject. Wilson portrayed the 'revolution' as a Kuhnian one. In announcing it he pointed out that new textbooks and new editions of textbooks incorporating the new ideas would soon be rolling off the presses (Wilson, 1968a: 15). This would in Kuhnian terms set the final seal on the 'revolution': Drift would be enshrined as the new orthodoxy and the history of the discipline as related in texts would be rewritten accordingly. His 'revolution' was not simply a change from static continents to moving ones. The new view of the earth was revolutionary also because it was a global theory in the dual sense of being a whole-earth theory and a theory which drew upon and drew together all fields of the science. Theory rather than description would become the centerpiece of a new discipline (Wilson, 1968a: 12). In the old geology each specialty and subject 'has been treated as an isolated set of methods of data collecting and as a package of accumulated facts'; earth science based on Drift would be radically different: '[it] can be put on a similar basis as other sciences, with a discussion of the principles of the behavior of the Earth as the framework and the essential parts of the traditional subjects used to illustrate this' (Wilson, 1968a: 13–4). For Wilson the revolution involved not only a change of theory but also the change from a descriptive historical 'Geology' to a predictive causal 'Earth Science' in which such once-central specialties as palaeontology would be moved to the periphery.

Drift in its incarnation as plate tectonics, which was just being adumbrated when Wilson proclaimed the revolution, filled a role which had been created but not yet cast. From the late 1950s there had been sporadic attempts to reshape and redefine 'Geology' into 'Earth Science'.[9] Wegener had used similar rhetoric in pleading for a broad assessment of his theory. As he had put it, 'all *earth sciences* must contribute evidence towards unveiling the state of our planet in earlier times' and, further, 'It is only by combining the information furnished by all the *earth sciences* that we can hope to determine truth here . . .' (Wegener, 1929t: 'Preface', emphasis added). The Central European tradition of 'earth sciences' and 'global geology' was in his day shared by few geologists in the Anglo-American world; e.g., T.C. Chamberlin and Daly. It was recreated or reinvigorated in Britain and North America in the late 1950s and the 1960s.

Enthusiasm for an inter-disciplinary study labeled 'Earth Science' preceded the general acceptance of Drift and seems to have owed little to theoretical developments *per se*. One impulse was the increasing prominence of geophysics and geochemistry within the 'old' geology. This had several effects. First, researchers with backgrounds in physics or chemistry and who were oriented towards theory rather than to descriptive field work were attracted into the discipline. Second, the growth of geophysics and geochemistry generated pressure for a stronger component of mathematics and physical science in geological training. The result was a tipping of the balance from geology as a historical science toward geology as an explanatory science. These changes were partly due to trends within the discipline; e.g., the development of more sophisticated techniques in petroleum exploration, the increased use of geophysical instrumentation and the growth of marine geology and geophysics. These trends were likely accelerated by the effects of the 'Sputnik' launch on the US political, educational and scientific communities and by the International Geophysical Year (IGY).

In the name of national defense and of maintaining the US's putative lead in science and technology, new school curricula were designed and implemented, programs to attract and retain students in chemistry, physics, biology and mathematics were introduced and government funding of scientific research was stepped up. The irony that vast sums were being poured into the space program while over two-thirds of the earth was as yet almost entirely unknown was not lost on geologists, nor was the fact that the space program was generating enormous public interest and support. IGY was intended to be a study of the whole earth and of the planetary earth from mid-1957 to the end of 1958. It incorporated studies of the oceans, the atmosphere, terrestrial magnetism and electrical fields surrounding the earth. In the aftermath of the IGY and in the context of the 'space race' which it helped spark there was a pragmatic concern: to attract to the study of the earth students dazzled by rockets, high-energy physics and medical and biological research. 'Project Mohole' would, it was hoped, attract to geology large-scale government funding and, in the words of its originator, 'would arouse the imagination of the public, and . . . would attract more young men into our science' (Quoted from Wood, 1985: 186; cf. Greenberg, 1967: 171–208). For many geologists this concern did not yet entail theoretical changes but rather discarding a fusty image associated with rock collections and maps and with fossils both in display cases and behind lecterns. The oft-reproduced photographs of the earth floating in space or of the 'earth rise' seen from the moon are symbolic of this reorientation away from geology as a feudal collection of specialties and toward geophysics and physical science, toward an 'Earth Science'. This reorientation engaged teachers and textbook authors in universities and schools. Initially, at least, much of it seems to have been cosmetic rather than cosmic.

One expression of the reorientation in North America came in the renaming of some departments of geology as departments of earth science or geoscience. J. Tuzo Wilson's in Toronto was one of the first to be rechristened; several in the US soon followed (Wood, 1985: 181). Curricula changed more slowly: Wilson commented in 1968 that for most departments little had altered except for a somewhat greater stress on geophysics instrumentation, mathematics, physics and chemistry (J.T. Wilson, 1968a: 12–3). A second expression was the restructuring and renaming of some North American textbooks. Of 19 new, or new editions of introductory university texts published between 1958 and 1966, four in the title or subtitle purported to be texts on 'Earth Science' compared with 15 labeled as 'Geology'. The process was further advanced in schools: of five secondary-school texts, four claimed to be texts on 'Earth Science' and only one, 'Geology'. Earth science texts included material on astronomy, meteorology and oceanography and gave greater emphasis to geophysics. Unlike geology which built up from rocks to strata toward, but never quite reaching, a vision of the earth as a system, the more glamorous earth science moved down in scale and generality from the planet earth to specific geological processes and structures. At least, this was the rhetoric to justify a place for earth science in general education and kindle student interest. Wilson described the expected appeal of earth science *vis-à-vis* geology in the following terms (Wilson, 1968a: 14):

> This new approach is necessary if earth sciences are to attract students. Why should bright young scientists elect to study subjects mulled over for a century, shown to be ineffective in coping with major problems of the Earth and often of only local interest when a more modern undergraduate curriculum can intro-duce him to the whole spectrum of Earth and planetary science? The full scope and the greatest appeal of the subject lie in pointing out that now is the time when other planets are being explored, when the sea floor is yielding exciting discoveries, when the interior of the planets is becoming understood. . . .

However, in the mid-1960s earth science did not yet espouse nor was it based upon a general theory of the earth.

Textbooks in the late 1960s and early 1970s began to mirror the growing acceptance of the newer versions of Drift and a role for this theory as the organizing principle of earth science. Foster's *General Geology* (1969) and Longwell's *Physical Geology* (Longwell, Flint & Saunders, 1969) both contained sections on seafloor spreading and showed partiality toward Drift as a general theory but both these traditionally-titled texts followed an otherwise conventional pattern of presentation. Neither made Drift the focus of the discipline. Verhoogen's *The Earth* explicitly aimed at an analytic rather than a descriptive geology and made no apology for insisting upon a solid foundation of physics and chemistry (Verhoogen *et al.*, 1970: v). Verhoogen toward the end of the book summarized seafloor spreading and the lines of evidence on which it was based. He and his collaborators judged this to be the most plausible program but in no sense is their text or the discipline as they construed it centered on or ordered by Drift (Verhoogen *et al.*, 1970: 670–94). A similar exposition, though with the Kuhnian gloss, was given by Mears (1970: 374–86). Arguably the first text to make the new versions of Drift the foundation of earth science was a British production. Staff of the new Open University were charged with finding a text which reflected the recent 'revolution' in the earth sciences; which employed plate tectonics to integrate the components of earth science and to make it an explanatory and predictive science rather than a historical one; and which portrayed 'present-day Earth scientists' as 'very far removed from the all too prevalent mental picture of elderly, bewhiskered gentlemen diligently brushing the dust from museum collections of rocks, minerals and fossils' (Gass, Smith & Wilson, 1970: 9). They found no such text and in short order commissioned and published one which consisted of contributions from many of the major participants in the 'revolution'. The produce was a collection of readings rather than a comprehensive systematic introduction to the discipline. However, its use of plate tectonics as the global theory and organizing principle of the earth sciences in some measure met J. Tuzo Wilson's call for a theory-centered earth science in which the traditional specialties would merely illustrate a central general theory.

Enshrinement of the new orthodoxy

In North America the 'new global theory' came to dominate earth science and geology by 1975. A plethora of texts either based upon Drift or which took a firmly positive stance toward it flooded classrooms (see Table 10.1). Other publications were emblematic of the now orthodox status of Drift: collections of 'classic' papers, histories of the 'revolution', bibliographies and popularizations.[10] This contrasted strikingly with the climate of a few years earlier. Mather's (1967) selection of 'classical papers' from the period 1900– 50 included no extract from Wegener nor any direct reference to Drift. He later commented that this was an 'oversight' (Marvin, 1973: 205) but more likely it represented a consensus amongst many American continent geologists – as late as the mid-1960s – that Drift was still not to be taken seriously or had no bearing on their research.

By the mid-1970s plate tectonics had, in North America, become the standard view. So banal was Drift that when in 1972 the Golden Press, a major publisher of inexpensive children's books which often featured Disney characters, produced a 'Golden Science Guide' to geology, it included a richly illustrated discussion of seafloor spreading, palaeomagnetism, earthquake plots and plate tectonics (Rhodes, 1972)! Further evidence that Drift had become the new orthodoxy is given by the results of a 1977 survey of senior professional geologists. Nearly 90% of the 205 respondents judged Drift either 'essentially established' (40%) or 'fairly well established' (46%). None judged Drift 'erroneous'. When asked to predict the status of Drift 10 years hence (i.e., in 1987), nine out of 10 predicted that it would be 'essentially established' (47%) or 'accepted with modifications' (43%). Only 6% believed that it would still be in doubt and only one thought it would be 'rejected' by 1987 (Nitecki *et al.*, 1978).

Geologists elsewhere were more restrained than their American counterparts in espousing plate tectonics. This may have been because versions of crustal mobilism had gained currency before the mid-1960s. The British geological community is an instructive example. With the exception of the various editions of Jeffreys' monograph, most British texts through the mid-1960s either endorsed Drift or at least provided a brief sketch of it as an alternative to Fixist interpretations. This likely indicates a greater acceptance or at least tolerance of Drift than in the US (See Table 10.2).[11]

British geologists may have held a more positive opinion of Drift than was apparent in textbooks. D.V. Ager, for example, was hostile to Drift but implied that a majority of his colleagues favored it, 'though American geologists appear to regard the Declaration of Independence [the separation

Table 10.1. *Drift in US textbooks*

(new or new editions of introductory textbooks; reprintings excluded)

Period	Drift-based	Pro-ish	Neutral	Anti	Silent	Total
1958–62	0	4	10	2	6	22
1963–67	0	6	6	3	2	17
1968–71	3	7	3	0	1	14
1972–75	11	8	2	0	0	21
Total	14	25	21	5	9	74

of North America from Europe] as retroactive to the Palaeozoic' (Ager, 1961: 138). Keith Bullen, a seismologist and former student of Jeffreys, was another hostile witness who testified to the popularity of Drift. He concluded a biographical entry on Wegener, written in the early 1960s (Wood, 1985: 204) with the observation that 'the enthusiasms of a considerable number of earth scientists lead them to assert, sometimes with religious fervor, that continental drift is now established' (Bullen, 1976: 216). His use of such rhetoric as 'enthusiasms', 'assert' and 'religious fervor' indicate defensiveness and, with the reference to 'considerable number', imply significant support for Drift among his colleagues.

British scientists were in the forefront of directionalist studies in the 1950s and in the elaboration of seafloor spreading in the mid-1960s. In the late 1960s and early 1970s McKenzie, Dewey and others were instrumental in the construction of plate tectonics. Still, there was no sense of a sudden collapse of the empire founded by Lyell. Perhaps many geologists tended their own specialist gardens instead of worrying about global geological issue and were either oblivious to the resurgence of Drift in the 1960s or at least not adamant in their opposition to it. Perhaps it was also due to the institutional divisions in Britain between geophysics and other geological specialties. Geophysicists were represented in the Royal Astronomical Society and the Royal Society; both institutions served as forums for discussions of Drift and of current geophysical research. However geophysicists, who formed the cutting edge of Drift research in Britain in the 1950s and 1960s, seem to have taken little part in the Geological Society which was the preserve of more traditional specialists. Within the Geological Society there were no noisy confrontations, there were no addresses by the Presidents nor special symposia on Drift and its implications for geologists. Indeed, this was

Table 10.2. *Drift in British textbooks*

(reprintings excluded)

Period	Drift-Based	Pro-ish	Neutral	Anti	Silent	Total
1940–50	0	3	2	1	0	6
1950–57	0	4	0	0	4	0
1958–62	0	1	5	2	0	8
1963–67	0	4	1	0	0	5
1968–71	2	2	0	1	1	6
1972–75	1	1	0	0	0	2
Total	3	15	8	4	5	27

true for most of the decade after Drift had been generally accepted. The 'revolution' in the US occasioned by the adoption of the seafloor-spreading or plate tectonics versions of Drift took the form in Britain of a gradual evolution in which those who had adopted Drift long before or during the mid-1960s were confirmed in their convictions and the opponents of Drift either muted their opposition (with the exception of Jeffreys) or gradually adopted some elements of the new global theory. In Switzerland, Austria, Germany, South Africa and Australia too, the newer versions of Drift produced no convulsions and were gradually assimilated into research and teaching.

The situation in France was more complex. Many of the senior members of the French geological community in the late 1950s and early 1960s had taken stands against Drift in the 1920s and early 1930s. Many in the geological 'establishment', and certainly most Parisian professors, opposed Wegener's version of Drift (Buffetaut, 1985: cf. Carozzi, 1985: 126–9). This was not unanimous. Marcel Roubault, a geochemist, gave an extensive summary of Wegener's theory in a widely cited monograph on orogeny and criticized the cool response Drift had received (Roubault, 1949: 126–78). Although he favored his own geochemical theory, he believed that 'the theory of Wegener contains a large measure of truth' (Roubault, 1949: 175). He was also an early partisan of 'Earth Science' as opposed to 'Geology': in 1953 he changed the name of the *Annales* of the geology and mining school at Nancy to *Sciences de la Terre* and campaigned for a broader conception of the discipline which would give prominence to such new specialties as geochemistry (Wood, 1985: 182, 189).[12] Opposition to Wegenerian Drift did not entail opposition to large lateral movement of pieces of the crust. Jean Goguel was hostile to Wegener's version. Nonetheless, he believed that large-scale horizontal movements were necessary to explain the formation of mountain chains and favored convection currents as the mechanism for shifting the crust on even a continental scale (Goguel, 1952t: 358–69). This attitude was widely shared in France (e.g., Moret, 1955: 451–6; Chanton, 1961: fasc. 2, 122, 142–3). By the 1960s even some geologists who had earlier adamantly upheld the permanence of continents and ocean basins (Theobald & Gama, 1956: 31–2; Furon, 1959: 347–50) allowed for horizontal motion produced by convection currents (Theobald & Gama, 1961: 485; Furon, 1967: 106–9). Finally, almost all of the textbooks in the 1950s and 1960s made at least a brief reference to Wegener and Drift (see Table 10.3).

Both seafloor spreading and plate tectonics incorporated horizontal crustal mobility and convection currents: ideas already endorsed by many

French geologists. The restrained but moderately favorable response to the new global tectonics in France is not unexpected. There was, however, a tone of disdain toward theorization and speculation: geology should be marshalled presentation of facts; theories and *hypothèses de travail* were of value primarily for directing further research and suggesting new observations. Accordingly, one should be wary of accepting the new versions of Drift uncritically or enthusiastically. Le Pichon in 1973, for example, was careful to categorize plate tectonics as a 'working hypothesis' (Le Pichon, Francheteau & Bonnin, 1973: e.g., vii, 1) and deplored the uncritical use of 'plate tectonics . . . to justify wild extrapolations from poor data with little rigor' (Le Pichon, *et al.*, 1973: vii). The hypothesis should be confined to present processes and structures; it was improper or at least led to many 'abuses' for one to extrapolate into the distant past either to explain particular structures or to construct general orogenic models. Dewey's work was criticized on both counts. This position, rigidly construed, could rule off most of the geological past as an object of study by plate tectonicians and could seriously undermine such efforts as Dewey's to render plate tectonics global in time as well as space or to make it the central theory for all geosciences (Le Pichon *et al.*, 1973: 270–5). There was certainly no clamorous plate tectonics bandwagon in France.

Geologists in the Soviet Union were, at least officially, very resistant in the late 1960s and early 1970s to all versions of Drift. Explanations for this opposition may invoke such 'social' aspects as the domination of the society by a gerontocracy (in the case of the geological community, by a gerontocracy devoted to older Fixist conceptions); the isolation of Soviet geologists from their Western counterparts and Western research literature; resistance to change on the part of the political and scientific bureaucracy; and a presumed consilience between Marxist theory and geological theories of rhythmic vertical tectonics (cf. Wood, 1985: 210–23). Vladimir V.

Table 10.3. *Drift in French textbooks*

Period	Drift-based	Pro-ish	Neutral	Anti	Silent	Total
1950–57	0	1	1	0	0	2
1958–62	0	2	3	3	1	9
1963–67	0	2	3	0	1	6
1968–71	1	3	2	0	0	6
1972–75	1	2	0	0	0	3
Total	2	10	9	3	2	26

Beloussov was the most prominent Russian opponent of Drift (e.g., Beloussov, 1968; 1970; 1976; 1979). He occupied positions of great prestige and authority in the crucial period: Vice-President of the IGY, Head of the Department of Geodynamics attached to the Academy of Sciences in Moscow, President in 1960 of the International Union of Geodesy and Geophysics, Chairman in 1972 of the International Upper Mantle Committee of the IUGG, and a participant in international conferences and research. We could analyze his opposition to Drift in terms of such 'social' factors as those mentioned above; it is plausible that the influence of his opposition may have been magnified by those factors. However, to draw a neat distinction between the 'social' and the 'scientific' or 'technical' and to use only the former to account for the negative attitude of Beloussov or of other Soviet geologists is misleading. Soviet geologists concentrated on continent, not oceanic, geology: after all, the USSR covers about one-sixth of all the dry land on the globe. There was an emphasis on economic geology, on the discovery and exploitation of oil and mineral resources. Beloussov had dismissed Wegener's Drift as 'fantastic and having nothing to do with science' (Beloussov, 1954t: 750). He had criticized it as 'overtly formalistic', as displaying 'a total and consistent disregard of the basic geotectonic data' and perhaps most importantly as 'explaining nothing of what must be explained in the first place' (Beloussov, 1954t: 753). 'First place' was held by the vast body of information about the geology of the USSR laboriously gathered by battalions of field geologists (Beloussov, 1954t: vi). This was better explained by the rhythmic vertical tectonics pioneered by Soviet theoreticians and refined by Beloussov himself than by Wegener's 'schematic' view. For Beloussov the explanation of regional details was preferable to vague global generalizations. He maintained this position with respect to newer versions of Drift: they too shed little light on the problems of special interest to Soviet geologists. These new versions showed 'excessive schematization' because they focussed on the oceans and the small body of geophysical data collected by marine geologists; data on continent geology were ignored, indeed 'the geology of continents is simply and completely annihilated' (Beloussov, 1968: 18; cf. Beloussov, 1979). Neither older nor newer versions of Drift seemed useful for Soviet problem-fields. The opposition was not monolithic. G. Gorshkov and A. Yakushova published a text which included favorable references to Drift (Gorshkov & Yakushova, 1973t: 672–5) and other Soviet geologists have published Drift-related research (see Wood, 1980; 1985: 217–20). We might suppose, nonetheless, that the opposition of Beloussov and others delayed the attainment of dominant status that Drift now possessed in other countries and that the incorporation of Drift into Soviet geology may come in a form different from that elsewhere; e.g., as a compromise between vertical and horizontal tectonics.

Life after the revolution

J. Tuzo Wilson (1976: vii) contended that life after the 'revolution' was profoundly different from what it had been before:

> Today many earth scientists believe that . . . a scientific revolution has occurred in their own subject. As in other cases, the new belief demands a reinterpretation of the observations of both geologists and geophysicists, ... The acceptance of continental drift has transformed the earth sciences from a group of rather unimaginative studies based upon pedestrian interpretations of natural phenomena into a unified science that holds the promise of great intellectual and practical advances.

An impressionistic survey two decades after the revolution lends some support to his contention. There are dissenters from the plate tectonics version of Drift but these proponents of Expansionism or other competing programs seem to represent very much a minority view. One characteristic of 'post-revolutionary' geology is that there is a high degree of consensus, though not unanimity, on the preferred research program. In this sense there has been a decline, though not a disappearance, of theoretical pluralism. It also seems that earth scientists are somewhat more conscious and make more frequent use of global theories in their practice and publications than in the three or four decades before the 'revolution'. Plate tectonics constitutes the general framework for most geologists and geophysicists in university research and teaching, in research and applied research in industrial and natural resource geology, and in mapping and surveying.[13]

The adoption of Drift gave rise to new research agendas. Some were 'mopping-up' operations (cf. Kuhn, 1962; 1970: 23–34); that is, reworking traditional research fields in terms of the new global tectonics. This included such geological specialties as structural geology, geomorphology, petrology, economic geology and palaeontology and also other disciplines on which geology impinged. For example, the 1976 Annual Biology Colloquium at Oregon State University had as its theme plate tectonics and biogeography. Virtually all of the participants, despite disagreements on other issues, were 'united' in casting their interpretations in terms of 'an unquestioned acceptance' of this program (Gray & Boucot, 1976: v). In line with Wilson's claim, most of the papers amounted to reinterpretations of old data and old solutions in terms of plate tectonics; a 'cashing-in' of promissory notes issued by earlier advocates of Drift solutions to biogeographical problems, a repackaging of previous work.

Similarly, it could be argued that the development of the notion of 'suspect' or 'exotic' 'terranes' within plate tectonics is in some ways a reinterpretation of the older geosynclinal model of continental accretion. The latter was intended to explain what appeared to be ancient 'shields' or 'cratons' around which were gathered more recent chunks of continental crust. In the early plate tectonics model, these assemblages were explained in terms of collisions of continents which then separated along different lines or of sediments scraped off an oceanic plate onto a continent as the former was subducted under a continental plate. Wilson (1968c: 316) had made passing reference to these processes in connection with the formation of mountains along the west coast of North America. This was supplanted in the late 1970s by the view that these mountain ranges, as well as other mountain ranges and parts of continents around the world, were made up of 'suspect', 'exotic' or 'allochthonous' 'terranes'. These were claimed to be pieces of continents or island chains or oceanic plateaus (the term 'terrane' rather than 'terrain' emphasizes their three-dimensional block-like nature) ranging from a few kilometers to hundreds or even thousands of kilometers across. These had been carried on plates for perhaps thousands of kilometers, sliced by faults, perhaps rotated, sometimes amalgamated with other terranes and then plastered against the cores of the present continents. The west coast of North America is supposedly a mosaic of over 50 terranes, each with a distinctive history as determined through stratigraphy, palaeontology and palaeomagnetics. Drift has thus been extended to resolve a problem previously solved by Permanentists and in a manner different from earlier versions of Drift.[14]

Other agendas involved 'mopping-up' of a different nature. These were directed toward resolving residual ambiguities within plate tectonics theory itself: the solving of problems and the gathering of data which had significance only within the context of Drift. These areas of research have been mostly the province of geophysicists.[15] A few examples should suffice. The problem of the detailed mechanisms of plate motions (establishing specific causal connections between convection currents and the movements of particular plates) has attracted much attention but thus far proposed solutions have not been judged satisfactory. Considerable effort has been put into more precise and detailed global reconstructions, particularly for the period prior to the breakup of Gondwana. This is connected with a third area, making more precise determinations of past plate movements. Finally, there are continued efforts using laser-ranging, satellites and even stellar interferometry (e.g., Carter & Robertson, 1986) to measure and to demonstrate present-day plate motion.

Still other agendas have led to modifications of the plate tectonics version, especially as it applies to continental structures. For example, the initial explanation of orogeny in terms of plate collisions has been challenged. On such 'active margins' as the west coasts of North and South America, orogeny is attributed not simply to the overriding of one plate by another but also to the plastering of terranes onto the North and South American plates. More recently, there has been much interest in explaining the formation of sedimentary basins adjacent to mountain ranges (i.e., 'basin-range provinces' such as the Great Basin in western North America or the Ross Embayment and Trans-Antarctic Mountains) and orogeny along 'passive margins' where the crust is pulling apart rather than converging or colliding. Different models have been suggested (cf. Kent *et al.*, 1982). McKenzie (1978) proposed that basins could be formed through the stretching, thinning and subsidence of a portion of a plate. This simple model formed the basis for more complex ones in which the plate might fracture and pull apart asymmetrically (Wernicke, 1981) or in which the upper more brittle part of a plate might 'delaminate' (i.e., separate from) and slide over the lower less brittle portion (Lister, Ethridge & Symonds, 1986). Proponents claim that these can explain not only subsidence on one edge of extensional rifting but also uplift on the opposing edge.[16]

These developments thus far amount to minor revisions of the dominant program: a more complicated account of orogeny, new ideas on rifting and passive margins, qualifications of the previous assumption that plates were rigid and undeformable, and so on. However, their possible importance in strengthening geologists' allegiance to and reliance upon plate tectonics should not be underestimated. Anita Harris, a well-known field geologist with the USGS, once complained that 'Geology often refutes plate tectonics. So the plate-tectonics boys tend to ignore data. The horror is the ignoring of basic facts, not bothering to be constrained by data' (McPhee, 1983: 209). She did not reject the version outright, her point was that the abstract model developed for the seafloors and extended to mountains by Dewey and Bird did not seem directly applicable to more complex continent structures and, moreover, that it did not address the wealth of regional data and their particular problems (cf. Scherman, Howell & Jones, 1984: 107). There was a vast chasm between an idealized global theory and the stones and strata underfoot (McPhee, 1983: 122). We might suppose that explanations of regional features in terms of terranes and extensional faulting and rifting go some way toward bridging this perceived chasm: as one puts it, these versions allow for 'fine-grained' explanations instead of 'arm-waving'.[17] This could be seen as a revival of localism compared with the grand generalities of early plate tectonics and as an adjustment of the simple tectonics of the seafloors to the variegated histories of continents. Terrane studies especially have enlisted the social and technical interests of geologists from more traditional specialties. A crucial part of the agenda is the identification and characterization of suspect terranes; stratigraphy and palaeontology are major tools (Nur, 1983; Scherman *et al.*, 1984: 112–5) for 'each terrane records a unique sequence of historical events' (Jones *et al.*, 1982: 57). It also represents a reassertion of the interests of continent geologists in vertical movements – uplift, subsidence, orogeny, faulting – which had been neglected in favor of the horizontal motions so characteristic of Drift (Kent *et al.*, 1982: 3).

It is certainly not the case that every research paper draws explicitly on plate tectonics. Plate tectonics may be the 'umbrella' program for most geologists. However, a geologist attacking a particular problem, mapping a particular region or constructing a research report may often have no need to refer to this high-level theory. Geologists may work on many problems which have no obvious connection with plate tectonics. One example is the current controversy involving palaeontologists and other specialists (including physicists, statisticians and astronomers) over putative mass extinctions, especially the so-called 'death of the dinosaurs' and whether or not mass extinctions, if they did occur, were periodic and had an extraterrestrial cause.[18] The controversy has generated hundreds of publications but plate tectonics does not figure in these except perhaps for an occasional appeal to subduction to remove evidence of meteoritic or bolide impacts.[19]

There are significant changes in post-revolutionary textbooks. All of the introductory geology or earth science texts published since the late 1970s give central place to Drift. This is, of course, quite obvious and expected. A more subtle difference from pre-revolutionary texts is that those produced in the past ten years are often structured around a global theory; i.e., the discipline is explicated in terms of plate tectonics, and this theory permeates much of the text instead of being banished to the end of the last chapter. This is even more the case with advanced texts, particularly in geophysics or physical geology. Moreover, although there may be a passing reference to 'sincere dissenters' (Strahler, 1981: 19), plate tectonics is often the only global theory presented.[20] There are a few exceptions. Clarke & Cook (1983: 470–4) summarize Expansionism but state that plate tectonics is the 'accepted' program whereas Expansionism is problematic and it 'not popularly accepted'.[21] This is a change from pre-revolutionary days when it was common for textbook authors to give summaries of rival programs and of the evidence and arguments for and against. Although plate tectonics is not usually depicted as unalterable, unproblematic and *true* but as the best available program; nor as fact but as generally pursued; not as completed but still in the process of construction, nonetheless, the clear impression received from textbooks is that pluralism in global theories is almost non-existent. In so far as textbooks serve as an indication of the training future geologists are receiving, that training could be arguably described as indoctrination into the dominant program. If this be the case, it ignores the continued existence of other, less widely-shared programs.

The dissenters

Drift's competitors were not abandoned by the mid-1970s. Drift had gained ascendancy in most quarters; it had won a high degree of acceptance not only among geophysicists but also in the geosciences community more generally. However, Permanentism, Expansionism, and even Contractionism continued to be pursued by some. These criticized what they judged the deficiences of plate tectonics and deplored what they thought to be a bandwagon effect; that is, an unthinking and uninformed stampede toward the currently fashionable theory.

Francis G. Stehli and Howard and Arthur Meyerhoff remained stout defenders of the permanence of continents and ocean basins. Stehli argued from the specialist perspective of palaeontology that the present distribution of brachiopods corresponded well to that in the Permian; since this distribution was temperature-dependent, the Permian equator must, despite palaeomagnetic evidence, have corresponded roughly to the present equator. Therefore, massive Drift has not taken place and the assumptions built into palaeomagnetic studies must be reassessed (Stehli, 1970). The Meyerhoffs, father and son, attacked plate tectonics largely by amassing specific observations which seemed to conflict with the theory; for example, the presence of some old rocks near active ridges.[22] Other criticism focussed on methodology: they objected that the process was viciously circular whereby plate margins were identified by earthquake plots and then the correspondence between plate margins and earthquake activity put as evidence for plate tectonics.[23] Their contribution of Permanentism was mainly defensive; e.g., arguing that some seafloor magnetic striping may be due to the earth's magnetic field but that much of it could be attributed to other causes such as the varying chemical composition of adjacent bands of rock (Meyerhoff & Meyerhoff, 1972a,b; A.A. Meyerhoff, 1978). The

Meyerhoffs perhaps made the point that the new global tectonics incorporated neither all of the data nor all the successes of the Permanentist program. However, their attempted resuscitation of the latter did not attract much support.[24]

Contractionism was defended by Jeffreys. He was joined by R.A. Lyttleton (1983), like Jeffreys an Emeritus Professor of Astronomy at Cambridge. Neither took much notice of plate tectonics or the growing body of data claimed to be in accord with that version: their earth 'has been found lying in the dusty attic of the imagination where it had been abandoned by the German geologists of the 1920s' (Wood, 1985: 209). One might conclude that Permanentism and Contractionism if not moribund are advocated primarily by theoretical dinosaurs who have not adapted to the changed environment of geology and who, producing no intellectual offspring, may soon become extinct.

Even if two rivals to Drift can be summarily dismissed thus, the third, Expansionism, remains more robust. Carey, from his first proposal of Expansionism to the present, has been a tireless itinerant evangelist for Expansionism. In brief, he contends that 'continents are permanent and fixed to the mantle below them (as Dana believed), and that oceans are formed by rifting apart of continents and once formed are fixed and permanent. This implies earth expansion' (Carey, 1981b:6). He argues that his theory can incorporate much of the pattern-matching, palaeoclimatic and palaeobiogeographical evidence used to support Drift, that seafloor spreading is a consequence of the expansion of the earth and that most of the palaeomagnetic results are consistent with his view (e.g., Carey, 1976: 9–13). He objects strenuously to what he terms 'the subduction myth'. He himself had thought in the early 1950s that the sub-crust, the conveyor belt which carried along the continents, was consumed in trenches; however, he soon discarded that view. He regarded subduction as both improbable and, on an expanding earth, unnecessary: the problem of disposing of old seafloor does not arise. Moreover, he presented specific objections to explanations of crustal features related to trenches and subduction and brought forward other features which he believed were more satisfactorily explained by Expansionism (Carey, 1976: 54–79). For a time Expansionism attained some standing as an alternative to both Fixist and older Drift theories. Many of Carey's analyses of structures, much of his terminology and some of his reconstructions of palaeocontinents were subsumed by seafloor spreading and plate tectonics.

Carey's analysis of the situation confronting Expansionism is illuminating. He was certainly aware of major empirical and conceptual problems for

Expansionism but as a first response suggested that it was a case of 'the pot calling the kettle black': Drift had problems too but these were overlooked by those who jumped blindly on the plate tectonics bandwagon and have since worn blinkers imposed by that dogma (Carey, 1976: 14; MS, Chapter, 10). As a second response, he has argued that the empirical objections to Expansionism are ill-founded (Carey, 1976: 16–18) and has developed his theory to meet some of the conceptual objections. To fit rigid crustal sections to a swelling earth would require extensive fracturing; he proposed that the crust is divided into a hierarchy of polygons ranging from 'first order' ones of thousand kilometers across down to 'sixth order' polygons a few tens of meters across. These permit the fitting of a rigid crust to an expanding earth (Carey, 1963; 1976: 42–50). As for a mechanism, he has examined and eliminated several possibilities, has spelled out certain constraints but in the end has self-consciously adopted the position of early Drifters: the lack of an acceptable mechanism should not constitute grounds for the rejection of the theory should it seem empirically adequate; i.e., should there be sufficient evidence that the earth has and is expanding (Carey, 1976: 446–60; MS, Chapter 13).

Irrespective of the merits of Carey's version of Expansionism or of the program in general, it is a program which had a following, albeit often a very small one, from the 1920s and 1930s up to the early 1970s.[25] The triumph of plate tectonics in the early 1970s did not obliterate Expansionism; it may have more adherents now than in the late 1960s. It seems to have garnered support especially from palaeontologists. This may reflect a specialist concern: reconstructions using plate tectonics do not in the opinion of some palaeontologists fit some palaeontological evidence as well as reconstructions using an expanding earth (H.G. Owen, 1976; 1981; Neil Archbold, private communication; Saxena *et al.*, 1985). The central claim of Expansionists however, continues to be that the subduction on the scale demanded by plate tectonics has not been established and that reconstructions of palaeocontinents without 'gaps' or 'gores' cannot be done on an earth of the present size (Carey, 1976; MS; H.G. Owen, 1976; 1981; Steiner, 1977; Ahmad, 1978; 1982). Three symposia (London, 1979, sponsored by the Geological Society; Sydney, 1981; and Moscow, 1981) have been devoted to Expansionism, and geologists in the USSR and elsewhere have worked to refine and apply the program (see Milanovskiy, 1983). Nonetheless, Carey may be over-optimistic to suggest that another revolution is at hand, that 'a Kuhnian orgasm is already quickening' (Carey, 1981b: 7). Plate tectonics seems firmly entrenched as a result of the modern revolution in geology, perhaps more firmly entrenched than Fixist theories in the 1920s.

Voice-Over

The story of how Drift won the hearts and minds of geologists is the stuff of which myths and legends can be made. This does not, however, obscure a number of issues concerning this case of theory acceptance and scientific change. What was the relationship between plate tectonics which became generally accepted and the earlier versions of Drift which had so often been rejected? Was there a sudden volte-face of geologists in the late 1960s and early 1970s which could be ascribed to a 'bandwagon' effect? These two questions are related and have obvious bearing on the evaluation of theories of theory change. We shall discuss them and then sketch other issues arising from this section of the narrative.

Drift and plate tectonics

Geologists sometimes seem uneasy in expressing the relationship between plate tectonics and the versions of Drift which preceded it. This is apparent in the use of such circumlocutions as 'continental drift–plate tectonics theory' (e.g., McArthur & Pestana, 1974: 105; Nitecki *et al.*, 1978: 664) or in characterizing plate tectonics as a *renamed* but more complete theory of Drift (Glen, 1975: 166) or as incorporating continental drift and seafloor spreading (Wyllie, 1976: 17, 21). Those who accepted versions of Drift prior to the early 1970s regarded the terms 'continental drift' and 'plate tectonics' as synonymous. Those who embraced plate tectonics after 1970 were more likely to perceive these terms as only loosely related (Nitecki *et al.*, 1978: 664). The perception by 'recent converts' of some disjunction between Drift and plate tectonics is explicable in several ways. At face value it is evidence for a claim that the history of Drift was quite irrelevant to the formulation and acceptance of plate tectonics and that geology only had a firm basis, a true paradigm, from about 1970.[26] A distinction is suggested between the science of plate tectonics and the ill-founded speculation which preceded it. This may serve the ends of textbook writers but it is a naive analysis of a complex issue. More plausibly, perceptions of 'recent converts' could be interpreted as instances of the ahistoricity of scientists, that 'A science that hesitates to forget its founders is lost' (Whitehead quoted in Kuhn, 1970: 137; cf. Kuhn, 1970: 135–40).

Alternatively, one might analyze the differing perceptions of 'old believers' and 'recent converts' in terms of differing intellectual and social interests. By stressing a *continuity* between Drift and plate tectonics the 'old believers', especially those who authored Drift publications, benefit in at least four ways. First, their earlier work is validated by association with the now accepted theory. Second, the acuity of their judgment is confirmed. Third, they establish some share in any credit which might be awarded for the triumph of the prevailing orthodoxy. Fourth, these three taken together would add to the authority which the 'old believers' might expect to wield in the discipline. We can attribute to 'recent converts' a different set of interests which would be served by their stressing of a *discontinuity* between Drift and plate tectonics. This might take the form of emphasizing the differences between Drift and plate tectonics or even ignoring the former; of claiming that plate tectonics is much more rigorous and precise; of stressing the cruciality of such newer lines of evidence as earthquake mechanism solutions and magnetic striping; of adopting new standards and methodologies; or of portraying plate tectonics as having subsumed earlier theoretical and practical work instead of being an outgrowth of it. Plate tectonicians, a group comprised mostly of recent converts and younger scientists, would thereby effectively appropriate the field. The 'modern revolution' would be seen in terms of the articulation of a 'new global tectonics', not the culmination of a half-century battle by proponents of Drift. The achievements and the intellectual and social authority of the 'old believers' would be undermined and devalued and that of plate tectonicians correspondingly enhanced.

My view is that plate tectonics is the most recent version of a succession of Drift theories and that it is closely related to those theories. First, many of the lines of evidence and argument used by exponents of the older versions of Drift are consonant with more recent ones. The earlier versions differ from the later versions and some of these differences were of great importance in the varying appraisals these theories received from geologists. It is reasonable for geologists to stress the great differences between the evidence for and models of Drift put forward by Wegener and plate tectonicians. The differences between the two are not unlike those between the Copernican and Newtonian astronomies. However, if one look at the gradual development from Wegener to Holmes to Hess to Vine–Matthews to Vine–Wilson to McKenzie, Morgan and others – rather than comparing the 'beginning' and the 'end' – this development could be better characterized as incremental (uniformitarian!) rather than disjunctive (catastrophic!). Second, most of the geologists and geophysicists who actually engaged in the 'revolution' did regard plate tectonics as an improved formulation and a lineal descendant of earlier Drift theories. It could be argued that the 'revolution' occurred for marine geologists and geophysicists in 1966–8 but that it was not completed for the larger disciplinary community until the early 1970s. If this is accepted, then we could talk about a 'revolution' associated with seafloor spreading blending into a 'revolution' associated with plate tectonics.[27] Finally, most of today's textbooks which include a historical sketch depict plate tectonics as the end point of a sequence of theoretical developments beginning with Wegener. In so far as texts may be taken to represent consensus, this is evidence for continuity; against this must be set the possible rewriting of history as described by Kuhn to present the current theory as the cumulative product of previous achievements.

Bandwagons

Geologists have themselves used the term 'bandwagon' with respect to the American adoption of Drift in the late 1960s and early 1970s.[28] This term can, however, have different connotations. Bullard in 1967 remarked that he had seen his subject, geophysics, 'transformed from a backwater into a bandwagon'.[29] Part of that transformation was due to the connection between geophysics and the construction of new versions of Drift. Part was due to the recent confluence of 'many streams of knowledge and many ways of looking at the earth' in the new earth science. He probably intended by the use of the word 'bandwagon' to refer not only to the meteoric rise of geophysics in the 1960s but also to the fact that Drift was rapidly gaining momentum, that it was attracting large numbers of adherents and that it seemed assured of general acceptance. Bullard, Marvin (1973: 187), Wyllie (1976: 186) and others use the term in this relatively neutral descriptive sense. It has also been used pejoratively: to suggest that many of the new Drifters had climbed aboard simply because it had become the fashionable theory and that they had done so uncritically, unthinkingly, heedless of any criticisms of or flaws in that theory.

The pejorative sense of 'bandwagon' features in A.O. Kelly's 1970 letter to *Science News* (Quoted in Wyllie, 1976: 21–22):

> Now when some authority proposes a hypothesis like ocean floor spreading, every junior scientist in the country jumps on the bandwagon. Criticism, it seems, is rude, egotistical and out of style. . . . These me-too scientists are piling hypothesis upon hypothesis. . . . Reasons, it seems, are no longer necessary; one simply backs up one speculation with another.[30]

This is an excellent example of social accounting for 'error': the implication is that if younger scientists had followed appropriate procedures and standards in assessing the new theory, they would have come to share Kelly's skepticism; since they reached a different verdict, this must be due to some distorting 'social' element (cf. Mulkay & Gilbert, 1982a). Kelly overlooks several key issues. How did 'some authority' acquire that authority in the first place? If it be on the basis of the important contributions of that individual to the discipline as judged by other members of the disciplinary community, then there may be good reason to take note of the pronouncements of that individual. The 'Matthew Effect' is perhaps applicable here as a guide to action, if not necessarily to belief, and it seems to be an efficient procedure. Further, if we take Kelly's description literally, there must be a 'bandwagon' already rolling along for junior scientists to jump on. This is not inconsistent with the idea that theory change in science depends crucially on the adoption of a novel theory by small groups of researchers who see value in the new theory for their research concerns and who for this reason pursue or accept it in their own research. The opinion of the larger community shifts more slowly.

Did American geologists mindlessly stampede to Drift or did they accept it for good reasons? This is the crucial point in Kelly's remarks. There was a tranformation in the majority view in the decade or so between the mid-1960s and the mid-1970s. We should, however, not exaggerate its breadth and depth. Many geologists would have received some exposure to Drift through their education and professional activities: Drift did not suddenly appear as a competitor theory, rather it was a familiar research program which perhaps had become progressively more competitive. Many geologists would have had no direct stake in the debate over global theories. Drift and its competitors impinged only marginally if at all on what could be termed descriptive or classificatory geology or even economic geology. A vague preference for Permanentism might be due to inertia, to an acceptance of the judgment of specialists more directly concerned, than to an active informed hostility to Drift. Finally, we ought to remember that though Drift was not a popular view in North America prior to the mid-1960s, it was a view held by a sizable minority. Nonetheless, there was certainly a change of opinion beginning in the mid-1960s. Scientists pursued or accepted Drift for many reasons and for reasons which may have varied from scientist to scientist (as Kuhn, 1977: 329–33 and L. Laudan, 1984: 68 suggest). The change to Drift was a gradual one, spreading from specialty to specialty, and though social reasons played some role, so too did cognitive and technical ones. There was not an irrational stampede as some readers of Kuhn might infer (e.g., Lakatos, 1978: 9–10, 91).

There were several periods in this transformation. First, there were those who accepted Drift prior to the mid-1960s. Many of these 'old believers' were motivated by local interests and their judgments of the merits of Drift compared with its competitors would likely be reinforced by the new lines of evidence and argument developed in the 1960s and early 1970s even though they may not have had a great familiarity with the lastest research literature.[31] Next, there were those who adopted Drift in the mid- to late 1960s. Many of these were geophysicists, marine geologists and other specialists whose research, teaching or employment interests were directly addressed by the newer versions of Drift and by the wealth of geophysical and ocean-floor data used to construct and underpin those theories. As these versions of Drift encroached upon and transcended more and more specialist or local interests, so more and more members of the community may have been impelled to reassess the relative merits of Drift and its competitors. This would have been very much the case in the late 1960s and early 1970s when Drift emerged from the oceans to invade the continents. It is plausible that many, though not necessarily all geologists who adopted Drift in this period did so from a conviction either that it was generally superior to its competitors in making sense of a steadily expanding database or, more narrowly, that it was superior with respect to specific problems and research fields.[32] Menard recounts how this occurred in 1966 and 1967 among marine geologists who had previously opposed or at least been skeptical of Drift. Their change of opinion came not through being swept along by a tide of sudden enthusiasm or by appeal to authority (Ewing, after all, was an authority but no partisan of Drift), but through a comparison of rival theories with one another and with the data (Menard, 1971: 123–4):

> Most said, 'My observations are not compatible with sea-floor spreading, and I shall prepare a critical demonstration that this is so and thus demolish this nutty idea and we can all get back to work.' One by one they found, for they were honest scientists, that in fact their data, regardless of the subject, were compatible with sea-floor spreading. An elaborate network of confirmations appeared, and most marine geologists got back to work but with a new paradigm.

What of those who accepted Drift in the 1970s? It seems wrong-headed to appeal to an irrational 'bandwagon' effect to explain the allegiance to Drift of younger geologists trained in the late 1960s or early 1970s when Drift was being incorporated in textbooks and curricula. It is more reasonable to regard their adherence to the now-prevailing theory as the normal, expected outcome of science education. Some more senior geologists, especially those concerned with continents rather than the oceans, may well have accepted Drift in the 1970s on the basis of local interests. For example, Drift might have offered little aid to Appalachian geologists till the early 1970s. Others, perhaps even many others, among the 'recent converts' may have relied upon the judgment of experts in other specialties if the issue of 'fixism' versus 'mobilism' was not particularly relevant to their own work. A similar phenomenon occurred in the early debates over Drift; now, however, the opinion of the key specialists favored Drift. Geologists involved in off-shore oil exploration may have taken their cue from marine geologists; regional geologists, from geophysicists or 'global' geologists, and so forth.[33] Perhaps the term 'bandwagon' in a very restricted sense would describe the change in attitude of these geologists, and surely there were some geologists who did opt for Drift simply because it was fashionable or to enhance their chances of getting research money. However it would be misleading simply to wave away the 'revolution' as a 'bandwagon effect', for this would be to confuse the core of this episode of theory change with an epiphenomenon.

Informal networks

Developments in marine geology, reversal chronologies, seafloor spreading and plate tectonics were all for the most part carried out by small groups of earth scientists from a few institutions. These groups were comprised of individuals with similar or complementary research interests who often shared theoretical convictions and aims. These informal networks are the counterparts in the social organization of science of what we have labeled 'problem-fields'.[34] In so far as problem fields cut across different specialties, these networks knit together researchers from different institutions, specialties and disciplines. In these groups social and cognitive interests or social and cognitive interactions cannot be easily distinguished: members might exchange data and preprints of articles, attend the same conferences, receive invitations to the same symposia and colloquia, engage in collaborative research and publication, referee each other's papers for journals and serve as referees for one another's grant proposals. These networks can shift in terms of both membership and research aims and problems as consensus on conceptual and empirical issues forms, evolves or dissolves. During periods of rapid theoretical or empirical growth, membership could be highly advantageous. Recall, for example, the importance of Hess's and Wilson's sojourns at Cambridge or the circulation of the latest information on the seafloors among a select few long before its appearance in print. Members who received advance notice of the latest empirical or theoretical work or who had a ready pool of sympathetic critics for their own latest results had an edge over competitors outside the network pursuing the same or another program. With reference to the transformations in geology over the last three decades, these informal networks formed their problem fields the spearhead of the larger geological community; to use Menard's categories, they were 'participants' in the revolution rather than 'onlookers' (cf. L. Laudan, 1977: 137).

Illustrating the earth

Striking features of much of the key literature of the revolution in the 1960s and early 1970s were visual models and evidence: anomaly profiles, the Bullard Fit, polar-wandering curves and depictions of plates, to mention some illustrations reproduced in this text. None of the models of scientific change under consideration deal in detail with the roles of visual materials. Bruno Latour in a suggestive discussion proposes that a scientist in his research may be likened to an explorer who, to document his travels and to persuade his audience of his veracity, must return from his travels with material objects. These 'objects' should be portable, durable, capable of being shown to others, legible and combinable with other 'objects' (Latour, 1985: 11). Geological graphs, charts, maps, pictures and diagrams fit this definition well; e.g., the *Eltanin*-19 profile. During the revolution, debate and discussion often centered on such 'objects' – on diagrams, graphs and other graphic representations of theories and data– rather than on the 'real' earth.[35]

In geology a 'picture' may be worth far more than a thousand words. With reference to data, symbolic maps (i.e., 'maps' of ridge systems, polar-wandering curves, plates, earthquake centers or magnetic patterns in which the earth is reduced to a schematic outline of continents) summarize vast amounts of data in a form which is concise, manipulable, easily learned and easily remembered. These summaries are of obvious value in pedagogy: at a glance, the student can assimilate the contents of innumerable research reports and papers. Similarly, with reference to theories, symbolic diagrams illustrating seafloor spreading (Figures 9.2, 9.4, 9.9) or plate tectonics (Figure 10.3) summarize complex geophysical theories. Theory could be literally superimposed on data. Researchers and students, confronted with these representations, did not have to read vast quantities of research literature or master arcane physics and mathematics in order to grasp the essentials of the new theories.

In proposing, defending and evaluating theories, these summaries can be manipulated without necessarily mastering the literature of the relevant specialties. The theories, evidence and arguments had been pre-digested: geologists need only see and remember a series of cartoons. This may help explain how knowledge of the new versions of Drift could spread rapidly over diverse specialties in the late 1960s and early 1970s and, perhaps, why many of the new adherents had not or did not read in full many of the key papers. These images of the earth and of theories of the earth were powerful seductive mnemonic devices which were instrumental in the *diffusion* of geological knowledge across specialty boundaries. Moreover, they facilitated the drawing together, the *concentration* of knowledge from different specialties with respect to specific problems or general global theories. They were incorporated into textbooks so that a text was not only illustrated by these images but was in a sense a development of them (e.g., Vine, 1965; Wyllie, 1976). Finally, these summaries were combined or juxtaposed with one another or superimposed on one another: images of such evidence as earthquake centers and ridges and trenches could be combined with one another and then superimposed on such images of theory as plate and transform faults to yield a unified 'geophysicist's earth' (cf. Latour, 1985: esp. 21–2).

'Pictures' of the earth served a *heuristic* function. Summaries of vast accumulations of detailed observations when plotted onto a paper earth yielded dramatic patterns of apparent and sometimes unexpected simplicity. The contours of the forest, as it were, were not obscured by individual trees. This heuristic element of symbolic maps and other graphic material is evident in Wegener's original inspiration and in the Drift interpretation of polar-wandering curves. Symbolic maps of earthquakes, volcanic activity, mountains, ridges, trenches and faults led Wilson to his 1959 hypothesis of a world-wide fracture system and thence to his 1965 theory of transform faults, mobile belts and rigid plates. Carey's work on maps led him to the orocline concept and to Expansionism. The methods used to construct the Bullard Fit were instrumental in the development of plate tectonics and the resolution of such problems as 'triple junctions' (where the edges of three plates meet) the solutions to which were in turn represented by diagrams and schematic maps.[36]

More speculatively, 'pictures' not merely *represent* but *constitute* the world for geologists. The meticulous geological surveys in the nineteenth and twentieth centuries had as their aim highly detailed maps: to capture on paper the earth in all of its detail. Geological maps, drawings of outcrops, stratigraphic columns, traverse sections and other forms of illustration formed part of an integrated discourse about the earth.[37] In order to color in on maps the different strata visible on the surface what was required was systematic field work; theory was secondary (cf. Secord, 1986: 29). Textbooks on the geology of this earth are filled with such maps and, notably, with photographs. Why photographs? These capture the earth in all its messy but treasured detail; they ensure that speculative theorizing does not stray too far from the 'real' earth, the earth underfoot; they are proof to the student that the author is describing and cataloguing the 'real' earth; they may substitute for 'being there', for personal observation in the course of fieldwork. By the conclusion of the 'revolution' photographs disappear, symbolic maps and diagrams dominate and these latter representations constitute the world which geologists are to explain. Holmes's text (1944) – even though it more than other older texts had a strong theoretical component – was filled with illustrations of the old 'real' world of pre-revolutionary geology: no fewer than 24 full-page and 173 half-page photographs of actual landforms and geological features. Wyllie's (1976) post-revolutionary text is filled with symbolic maps, symbolized theories, graphs and charts: there is not a single photograph of an actual landform. These two examples may be extreme but they are symbolic of the change in images of the earth. At the least, this use of abstracted images reflects the increasing importance of geophysics as a specialty and the increasing emphasis on geology as an explanatory rather than a descriptive science. For some of the generation of geologists and especially geophysicists who matured during the revolution, the 'real' earth was the paper earth – the earth of symbolic maps, theories, graphs, and diagrams – bound into journals and monographs. The problems to be solved and the solutions proposed emerged from this paper earth. In this sense their earth was constituted by these images.

Some characteristics of images of the earth fit well with Kuhn's discussion of 'visual gestalts' (Kuhn, 1970: 111–35, esp. 118–25). If one look at Gutenberg and Richter's plots of earthquake distributions (Figure 8.4), one sees a puzzling pattern; if one look at the same symbolic map after looking at a map of the plates (Figure 10.1), one now 'sees' the boundaries of the plates. This indicates the power of these images for teaching, learning and persuasion. However, Kuhn's analogy between 'visual gestalts' and 'conceptual breaks' and especially his contention that scientists holding different programs actually experience and perceive 'the world' differently is not borne out by our study. Nonetheless, Kuhn has drawn attention to a phenomenon which has been neglected by post-Kuhnian theorists of science who have rejected both his thesis of monolithic paradigms and a strong version of incommensurability.

Heretics and schismatics

Plate tectonics became the orthodox global theory, the conventional wisdom in the earth sciences. Most geologists accept and base their work on it or at least give it nominal allegiance. There may be among geologists worldwide a greater degree of consensus on the 'best' global theory than ever before. Nonetheless, the verdict is not unanimous. It is tempting to suggest that this is only because some dinosaurs are very long-lived. Carey, Jeffreys, Beloussov and many of the other heretics and schismatics are survivors from the pre-revolutionary era: with their passing so too will pass the last vestiges of the older, pre-plate tectonics geology. This is the spirit of Max Planck's oft-quoted remark that 'a new scientific truth does not triumph by convincing its opponents and making them see the light, but rather because its opponents eventually die, and a new generation grows up that is familiar with it' (Planck quoted in Kuhn, 1970: 150). The temptation should be resisted. The more prominent dissenters do champion versions of programs similar to those they proposed and defended prior to the 1960s. Perhaps we can set aside Jeffreys and Contractionism as vestigial; perhaps we can pass over 'localism' and claim that Beloussov and his vertical tectonics version of Permanentism continues to attract some Soviet adherents because of the social structure of Soviet science and ignore the local structure of the Soviet landmass. However, Expansionism is not so easily dismissed. Expansionists are not ruled out as 'cranks' even by their opponents and their articles continue to be published. Expansionism continues to gain recruits, though in very small numbers: it is at least in this sense a dynamic program, not a dying one. For the moment Expansionism seems to be a legitimate alternative though one which faces conceptual and empirical difficulties and which is accepted or pursued by only a tiny fraction of the discipline. To this extent there is still theoretical pluralism in the discipline though probably much less than before the 'modern revolution'.

Notes

1 Sykes's paper as published was revised from the version read; his conclusion may have been expressed more tentatively at the symposium.

2 *The Geological Society of America. Abstracts for 1966*, pp. 3, 44, 100–1, 229–30.

3 *The Geological Society of America. Abstracts for 1967*

4 Vol. 48 of the *Transactions* of the AGU contains the program and the abstracts for the 48th annual meeting. Phinney (1968b: 9n) claims 70 abstracts on seafloor spreading for this meeting; my count of 52 does not undercut his contention of a dramatic increase in interest in the months between the Goddard symposium and the 1967 AGU.

5 Carey (MS: Chapter 15) remarks on the range in size and time of the phenomena which different specialties take as their province. In size the scale extends from rock mechanics (1 ångström to 10 micrometers) to structural geology (10 meters to 10 kilometers) and tectonics (10–10 000 kilometers).

6 Nobel Prizes are not awarded for geology.

7 One exception was W.B. Harland of the Sedgwick Museum at Cambridge. He gave a sweeping account of the tectonic history of Spitsbergen in terms of seafloor spreading and sprinkled his paper with references to Vine, J. Tuzo Wilson, and other 'spreaders' (Harland, 1969).

8 J. Tuzo Wilson had conjectured this proto-Atlantic on the grounds that there seemed to be fragments of an American fossil assemblage stuck to parts of the

coastline of northern Europe and the British Isles and, conversely, fragments of an apparently European assemblage stuck to parts of eastern Canada and New England. In the same year as the Gander symposium, he cited in support of his speculation the discovery of 'European' trilobites in Florida (J.T. Wilson, 1967).

9 I here draw freely from Wood (1985: 179–93).

10 For some examples, see Takeuchi *et al.*, (1967; and four more printings by 1970), Tarling & Tarling (1971), Kasbeer (1972), J. Tuzo Wilson (1972; 1976), Cox (1973), Hallam (1973), Marvin (1973), and Sullivan (1974).

11 These figures exclude more specialized monographs such as Zeuner (1946; 1950; 1952; 1958) and texts in palaeobiogeography which often expressed approbation of Drift. The 'anti-Drift' group is comprised of editions of Jeffreys's *The Earth* which was not truly an introductory work.

12 There is an element of interest in his promotion of a definition of the discipline in which his own specialty would be more central.

13 I am perhaps rash to make such a sweeping generalization, for I have not made a systematic analysis of all relevant publications nor have I conducted an 'opinion poll' of a representative sample of geologists around the world. There are, however, several pieces of evidence apart from informal discussions with a small number of geologists in Australia, the US and France. One indication of the rise to prominence of Drift in the early 1970s is abstracts submitted for the annual meetings of the Geological Society of America: there was a rise in Drift-related papers from less than 2% of the total in 1966 and 1967 to 10% in 1971 and 15% in 1972, counting in the latter two years only those *abstracts* which made explicit reference to plate tectonics. In 1986 the figure was just under 30%. Further evidence for the mid-1970s comes from Nitecki *et al.*'s (1978) survey: of the 133 geologists who reported having produced publications since 1973, 42% had published at least one item explicitly dealing with plate tectonics. Nitecki and his co-workers subsequently completed a more extensive survey but as yet their findings, if any, have not been made public. A cursory examination of some major journals published in 1986: *Geology, Bulletin of the Geological Society of America* and *Nature* does not suggest any change in the dominant status of the plate tectonics version of Drift. Finally, introductory and advanced textbooks published since 1975 speak univocally for Drift as the preferred program.

14 The story of how some geologists were led to propose this modification of theory, the theoretical and empirical arguments for and against, the initial rejection and eventual acceptance, and the role in this story of local and specialist interests would be fruitful for analyses of the nature of scientific change. Some of the seminal papers are Monger & Ross (1971), Irwin (1972) and Jones, Irwing & Ovenshine (1972). It is notably that these researchers were employed by national geological surveys. An accessible introduction is Jones *et al.* (1982). Two useful overviews are given by Nur (1983) and Scherman *et al.* (1984); the latter includes a brief historical introduction.

15 It might be expected that geophysics, because if its close association with the new evidence and arguments for Drift, would become more central in the post-revolutionary period. Some support for such a change is found in the increase in size of older geophysics journals (e.g., between 1965 and 1985 the annual size of the *Journal of Geophysical Research* doubled to over 13 000 pages), the apparent

increase in prestige of geophysics journals founded in the mid-1960s (e.g., *Tectonophysics* and the *Review of Geophysics and Space Physics*), the inauguration of such new journals as the *Annual Review of Earth and Planetary Sciences* (1973) and *Tectonics* (1982) and increased coverage of geophysics in general textbooks. However much further research is required before such an argument could be sustained.

16 My remarks do scant justice to this recent rapidly-developing problem-field. Investigators of scientific change may in future give both this topic and the 'suspect terrane' story the careful examination which they merit.

17 Dr. A.J.W. Gleadow, Geotrack International, private communication.

18 Raup (1986) gives an account for the general reader of the theory and its fortunes as seen by one of its partisans.

19 The journal *Geology* is published by the GSA as a forum for recent research across the discipline. The volume for 1986 (vol. 14) contains 226 papers by 571 different authors; 400 affiliated with US institutions, 171 (including 57 Canadian) affiliated with institutions in 20 other countries. Of these papers, 44 depend crucially and explicitly upon plate tectonics theory, another 48 use plate tectonics, another 24 make passing reference to the program but 110 make no explicit reference to plate tectonics. Of these papers, incidentally, 20 relate to passive margin or extensional faulting, 19 to exotic terranes and 17 to mass extinction. Oldroyd (1987) gives a highly detailed historical case study of one recent attempt to give geological underpinning to the biological theory of punctuated equilibrium; here again, plate tectonics is of marginal relevance.

20 In 13 of 15 English-language texts published since 1979 which were randomly chosen.

21 It may be relevant that Carey is the former Professor of Geology at the University of Tasmania and that Clarke & Cook (1983) was prepared as a project of the Australian Academy of Sciences.

22 Drifters explain these as continental rocks deposited by melting icebergs.

23 The negative tenor of their response is exemplified in Arthur Meyerhoff's review (1972) of a popular account of the new theory (Tarling & Tarling, 1971) in which he implied that the authors perpetrated a 'snow job' and he suggested the book was better categorized as fiction than non-fiction. His review elicited sharp rejoinders from the Tarlings and others (*Geotimes*, July, 1972: 9–10).

24 The Meyerhoffs seem to have withdrawn from the main theoretical arenas – though A. Meyerhoff has sniped at Drift in more obscure venues (e.g., A. Meyerhoff, 1980; Saxena *et al.*, 1985) – and concentrated on their specialty of natural resource and economic geology.

25 Carey (1976: 23–38) gives a brief historical sketch of the program.

26 Kitts (1974) attempts to draw a distinction between continental drift as a 'historical' hypothesis and plate tectonics as a 'theoretical' hypothesis in which the content is not reducible to simple description (cf. R. Laudan, 1980a).

27 This might lead one to question the meaning of the term 'revolution' as it is used by participants as opposed to historians and philosophers of science.

28 If we accept the common contention that this geological community was the most adamantly opposed to Drift before 1960 and the most enthusiastic in its acceptance of Drift in the 1970s, then we should look there for a presumed 'bandwagon' effect.

29 *Proceedings of the Geological Society of London*, 1968: 280.

30 The same tone is apparent in the 'Plate Tectonics Creed' (Scharnberger in Kern, 1972): 'I believe in Plate Tectonics Almighty, Unifier of the Earth Sciences, and explanation of all things geological and geophysical; . . . Hypothesis of Hypothesis, theory of theory, Very Fact of Very Fact; deduced not assumed. . . .'

31 Nitecki *et al.* (1978) note that as a group, 'old believers' were less familiar with many of the benchmark publications of the 1960s and early 1970s than later 'converts'.

32 There is some evidence (Nitecki *et al.*, 1978) that geologists who opted for Drift in this period had a markedly greater familiarity with the Drift literature. This suggests but does not establish that their decision in favor of Drift was based on knowledge and informed evaluation of the rival theories. Nitecki puts forward as a hypothesis that 'there was at some period – perhaps during the mid-1960s – a "chain reaction" or other general shift in opinion toward the theory. . .'. This chain reaction or ripple effect is at least in part explicable in terms of the gradually expanding domain of the successive versions of Drift: a chain reaction or ripple from ridges to ocean floors to continents to the whole globe.

33 Nitecki *et al.* (1978) report that both those who had not accepted the theory by 1978 and those who had accepted it since 1970 had significantly less familiarity with the literature than those who had opted for Drift before 1970; this is consistent with the analysis suggested.

34 See Rudwick (1985: 418–28) for a discussion of what he terms 'the field of competence' and 'the core set' in the Devonian Controversy. He claims that a map of the social field (which we could roughly equate to the research network and the positions of individuals within that network) is at the same time a map of its 'cognitive topography' (which we could roughly equate to the set problems, solutions, theories, techniques and so forth comprising the focus of that network).

35 Latour & Woolgar (1979: 49–51, 60, 63–4) give a provocative discussion of the complex process whereby experiments and observations in a biological research laboratory are 'transformed' into curves, graphs and charts which, rather than the raw data, become the focus of attention.

36 The illustrations in our book serve many functions. They are pedagogical, at least for readers unfamiliar with the subject matter. They amplify and are amplified by the discussion in which they are embedded. They help 'signpost' the empirical and theoretical evolution of the rival programs, especially Drift. In this section, they are objects of discourse.

37 For a study of the development of these representations in the 19th century, see Rudwick (1976); cf. Secord (1986).

11

Theories of scientific change and the modern revolution in geology

In 1962 Hess's 'History of Ocean Basins' outlined seafloor spreading. Hess's paper could be said to have marked the beginning of an establishment of consensus in geology. The 'modern revolution' in geology was complete by the early 1970s: a high degree of consensus on plate tectonics existed. In 1962 Kuhn's *Structure of Scientific Revolutions* appeared. In the 1950s there had been a high degree of consensus on the nature of science among philosophers, historians and sociologists of science. Kuhn's book was symptomatic of a dissolution of that consensus. No consensus on a general theory of the development of science existed in the early 1970s. There is still no consensus. For extreme rationalists, scientists do not exist, only disembodied theories and facts; for extreme social analysts, 'nature' does not exist except as it is constructed as an expression of social interests. These extremes are caricatures but disputes over the nature of scientific change seem far from resolution. These may be intensified by the conflicting cognitive and social interests of protagonists. Philosophers might be expected to claim that the development of science is best understood and analyzed in philosophical terms; sociologists, in sociological terms.[1] Often it seems that positions abandoned – or never held – continue to be attacked. This may be effective rhetoric but generates more heat than light. Kuhn's *Structure* marked a watershed, more so perhaps because of questions posed and problems raised than because of the answers he sketched. Just as research programs in geology changed between 1912 and 1970, so general theories of science have changed since 1962. Kuhn's later writings (1970; 1977) seem to qualify some of the more radical implications of his initial (1962) theory of science. Other models have been put forward; e.g., those of L. Laudan (1977) and Lakatos (1978) and the 'interests' and 'struggle for authority' approaches. Some of these have also undergone revision.[2] Also in the past 25 years, accounts of historical and contemporary episodes have increased in number, range and depth. Ironically, this growth of the 'data base' for developing and evaluating theories of scientific change may contribute to the difficulties of building consensus: advocates of different theories often appeal to different examples. Rarely do historians, philosophers and sociologists explicitly evaluate competing models with respect to the same episodes.[3] Consensus emerged among geologists partly because the newer versions of Drift embraced and were judged useful in more and more problem-fields and specialties. Perhaps one reason why there is little consensus among philosophers, historians and sociologists of science is that there is no program which embraces and is judged useful in the relevant specialties and problem-fields of those researchers. Philosophers seem for the most part to address only philosophers; historians, historians; and sociologists, sociologists.

In our exploration we adopted a procedure akin to the geological method of 'multiple working hypotheses'. We appraised aspects of several models with respect to *my* history of Drift. We must remember that my history has been shaped by my presuppositions: my beliefs that scientific fields are normally rife with contending research programs and theories, that analyses of science in either exclusively cognitive or exclusively social terms are inadequate, and that science can be usefully characterized as a problem-solving activity. My presuppositions have influenced the selection of rationalistic and social models we have examined. Philosophers of science will have noticed that I have ignored much of the literature in their field. I have not discussed, for example, 'falsificationism', 'logical positivism', 'confirmation theory', 'convergent epistemological realism' or 'evolutionary epistemology'. These omissions are deliberate: these approaches may have much importance within philosophical discourse but seem to me to have at present only remote connections with our study.[4] From the welter of 'social' analyses I chose the 'interests' and 'internal struggle' approaches to examine because I believed them to be the most promising.

Although my opinions of these general theories have altered as I prepared this study, I find none of them satisfactory in their current versions. No one model is vindicated. Proponents of any of these models can find parts of our story which 'confirm' features of their preferred schemes. But other parts conflict sharply with distinctive and central components of most schemes. Our explorations have been neither inconclusive nor fruitless. In the 'Voice-Overs' I made provisional judgments on many specific issues; others – now more sharply defined – remain on the agenda for future investigation. I shall summarize some of the general conclusions which we might draw from our study. I shall then indicate some lines for future research which could either reopen issues we have resolved or resolve ones we have left open.

Philosophical or rationalistic models

I have characterized science as a problem- or puzzle-solving activity. This accords with Kuhn's and Laudan's schemes and is not inconsistent with Lakatos's nor, for that matter, with the 'interests' or 'struggle' models. Others may describe science as a search for truth, building better theories, generating more accurate predictions, implementing social interests, mastering our environment, or imposing order on Nature. We could also describe science as a system of knowledge, as the process of acquiring or constructing that knowledge and as the occupation of people who possess that knowledge and engage in that process. All of these descriptions have advocates among scientists, historians, philosophers and sociologists. In our analysis of Drift, however, we cast these in terms of problem-solving. We were able to do this in part by treating conflicts involving aims, methods, standards, world-views, and beliefs held in different specialties and disciplines as 'conceptual problems'.[5] In this context what general conclusions might we reach about the models of Kuhn, Laudan and Lakatos?

Theories do not stand alone. They are part of complex structures. Kuhn terms these 'paradigms'; Lakatos, 'Research Programmes'; Laudan, 'Research Traditions'. These include not only theories but also aims, methods, techniques, standards and data. Socially, research programs may represent different mobilizations and deployments of resources, different hierarchies of specialties and specialists, different patterns of authority; in brief, different definitions of communities and community life.

Scientists were not unanimous in their choice of research program. There were usually several contenders of varying popularity. One program may, on occasion, have achieved a position of near dominance but rarely was there anything approaching complete consensus. This tells against Kuhn's (1962) claim that a mature scientific field is normally dominated by a single paradigm and in favor of pluralistic models of science (e.g., those of Lakatos, Feyerabend and Laudan as well as the 'interest' and 'struggle' models).

These programs were neither monolithic nor static. Scientists working in different programs may have had the same aims, methods, standards and techniques yet still disagreed about which of the available programs was preferred; there was also scope for disagreement on these issues among the proponents of the same program. In the Drift story there was sometimes a greater measure of consensus across programs or dispute within a program than Kuhn would seem to allow (cf. L. Laudan, 1984). These global theories were not static: older versions fell by the wayside, new versions were proposed, and whole new programs were sometimes put forward. Within the programs, the development of successive versions was not a chance affair. New versions were developed at least in part to meet perceived shortcomings of older ones. This accords with the views of Laudan and Lakatos. However, the Lakatosian ideal of a 'hard core' is problematic: it seems that the 'hard core' of Drift could be defined only retrospectively and in vague terms: there was no agreed, explicit core belief set down by Wegener which held for all subsequent versions of Drift or for all Drifters (see Frankel, 1979a).

Neither these structures nor the theories embedded within them were surrendered because of straightforwardly empirical problems; i.e., because of conflict between 'facts' and 'theory'. This accords with most of the models of science we have taken up. Both 'facts' and 'theories' were revisable; moreover, no sharp distinction could be made between theory-laden facts and fact-laden theories. This is not to say that empirical problems were ignored: opponents of Drift levelled empirical attacks against it and proponents replied in kind. Empirical anomalies do not appear to have been by themselves decisive. Geologists often expressed the view that the empirical data alone did not provide the basis for a firm decision on a given program or theory and held final judgment in abeyance pending further investigations. We should not lightly dismiss this as purely rhetorical.

Matters other than strictly empirical problems appear to have been crucial in the appraisal (and articulation) of rival programs and theories. From a rationalist's point of view, conceptual problems were ubiquitous: inter-disciplinary conflicts, methodological disputes, divergent opinions on the proper aims of geology, catastrophism versus uniformitarianism, and so on. This is in agreement with Laudan's model and to some extent with that of Kuhn but seems to go against Lakatos's scheme which emphasizes the prediction of 'novel facts'.[6] The roles of geological localism and of problem-fields and specialist interests may also be assimilable to a cognitive analysis as well as to several of the social models. The social analyst would likely argue that empirical and conceptual issues in appraisal were inseparable from intra-scientific social interests.

L. Laudan (1984) calls into question what he terms the 'hierarchical' approach: appealing to methodological rules to arbitrate theoretical disputes, appealling to shared aims and goals to arbitrate methodological disputes, accepting differences in aims as irresolvable matters of taste. He also rejects Kuhn's view that data, theories, methods and aims comes as discrete non-negotiable packages (paradigms). Instead, he proposes that there is interaction and the possibility of rational argument within and among all of these levels and that these elements of a Research Tradition can be revised in a piecemeal fashion. He suggests, for example, that methodological creeds and views on the proper aims of science are shaped to some extent by what are agreed to be 'successful' practices and acceptable theories and programs. Some support can be found in our narrative for this claim. The acceptance of Drift – at least in the English-speaking world – seems to have been accompanied by a shift in emphasis from data collection and simple inductivism to theories and speculative hypotheses and by a shift of emphasis in aim from comprehensive description to explanation. This tentative finding merits further research. In support of 'piecemeal' change, it could also be argued that even many scientists who continued to adhere to the former methods and aims nonetheless accepted Drift.

The judgments of scientists were comparative, not absolute. Drift, for example, was explicitly evaluated not simply on the basis of its internal coherence and consistency of its fit with the claimed facts of the matter but with respect to its competitors. Similarly, different versions of Drift were evaluated with respect to other versions of the program. However, a program or theory was not compared with all *possible* programs or theories but with *available* programs or theories. We noted also that discussions of those different theories and programs often evidenced disagreement and differences of interpretation but not the incommensurability which Kuhn (1962) invoked. Finally, scientists on occasion showed flexibility in their evaluations of newer programs and theories compared with long-established ones. Some who were disdainful of Drift nevertheless stated that it merited tolerance because it was relatively new, because its promise – though not yet its performance – in specific areas was significant, or perhaps because it evidenced early and rapid successes in solving some previously intractable problems.

These evaluations were often phrased in provisional terms. The programs, theories, database, aims, methods, problems, anomalies, techniques and so on were not static nor were they seen to be so. Furthermore, the process of scientific change was not fully cumulative; we might identify both empirical and theoretical gains when we compare later with earlier versions of Drift or Drift with its rivals, but there also were losses. If there were 'progress', it was not equivalent to a simple cumulation of facts, theories and solved problems. If we wish to talk about 'progress' within a program or in comparing different programs, we need to talk about progress with respect to some specified aim (cf. L. Laudan, 1984: 64–6); for example, increased problem-solving capability within a specialty or problem-field or the unification of specialties by an over-arching theoretical framework.

Few scientists directly and explicitly engaged in disputes and debates over programs and theories, gave public appraisals of them or actively participated in their development. Most geologists seem to have been content to leave aside such activities and to get on with their own work. In this sense, the brunt of effecting conceptual change fell upon an 'elite' rather than on the general community. Those scientists who did comment publicly on the competing theories and programs often did not reach simple 'yes or no' judgments. Confronted with a paradigm and a 'paradigm candidate', the Kuhnian scientists either 'converts' to the new view or remains firmly within the old: the choice is acceptance or rejection. However, as we have seen, geologists might accept or reject a particular theory or program but they might also adopt stances or entertainment, employment and pursuit.

My preferred rationalist account of scientific change is that of Laudan. More than its alternatives, it is consistent with the provisional conclusions given above. It is also compatible with our Drift story at a general level and in many matters of detail. However, there are a number of issues which that model as yet does not address. Among those which we have noted are localism, specialization and technical interests. Further, it is not clear to what extent his model is applicable to the decisions – and to the rationality – of individuals or perhaps small groups of specialists as opposed to an amorphous community. Moreover, any role for the social interests of scientists within a scientific field seems to be excluded if it be not subsumable as problem-solving. Laudan himself is blunt (1977: 208): 'the need for sociological analysis of a case only arises when we can show that the actual evaluation of a particular theory [or Research Tradition] in the past was radically at odds with the appraisal it should be accorded by the lights of the problem-solving model of rationality.' This does not entail a complete opposition between his model and some of the social models. He does not deny that both social and cognitive accounts of the same appraisal might be given; he merely asserts that a cognitive account is sufficient. Proponents of the 'struggle' model would argue that scientific beliefs are determined both cognitively and socially, and that these cannot be distinguished except analytically. Further, both the 'interest' and 'struggle' approaches imply that the weighting of problems, the choice of problems, criteria for acceptable solutions, aims, techniques, methods and so on, are negotiable and contingent upon the perceived interests of scientists. Within a problem-solving characterization of science, scientists aim not merely to put forward solutions, but to put forward solutions which are or which become solutions acceptable to the scientific community of which they are a part. For such solutions the successful problem-solver receives some measure of credibility, of standing within the community. This process is neither mechanical nor free from controversy; e.g., more than one possible theoretical construction can be placed on a set of data. Involved in this process are both competition and co-operation. Science is a problem-solving activity but it is also a social activity in which fellow scientists play crucial roles in the proposal and assessment of knowledge claims. Disputes in this sense are opened and closed by scientists, not by nature (cf. Latour, 1987).

Social models

The applicability of a grand, general, diffuse, 'external' social analysis is moot. We can sketch how such an analysis might proceed. For the period between the wars we might claim that Britain and the US experienced relative social, political and economic stability. Let us crudely equate this to the 'scientific' views of uniformitarianism and continental permanence. We could further add that in North America the geological parochialism engendered by specialization was undergirded by a strong political sentiment of isolationism. Turning to Europe, most of the continental powers, especially Germany, underwent considerable economic and political dislocation in the aftermath of the World War I. Let us crudely equate political and social instability with a willingness to entertain radical new theories, especially a theory incorporating the instability of the continents themselves. In this way we might try to link broader social interests with the varying fortunes of Drift in different countries, explaining away exceptions such as du Toit and Holmes with a greater or lesser degree of difficulty. We could then argue that the winds of change which whistled through the US and Britain in the 1960s and early 1970s swept away the earlier sense of order and stability. We might point especially to social unrest and incipient revolution: demonstrations against the Vietnam war, student activism, economic problems; in short, global economic and political instability. In this climate geologists of course abandoned Fixism in favor of Drift (or more modestly, were at last able to judge Drift on its merits). Whatever appeal such a sweeping social account may have, I think it extraordinarily difficult if not impossible to tie it to the actions and judgments of groups of scientists, much less individuals. This could perhaps be done by attributing 'class' interests to the participants in our story and explaining away any contradictions between their class interests and their actions or stated beliefs in terms of 'false consciousness' or some such manoeuvre. To me it seems far more plausible, if we wish to assign social interests, to do so in terms of the operation of social interests and the social order within the scientific field.

There were some 'external influences' which did affect geology but not in a direct and an obvious way to the detriment or the advantage of particular research programs. The differences in the organization of academic science in the English-speaking world and Europe may have made North American geologists more resistant to global synthetic theories but at the same time may have created specialist niches congenial to Drift. Narrow nationalism was not an obvious factor; the most that could be said about naked (or even subtly-clothed) nationalism is that whatever its contribution to criticisms of Wegener, such criticisms were 'overdetermined' by the way in which Wegener presented his case and by the prevailing views of the non-German geological community. Similarly, the injection of funding by the ONR had many repercussions but neither this funding nor the research that it purchased was perceived at the time by either the ONR or by geologists to be tied to particular global theories. It could, however, be argued that the lack of effective international co-operation in geology before the late 1950s did militate against Drift. The difficulties of establishing a global geology were acute for English-speaking workers, given their emphasis on examining specimens and structures first-hand and their distrust (sometimes well-founded) of published accounts. Only a handful would have been able to emulate du Toit and to compare South African and South American 'matches' with their own eyes (assuming that this handful would have been sufficiently impressed by Drift to commit the resources in time and money required for such an effort!). In the absence of global collaboration and co-operation, it is no surprise that Wegener (1929t: viii) lamented that 'Scientists still do not appear to understand sufficiently that all earth sciences must contribute evidence toward unveiling the state of our planet in earlier times, and that the truth of the matter can only be reached by combining all the evidence'. Leaving aside 'internal' obstacles to a 'global geology', there were 'external ones'. The 'Great Depression' created economic hurdles to co-operation and extensive travel and likely turned geologists' attention to such practical problems as securing or keeping employment. Then, political events culminating in World War II put paid to such attempts as the 1939 Frankfurt symposium to pool international expertise on problems of global geology. It was not until the 1950s that global research began to receive significant funding, that international co-operation was emphasized in the IGY, and that less 'local' specialties such as oceanography and palaeomagnetism began to acquire prominence. Even so, a claim that these developments, separately or together, *directly* produced a swing of opinion in favor of Drift seems very difficult to sustain. What could be said is that they made possible a more 'global' evaluation of 'global theories' but that such a procedure might have been favored or opposed for a variety of technical and social interests within geology.

I incline more toward the 'internal struggle' than to the 'interest' approach. First, 'internal' interests seem to have been far more ubiquitous and important than 'external' ones. While some proponents of the 'interest' approach accept this point, not all do. In the 'struggle' analyses primacy is given to internal interests although this does not rule out those interests being affected by 'external' influences; e.g., the funding of oceanography by the ONR. Second, it is axiomatic within the 'struggle' approach that intellectual and social interests cannot be distinguished in practice, that any distinction is only an analytic one, and that any account of scientific change must be simultaneously social and cognitive. This may be acceptable to some 'interest' proponents but thus far there has been not only a tendency to look almost exclusively at 'social interests', whether external or internal to science, but also to regard these as dominant if not determinant. From our story it seems plausible to conclude that the intellectual and social interests of a participant within the field were constantly interacting and mutually reinforcing and, since we are unable to enquire into the conscious and subconscious motivations of participants, impossible to separate. In this approach cognitive developments are situated in the human context of competition and co-operation.

A reasonably coherent cognitive account of the rejection of Drift by most geologists prior to the 1960s and its acceptance by most geologists since the 1960s can be given in terms of Laudan's model. But, to Laudan's stricture on social analyses, the social analysts could reply, 'There is no need to invoke a problem-solving model of rationality unless the actual evaluation of a particular theory in the past was radically at odds with the appraisal it should be accorded by the lights of the internal struggle model of scientific practice.' The modern revolution in geology was not a sudden 'paradigm switch'. The revolution occurred when the Drift research programme (or rather, a new version of this long-enunciated programme) had such success in problem-solving at both the conceptual and empirical levels and enough geologists saw their technical and social interests being best served by it that it could no longer be ignored or taken lightly. Many geologists began to employ or pursue or accept this program; others such as Carey, the Meyerhoffs, and, perhaps, some Soviet geologists, concluded that their intellectual and social interests were better served by continued refusal to embrace Drift. None the less they took into account in their own programs the successes with which Drift had been credited. That either a cognitive or a social account could be given of the revolution would delight, not distress, a follower of the 'struggle' interpretation: ideally, an intellectual or a social analysis should arrive at the same conclusions.

An unfinished study

I have painted my historical picture with very large brush-strokes. Often the result more or less resembles the pictures painted by others (e.g., Hallam 1973 or Wood, 1985). I have sometimes put in more detail (for example, the competition between the programs and especially the fate of Drift in the 1930s, 1940s and 1950s) where that has yielded interpretations different from those of previous writers. My framing of questions and my conclusions are of course in the same way controvertible or revisable. There are as yet no blow-by-blow commentaries on the history of Drift comparable to the painstaking and provocative reconstructions of nineteenth century geological controversies presented in the monographs of Rudwick (1985) or Secord (1986) or to Greene's (1982) account of tectonic theories. There are no comprehensive studies of any of the major actors nor of any of the major empirical or theoretical developments.[7] Future, more detailed historical studies of Drift may well result in substantial modification or qualification of our findings as well as perhaps resolving some outstanding issues. Leaving aside the question of rewriting history to reflect our changing interests, there are many histories of modern geology not yet written: institutional histories, biographies, social histories, histories of concepts and instruments. We have turned only a few spade-fulls of soil; the field is far from cultivated.

This historical task is inseparable from the task of constructing and evaluating frameworks for analyzing and understanding scientific change. I have singled out as worthy of pursuit with respect to the Drift case Laudan's model and the 'struggle' interpretation. It remains to be seen to what extent these two seemingly opposed programs, the first directed to rational evaluative procedures, the second to the 'coalface' of daily scientific practice, may be complementary or even convergent. Proponents of both need to elaborate these programs to give a systematic treatment of specialization, localism, problem-fields and research networks. As we have noted, these features of the social and cognitive organization of science can be important not only in the appraisal of competing research programs and theories and in the judgment of knowledge claims but also in such matters of practice as the selection of problems to be attacked, criteria for acceptable solutions, and decisions about appropriate standards, aims, methods, instruments and techniques. Laudan's model may be extended to include these elements. This would probably require modification or reinterpretation or perhaps the proposal in addition to his general model of a version specifically applicable to specialties and problem-fields. 'Strugglers' are well aware of the importance of these elements. But there seem to be no explicit agreed guidelines on how they are to be meshed with 'technical capacity' and 'social power' in analyzing developments within a scientific field. What are needed are clear formulations of specific arguments, rules or principles which could then be applied to and criticized on the basis of many episodes. Of course, the pursuit of either or both the Laudan and 'struggle' models or even of possible amalgamations of them does not foreclose the proposal of new general theories of scientific change which might subsume or provide alternatives to the more useful aspects of both, in the same way that Wegener propounded Drift as an alternative to both Permanentism and Contractionism or that Carey proposed Expansionism as an alternative to all three.

We have confined our evaluation of competing views of science to Drift. How much of what we have learned can be applied to other instances of scientific change in geology or, more ambitiously, to the growth of other sciences? Most of the matters we have discussed are not restricted to twentieth-century geology, they are not content-specific, and to this extent we could claim that our provisional conclusions should hold for other fields of science. Grander themes such as theoretical pluralism, specialization, conceptual problems, technical interests, the theory-ladeness of data, black boxes, patterns of authority, and so on are to be found in eighteenth-century chemistry, nineteenth-century physics and twentieth-century biology, to name just a few examples. Such matters as localism, the method of 'multiple working hypotheses', the limited scope for experimentation (in the usual sense) and the importance of visual representations may be peculiar to geology. These in retrospect were of less importance in our appraisals of which models merited pursuit than in suggesting topics which new versions of those models should address. A reasonable and pragmatic course of action is to assume that our more general findings are applicable to scientific change in other periods and other disciplines. With our set of preferred models, questions and unresolved issues as a starting point we could then explore other episodes with the aim of gaining a deeper understanding of this human activity which is indistinguishably cognitive and social.

Notes

1 See e.g., the exchange between L. Laudan (1981a) and Bloor (1981).
2 For examples, L. Laudan (1984) compared with (1977); Frankel's (1979a) proposed changes to the Lakatosian concepts of the 'hard core' and 'novel facts'.
3 For a bibliography of some case studies carried out by 'rationalists', see L. Laudan *et al.*, (1986: 218–23); by 'social analysts', Shapin (1982: 204–11).
4 To this extent I could be said to be working in the same problem-field or to share 'technical interests' with Kuhn, Lakatos, Laudan and others.
5 I have depicted scientists primarily as 'problem-solvers'. We have paid scant attention to the process of 'problem-making' or to scientists as 'problem-makers'. To secure agreement that a problem exists may be as arduous as securing agreement on a solution. Problems do not leap out from the collision of disembodied ideas, facts and theories: they are created and defined by scientists. It is open to proponents of problem-solving models to argue that rarely do scientists simply put forward problems; rather, the identification of a problem is almost always accompanied by a proposed solution.
6 In Lakatos's extensive writings on the philosophy of mathematics conceptual problems occupy a central place and some Lakatosians may feel that I have dealt unfairly with his views; however, his general model of science (1978) emphasizes empirical issues (but see 1978: 50).
7 Excepting perhaps Glen's (1982) account of the development of polarity-reversal scales and their relation to seafloor spreading.

Bibliography

- -

Ager, D.V. (1961). *Introducing Geology: The Earth's Crust Considered as History*. London: Faber & Faber.

Ahmad, Fakhruddin. (1966). An estimate of the rate of continental drift in the permian period. *Nature*, 210, 81–3.

Ahmad, F. (1968). Orogeny, geosyncline and continental drift. *Tectonophysics*, 5, 177–89.

Ahmad, F. (1978). *'Gondwanaland', The Concept that Failed*. Lucknow: Birbal Sahni Institute of Palaeobotany (27 pp.).

Ahmad, F. (1982). The myth of oceanic tethys. *Bollettino della Società Paleontologica Italiana*, 21, 153–68.

Andrée, Karl. (1914). *Über die Bedingungen der Gebirgsbildung*. Berlin: Gebrüder Bornträger.

Andrée, K. (1917). Alfred Wegeners hypothese von der horizontalverschiebung der kontinentalschollen und das permanenzproblem im lichte der paläogeographie und dynamischen geologie. *Petermanns Geographische Mitteilungen*, 63, 50–3; 77–81.

Andrews, Ernest C. (1926). Hypothesis of mountain-building. *Bulletin of the Geological Society of America*, 37, 439–54.

Andrews, E.C. (1938). Presidential address: Some major problems in structural geology. *The Proceedings of the Linnaean Society of New South Wales*. 63, iv–xl.

Argand, Emile. (1924). La tectonique de l'Asie. *Compte-rendu du Congrès géologique international (XIIIe session, Liège, 1922)*, 1, 171–372.

Arldt, Theodor K. (1918). Die frage der permanenz der kontinente und ozean. *Geographische Anzeiger*, 19, 2–12.

Bailey, Edward B. (1929). The palaeozoic mountain systems of Europe and America. In *Report of the B.A.A.S.*, pp. 57–76. Glasgow: B.A.A.S.

Bailey, E.B. (1935). *Tectonic Essays: Mainly Alpine*. Oxford: Clarendon Press.

Bailey, E.B. (1939). *Introduction to Geology*. London: Macmillan.

Baker, Howard B. (1912–14). The origin of continental forms. In *Annual Report of the Michigan Academy of Science*, 1912, pp. 116–41; 1913, pp. 26–32; 107–13; 1914, pp. 99–103. Michigan Academy of Science.

Baker, H.B. (1932). *The Atlantic Rift and its Meaning*. Detroit: Privately Printed.

Barnes, Barry. (1977). *Interests and the Growth of Knowledge*. London: Routledge & Kegan Paul.

Barnes, B. (1982). *T.S. Kuhn and Social Science*. London: Macmillan.

Barnes. B. & McKenzie, D. (1979). On the role of interests in scientific change. In *On the Margins of Science: The Social Construction of Rejected Knowledge (Sociological Review Monograph 27)*, ed. R. Wallis, pp. 49–66. Keele: University of Keele.

Barnes, B. & Shapin, S., eds. (1979). *Natural Order: Historical Studies of Scientific Change*. Beverly Hills: Sage Publications.

Barrell, Joseph, (1914). The status of hypotheses of polar wandering. *Science*, **40**, 333–40.

Barrell, J. (1917). Rhythms and the measurement of geologic time. *Bulletin of the Geological Society of America*, **28**, 745–904.

Barrell, J. (1927). On continental fragmentation, and the geologic bearing of the moon's surficial features. *American Journal of Science*, 5th series, **13**, 183–314.

Beloussov, Vladimir V. (1954t). *Basic Problems in Geotectonics*. trans. P.T. Broneer 1962. New York: McGraw-Hill.

Beloussov, V.V. (1968). An open letter to J.T. Wilson. *Geotimes*, Dec., 17–19.

Beloussov, V.V. (1970). Against the hypothesis of sea-floor spreading. *Tectonophysics*, **9**, 489–511.

Beloussov, V.V. (1976). Against continental drift. *Science Journal*, 56–61.

Beloussov, V.V. (1979). Why I do not accept plate tectonics. *EOS*, **60**, 207–11.

Ben David, Joseph. (1971). *The Scientists's Role in Society*. Englewood Cliffs: Prentice–Hall.

Berry, Edward W. (1928). Comments on the Wegener hypothesis. In van der Gracht (1928a), pp. 194–96.

Betim, Alberto. (1929). Etat des connaissances géologiques sur le Brésil (Rapport avec la théorie de Wegener sur la dérive des continents). *Bulletin de la Société Géologique de France*, 4th series, **29**, 35–87.

Blackett, Patrick M.S. (1947). The magnetic field of massive rotating bodies. *Nature*, **159**, 658–66.

Blackett, P.M.S. (1952). A negative experiment relating to magnetism and the earth's rotation. *Transactions of the Royal Society*, A**245**, 309–70.

Blackett, P.M.S. (1956). *Lectures on Rock Magnetism*. Jerusalem: Weizmann Science Books.

Blackett, P.M.S. (1961). Comparison of ancient climates with the ancient latitudes deduced from rock magnetic measurements. *Proceedings of the Royal Society*, A**263**, 1–30.

Blackett, P.M.S. (1965). Introduction. In Blackett, Bullard & Runcorn (1965), pp. vii–x.

Blackett, P.M.S., Bullard, E. & Runcorn, S.K. (1965). *A Symposium on Continental Drift*. London: The Royal Society.

Blackett, P.M.S., Clegg, J.A. & Stubbs, P.H.S. (1960). An analysis of rock magnetic data. *Proceedings of the Royal Society*, A**256**, 291–322.

Bloor, David. (1981). The strengths of the strong programme. *Philosophy of the Social Sciences*, **11**, 199–213.

Bourdieu, Pierre. (1975). The specificity of the scientific field and the social conditions of the progress of reason. *Social Science Information*, **14**, no. 6, 19–47.

Brannigan, Augustine, (1981). *The Social basis of Scientific Discoveries*. Cambridge: Cambridge University Press.

Brooks, Charles E.P. (1922). *The Evolution of Climate*. London: Benn Brothers Ltd.

Brooks, C.E.P. (1926). *Climate through the Ages*. London: Ernest Benn Ltd.

Brooks, C.E.P. (1949). *Climate through the Ages*. 2nd edn. London: Ernest Benn Ltd.

Brunnschweiler, Rudolf O. (1981). Evolution of geotectonic concepts in the past century. In Carey (1981). pp. 9–15.

Brush, Stephen G. (1979). Nineteenth-century debates about the inside of the earth: solid, liquid, or gas? *Annals of Science*, **36**, 225–54.

Bucher, Walter H. (1933). *The Deformation of the Earth's Crust: An Inductive Approach to the*

Problems of Diastrophism. Princeton: Princeton University Press.

Bucher, W.H. (1952). Continental drift versus land bridges. In Mayr (1952a). pp. 93–103.

Buffetaut, Eric. (1985). Annexe: Les géologues français. In Schwarzbach (1985t). 132–7.

Bull, Alfred J. (1921). A hypothesis of mountain building. *Geological Magazine*, **58**, 364–7.

Bull, A.J. (1927). Some aspects of the mountain building problem. *Proceedings of the Geological Association*, **38**, 144–56.

Bullard, Edward C. (1949). The magnetic field within the earth. *Proceedings of the Royal Society*, A**197**, 433–53.

Bullard, E.C. (1954). A comparison of oceans and continents. *Proceedings of the Royal Society*, A**222**, 403–7.

Bullard, E.C. (1961). Forces and processes at work in ocean basins. In Sears (1961), pp. 39–50.

Bullard, E.C. (1964). Continental drift. *Quarterly Journal of the Geological Society of London*, **120**, 1–33.

Bullard, E.C. (1968). Conference on the history of the earth's crust. In Phinney (1968a), pp. 231–5.

Bullard, E.C. (1975a). The emergence of plate tectonics: A personal view. *Annual Review of Earth and Planetary Science*, **3**, 1–30.

Bullard, E.C. (1975b). Overview of plate tectonics. In *Petroleum and Global Tectonics*, ed. A.G. Fischer & S. Judson, pp. 5–19. Princeton: Princeton University Press.

Bullard, E.C., Everett, J.E. & Gilbert Smith, A. (1965). The fit of the continents around the Atlantic. In Blackett, Bullard & Runcorn (1965), pp. 41–51.

Bullard, E.C. & Gellman, H. (1954). Homogeneous dynamos and terrestrial magnetism. *Philosophical Transactions*, A**247**, 213–78.

Bullen, Keith. (1976). Wegener, Alfred Lothar. *Dictionary of Scientific Biography*, **14**, 214–17.

Burchfield, Joe D. (1975). *Lord Kelvin and the Age of the Earth*. New York: Macmillan Press.

Burmeister, Friedrich. (1921). Die verschiebung Grönlands nach den astronomischen längenbestimmungen. *Petermanns Geographische Mitteilungen*, **67**, 225–7.

Cannon, W.F. (1960). The uniformitarian–catastrophist debate. *Isis*, **51**, 38–55.

Carey, S. Warren, ed. (1958a). *Continental Drift: A Symposium*. Hobart: Geology Department, University of Tasmania.

Carey, S.W. (1958b). The tectonic approach to continental drift. In Carey (1958a), pp. 177–355.

Carey, S.W. (1963). The asymmetry of the earth. *The Australian Journal of Science*, **25**, 479–88.

Carey, S.W. (1976). *The Expanding Earth*. Developments in Geotectonics Vol. 10. New York: Elsevier.

Carey, S.W., ed. (1981a). *The Expanding Earth: A Symposium*. Hobart: University of Tasmania.

Carey, S.W. (1981b). Evolution of beliefs on the nature and origin of the earth. In Carey (1981a), pp. 3–7.

Carey, S.W. (MS). *A Philosophy of the Earth and Universe*. Stanford: Stanford University Press.

Carozzi, Albert V. (1985). The reaction in continental Europe to Wegener's theory of continental drift. *Earth Sciences History*, **4**, 122–37.

Carter, William E. & Robertson, D.S. (1986). Studying the earth by very-long-baseline interferometry. *Scientific American*, **255**, no. 5, 44–52.

Caster, Kenneth E. (1952). Stratigraphic and paleontological data relevant to the problem of Afro-American ligation during the paleozoic and mesozoic. In Mayr (1952a), 105–52.

Chalmers, Alan F. (1982). *What Is This Thing Called Science?*, 2nd edn. St Lucia: Queensland University Press.

Chamberlin, Rollin T. (1928). Some of the objections to Wegener's theory. In van der Gracht (1928a), pp. 83–7.

Chaney, Ralph W. (1940). Bearing of forests on the theory of continental drift. *Scientific Monthly*, **51**, 489–99.

Chanton, Louis-Robert. (1961). *Géologie*, 2nd edn. Paris: Centre de Documentation Universitaire.

Chelikowsky, Joseph R. (1944). A new idea on continental drift. *American Journal of Science*, **242**, 673–85.

Chorley, R.J., Dunn, A.J. & Beckinsale, R.P. (1964). *The History of the Study of Landforms. The Development of Geomorphology*, Vol. 1. London: Methuen.

Choubert, Boris (1935). Recherches sur la genèse des chaines paléozoiques et antécambriennes. *Revue de géographie physique et de géologie dynamique*, **8**, 5–50.

Clarke, I.F. & Cook, B.J. (1983). *Geological Sciences: Perspectives of the Earth*. Canberra: Australian Academy of Sciences.

Clegg, J.A., Almond, M. & Stubbs, P.H.S. (1954a). The remanent magnetism of some sedimentary rocks in Britain. *Philosophical Magazine*, series 7, **45**, 583–98.

Clegg, J.A., Almond, M. & Stubbs, P.H.S. (1954b). Some recent studies of the pre-history of the earth's magnetic field. *Journal of Geomagnetism and Geoelectricity*, **6**, 194–9.

Clegg, J.A., Deutsch, E.R. & Griffiths, D.H. (1956). Rock magnetism in India. *The Philosophical Magazine*, ser. 8, **1**, 419–31.

Cole, Grenville A.J. (1915). Geology: Crustal movements. *Report of the British Association for the Advancement of Science, Manchester*, 403–19.

Coleman, Arthur P. (1916). Dry land in Geology. *Bulletin of the Geological Society of America*, **27**, 171–204.

Coleman, A.P. (1925). Permo-carboniferous glaciation and the Wegener hypothesis. *Nature*, **115**, 602.

Collet, Léon William (1927). *The Structure of the Alps*. London: Edward Arnold.

Collet, L.W. (1935). *The Structure of the Alps*, 2nd edn. London: Edward Arnold; rept. Huntington, N.Y.: Robert E. Kreiger.

Collinson, David W., Creer, K.M., Irving, E. & Runcorn, S.K. (1957). The measurement of the permanent magnetization of rocks. *Philosophical Transactions*, A**250**, 73–82.

Collinson, D.W. & Runcorn, S.K. (1960). Polar wandering and continental drift: Evidence from palaeomagnetic observations in the United States. *Bulletin of the Geological Society of America*, **71**, 915–58.

Cotton, Leo A. (1923). Some fundamental problems of diastrophism and their geological corollaries with special reference to polar wandering. *American Journal of Science*, 5th series, **6**, 453–503.

Cotton, L.A. (1929). Presidential Address: Causes of diastrophism and their status in current geological thought. *Report of the Nineteenth Meeting of ANZAAS (Hobart, 1928)*, Hobart: 1929, pp. 171–210.

Cox, Allan, ed. (1973). *Plate Tectonics and Geomagnetic Reversals.* San Francisco: W.H. Freeman.

Cox, A. & Doell, R.R. (1960). Review of palaeomagnetism. *Bulletin of the Geological Society of America,* **71**, 645–768.

Cox, A., Doell, R.R. & Dalrymple, G.B. (1963). Geomagnetic polarity epochs and pleistocene geochronometry. *Nature,* **198**, 1049–51.

Cox, A., Doell, R.R. & Dalrymple, G.B. (1964). Reversals of the earth's magnetic field. *Science,* **144**, 1537–43.

Creer, K.M. (1957a). The natural remanent magnetization of certain stable rocks from Great Britain. *Philosophical Transactions,* A**250**, 111–29.

Creer, K.M. (1957b). The remanent magnetization of unstable keuper marls. *Philosophical Transactions,* A**250**, 130–43.

Creer, K.M. (1965). An expanding earth? *Nature,* **205**, 539–44.

Creer, K.M., Irving, E. & Runcorn, S.K. (1954). The direction of the geomagnetic field in remote epochs in Great Britain. *Journal of Geomagnetism and Geoelectricity,* **6**, 163–8.

Creer, K.M., Irving, E. & Runcorn, S.K. (1957). Geophysical interpretations of palaeomagnetic directions from Great Britain. *Philosophical Transactions,* A**250**, 144–56.

Creer, K.M., Irving, E. & Runcorn, S.K. (1958). Palaeomagnetic results from different continents and their relation to the problem of continental drift. *Annales de Geophysique,* **14**, 492–501.

Croneis, Carey & Krumbein, W.C. (1966). *Down to Earth: An Introduction to Geology.* first published 1936, Chicago: University of Chicago Press.

Dacqué, Edgar (1915). *Grundlagen und Methoden der Paläogeographie.* Jena: G. Fischer.

Daly, Reginald A. (1923a). Earth's crust and its evolution (abstract only). *Bulletin of the Geological Society of America,* **34**, 61.

Daly, R.A. (1923b). The earth's crust and its stability. *American Journal of Science,* **205**, 349–71.

Daly, R.A. (1923c). Decrease of the earth's rotational velocity and its geological effects. *American Journal of Science,* **205**, 372–77.

Daly, R.A. (1926). *Our Mobile Earth.* New York: Scribner's.

Daly, R.A. (1933). *Igneous Rocks and the Depths of the Earth.* New York: McGraw–Hill.

Daly, R.A. (1942). *The Floor of the Ocean.* Chapel Hill: University of North Carolina Press.

Dana, James D. (1873). On some results of the earth's contraction from cooling; Part V, Formation of the continental plateaus and oceanic depressions. *American Journal of Science,* 3rd series, **6**, 161–72.

Dana, J.D. (1881). The continents always continents. *Nature,* **23**, 410.

David, Tannant W.E. (1928). (Notes on Lecture by David entitled:) Drifting continents: The Wegener hypothesis. *The Australian Geographer,* **1**, Part 1, 60–2.

David, T.W.E. & Browne, W.R. (1950). *The Geology of the Commonwealth of Australia; Edited and Much Supplemented by W.R. Browne,* vol. 1. London: Edward Arnold & Co.

Davies, Gordon L. (1969). *The Earth in Decay: A History of British Geomorphology.* New York: Elsevier.

Day, Deborah. (1985). *A Guide to the Roger Randall Douglas Revelle Papers 1928–1979.* La Jolla: Archives of the Scripps Institution of Oceanography.

Deutsch, Ernst R. (1958a). Value of rock magnetism to the earth sciences. In Raasch (1958a), 8–19.

Deutsch, E.R. (1958b). Recent palaeomagnetic evidence for northward movement of India. In Raasch (1958a), 19–26.

Deutsch, E.R. (1963a). Polar wandering and continental drift: An evaluation of recent evidence. In Munyan (1963a), 4–46.

Deutsch, E.R. (1963b). Polar wandering–a phantom event. *American Journal of Science*, 261, 194–9.

Dewey, John F. (1969a). Structure and sequence in paratectonic British Caledonides. In Kay (1969a), 309–35.

Dewey, J.F. (1969b). Evolution of the Appalachian/Caledonian orogen. *Nature*, 222, 124–9.

Dewey, J.F. & Bird, J.M. (1970). Mountain belts and the new global tectonics. *Journal of Geophysical Research*, 75, 2625–47.

Dewey, J.F. & Kay, M. (1968). Appalachian and Caledonian evidence for drift in the north Atlantic. In Phinney (1968a), 161–7.

Dicke, R.H. (1957). Principle of equivalence and the weak interactions. *Reviews of Modern Physics*, 29, 355–62.

Dicke, R.H. (1962). The earth and cosmology. *Science*, 138, 653–64.

Dickinson, Robert E. (1969). *The Makers of Modern Geography*. London: Routledge & Kegan Paul.

Diener, Carl. (1915). Die grossformen der Erdoberfläche. *Mitteilungen der K. K. Geographische Gesellschaft in Wein*, 58, 329–49.

Dietz, Robert S. (1961). Continental ocean basin evolution by spreading of the sea floor. *Nature*, 190, 854–7.

Dietz, R.S. (1968). Reply. *Journal of Geophysical Research*, 73, 6567.

Dixey, Frank. (1938). Some observations on the physiographical development of central and southern Africa. *Geological Society of South Africa: Proceedings and Transactions*, 41, 113–71; esp. 164–9.

Doell, Richard R. & Cox, A. (1961). Paleomagnetism. *Advances in Geophysics*, 8, 221–313.

Dott, R.H. (1974). The geosynclinal concept. In *Modern and Ancient Geosynclinal Sedimentation*, ed. R.H. Dott and R.H. Shaver, pp. 1–13. Tulsa: Society of Economic Paleontologists and Mineralogists.

Dott, R.H. (1978). Tectonics and sedimentation a century later. *Earth-Science Reviews*, 14, 1–34.

Drake, Charles L. & Girdler, R.W. (1964). A geophysical study of the Red Sea. *The Geophysical Journal of the Royal Astronomical Society*, 8, 473–95.

Drake, Ellen T. & Jordan, W.M., eds. (1985). *GSA Centennial Special Volume I*. Boulder, Colorado: Geological Society of America.

Dunn, E.R. (1925). The host–parasite method and the distribution of frogs. *The American Naturalist*, 59, 370–5.

Durham, J. Wyatt. (1950). Cenozoic marine climates of the Pacific coast. *Bulletin of the Geological Society of America*, 61, 1243–64.

Durham, J.W. (1952). Early tertiary marine faunas and continental drift. *American Journal of Science*, 250, 321–43.

Dury, George H. (1959). *The Face of the Earth*. Harmondsworth: Penguin Books.

du Toit, Alexander L. (1921). Land connections between the other continents and South Africa in the past. *South African Journal of Science*, 18, 120–40.

du Toit, A.L. (1922). The carboniferous glaciation of South Africa. (read 10 October

1921) *Transactions of the Geological Society of South Africa,* **24,** 188–227.

du Toit, A.L. (1927). *A Geological Comparison of South America with South Africa.* Washington: Carnegie Institution.

du Toit, A.L. (1929a). The continental displacement hypothesis as viewed by du Toit. *American Journal of Science,* **217,** 179–83.

du Toit, A.L. (1929b). Some reflections upon a geological comparison of South Africa with South America. *Proceedings of the Geological Society of South Africa,* **31,** xix–xxxviii.

du Toit, A.L. (1937). *Our Wandering Continents.* Edinburgh: Oliver & Boyd.

du Toit, A.L. (1939). The origin of the Atlantic-Arctic-Ocean. *Geologische Rundschau,* **30,** 138–47.

du Toit, A.L. (1944). Tertiary mammals and continental drift: A rejoinder to George G. Simpson. *American Journal of Science,* **242,** 145–63.

Edge, David (1976). *Astronomy Transformed: The Emergence of Radio Astronomy.* New York: Wiley.

Edge, D. (1979). Quantitative measures of communication in science: A critical review. *History of Science,* **17,** 102–34.

Egyed, Laszlo. (1956). The change in the earth's dimensions determined from palaeogeographical data. *Geofisica Pura e Applicata,* **33,** 42–8.

Egyed, L. (1957). A new dynamic conception of the internal constitution of the earth. *Geologische Rundschau,* **46,** 101–21.

Egyed, L. (1960). Some remarks on continental drift. *Geofisica Pura e Applicata,* **45,** 115–16.

Elsasser, Walter M. (1946). Induction effects in terrestrial magnetism. *Physical Review,* **69,** 106–16; 202–12.

Evans, John W. (1923). The Wegener hypothesis of continental drift. *Nature,* **111,** 393–94.

Evans, J.W. (1924). Introduction. In A. Wegener (1924t), pp. vii–xii.

Ewing, M. & Donn, W.L. (1963). Polar Wandering and Climate. In Munyan (1963a), pp. 94–9.

Ewing, M. & Heezen, B.C. (1956a). Mid-Atlantic Ridge seismic belt (Abstract). *Transactions American Geophysical Union,* **37,** 343.

Ewing, M. & Heezen, B.C. (1956b). Some problems of Antarctic submarine geology. *Geophysical Monographs,* **1,** 75–81.

Ewing, M., Hirschman, J. & Heezen, B.C. (1959). Magnetic Anomalies of the Mid-Oceanic Rift. In Sears (1959), p. 24.

Feyerabend, Paul (1975). *Against Method.* London: New Left Books.

Feyerabend, P. (1981). More clothes from the emperor's bargain basement (a review of L. Laudan: progress and its problems). *British Journal for the Philosophy of Science,* **32,** 57–71.

Fisher, Osmond (1882). On the physical cause of the ocean basins. *Nature,* **25,** 242–44.

Fisher, R.L. & Hess, H.H. (1963). Trenches. In *The Sea,* **3,** ed. M.N. Hill, pp. 411–36. John Wiley: New York.

Forman, Paul. (1971). Weimar culture, causality, and quantum theory, 1918–1927: Adaptation by German physicists and mathematicians to a hostile intellectual environment. *Historical Studies in the Physical Sciences,* **3,** 1–115.

Forman, P. (1973). Scientific internationalism and the Weimar physicists: The ideology and its manipulation in Germany after World War I. *Isis,* **64,** 151–80.

Foster, Robert J. (1966). *Geology*. Columbus: Charles E. Merrill.

Foster, R.J. (1969). *General Geology*. Columbus: Charles E. Merrill.

Frankel, Henry. (1976). Alfred Wegener and the specialists. *Centaurus*, **20**, 305–24.

Frankel, H. (1978). Arthur Holmes and continental drift. *British Journal for the History of Science*, **11**, 130–49.

Frankel, H. (1979a). The career of continental drift theory: An application of Imre Lakatos' analysis of scientific growth to the rise of drift theory. *Studies in History and Philosophy of Science*, **10**, 21–66.

Frankel, H. (1979b). The reception and acceptance of continental drift theory as a rational episode in the history of science. In *The Reception of Unconventional Science*, ed. S.H. Mauskopf, pp. 51–90. Boulder, Colorado: Westview Press for AAAS.

Frankel, H. (1979c). Why drift theory was accepted with the confirmation of Harry Hess' concept of sea floor spreading. In *Two Hundred Years of Geology in America*, ed. C.J. Schoeer, pp. 337–53. Hanover, New Hampshire: University Press of New England.

Frankel, H. (1980). Hess's development of his seafloor spreading hypothesis. In *Scientific Discovery: Case Studies*, vol. 60 of Boston Studies in the Philosophy of Science, ed. Thomas Nickles, pp. 345–66. Dordrecht: D. Reidel.

Frankel, H. (1981). The paleobiogeographical debate over the problem of disjunctively distributed life forms. *Studies in History and Philosophy of Science*, **12**, no. 3, 211–59.

Frankel, H. (1982). The development, reception, and acceptance of the Vine–Matthews–Morley hypothesis. *Historical Studies in the Physical Sciences*, **13**, 1–39.

Frankel, H. (1984a). Biogeography, before and after the rise of sea floor spreading. *Studies in History and Philosophy of Science*, **15**, 141–68.

Frankel, H. (1984b). The permo–carboniferous ice cap and continental drift. *Compte rendu de Neuviéme Congrés International de Stratigraphée et de Géologie du Carbonifère (1979)*, **1**, 113–20.

Frankel, H. (1985). The biogeographical aspect of the debate over continental drift. *Earth Sciences History*, **4**, 160–81.

Fulford, Margaret (1963). Continental drift and distribution patterns in the leafy Hepaticae. In Munyan (1963a), pp. 140–5.

Furon, Raymond. (1959). *La Paléogéographie: Essai sur l'Evolution des Continents et des Océans*, 2nd edn. Paris: Payot.

Furon, R. (1967). *Cours de Géologie*. Paris: Centre de documentation Universitaire.

Garland, George D., ed. (1966). *Continental Drift*. Royal Society of Canada Special Publications No. 9, Toronto: University of Toronto Press/Royal Society of Canada.

Garrels, Robert M. (1951). *A Textbook of Geology*. New York: Harper.

Gass, I.G., Smith, P.J. and Wilson, R.C.L., eds. (1970). *Understanding the Earth: A Reader in the Earth Sciences*. Milton Keynes: Open University Press.

Geikie, Archibald. (1901). *The Founders of Geology*. London: Macmillan.

Geikie, A. (1903). *Textbook of Geology*, 4th edn, 2 vols. London: Macmillan.

Gevers, Traugott W. (1949). The life and work of Dr Alex. L. du Toit. Annexure to: *Geological Society of South Africa: Transactions and Proceedings*, 52 (109 pp.).

Gheyselinck, Robert. (1939). *The Restless Earth: Geology for Everyman*. London: The Scientific Book Club, esp. pp. 189–229; 281–5.

Gicouate, Moisés (1945). *Manual de Geologia*. Sao Paulo: Ediçoes Melhoramentos.

Gillispie, Charles C. (1951). *Genesis and Geology*. Cambridge, Mass.: Harvard University Press.

Girdler, Ronald W. (1958). The relationship of the Red Sea to the East African rift system. *Quarterly Journal of the Geological Society of London,* **114,** 79–105.

Girdler, R.W. (1962). Initiation of continental drift. *Nature,* **194,** 521–4.

Glen, William. (1975). *Continental Drift and Plate Tectonics.* Columbus: Charles E. Merrill.

Glen, W. (1982). *The Road to Jaramillo: Critical Years of the Revolution in Earth Science.* Stanford: Stanford University Press.

Goguel, Jean. (1952t). *Tectonics.* trans. H.E. Thalman from *Traité de Tectonique* (Paris, 1952), 1962, San Francisco: W.H. Freeman.

Gold, T. (1955). Instability of the earth's axis of rotation. *Nature,* **175,** 526–9.

Gorshkov, G. & Yakushova, A. (1973t). *Physical Geology.* trans V.V. Shiffer 1977, Moscow: Mir Publishers.

Gould, Stephen J. (1965). Is uniformitarianism really necessary? *American Journal of Science,* **263,** 223–8.

Gould, S.J. (1977). *Ever Since Darwin: Reflections in Natural History.* New York: W.W. Norton.

Gould, S.J. (1980). *The Panda's Thumb: More Reflections on Natural History.* New York: W.W. Norton.

Gould, S.J. (1983). *Hen's Teeth and Horse's Toes: Further Reflections on Natural History.* New York: W.W. Norton.

Grabau, Amadeus W. (1939). Present status of the polar control theory of earth development. *Bulletin of the Geological Society of China,* **19,** no. 2, 189–205.

Graham, John W. (1949). The stability and significance of magnetism in sedimentary rocks. *Journal of Geophysical Research,* **54,** 131–67.

Graham, K.W.T. & Hales, A.L. (1957). Palaeomagnetic measurements on Karroo dolerites. *Advances in Physics,* **6,** 149–61.

Gray, Jane & Boucot, A.J., eds. (1976). *Historical Biogeography, Plate Tectonics, and the Changing Environment.* Corvallis: Oregon State University Press.

Greenberg, Daniel S. (1967). *The Politics of Pure Science.* New York: The New American Library.

Greene, Mott T. (1982). *Geology in the Nineteenth Century.* Ithaca: Cornell University Press.

Greene, M.T. (1984). Alfred Wegener. *Social Research,* **51,** 739–61.

Greene, M.T. (1985). Plate tectonics and biogeography in historical perspective. *Earth Sciences History,* **4,** 93–7.

Gregory, John W. (1915). *Geology of Today; a Popular Introduction in Simple Language.* London: Seeley, Service & Co.

Gregory, J.W. (1925). Review (of Wegener, 1924t). *Nature,* **115,** 255–7.

Gregory, J.W. (1928). Wegener's hypothesis. In van der Gracht (1928a), 93–7.

Gregory, J.W. (1930). The machinery of the earth. *Nature,* **126,** 959–63.

Griggs, David. (1939). A theory of mountain-building. *American Journal of Science,* **237,** 611–50.

Gussow, William C. (1958). Metastasy or crustal shift?. In Raasch (1958a). 44–8.

Gutenberg, Beno. (1936). Structure of the earth's crust and the spreading of the continents. *Bulletin of the Geological Society of America,* **47,** 1587–610.

Gutenberg, B., Benioff, H., Burgers, J.M. & Griggs, D., eds. (1951). Colloquium on plastic flow and deformation within the earth. *Transactions American Geophysical Union,* **32,** 497–543.

Gutenberg, B. & Richter, C.F. (1949). *Seismicity of the Earth and Associated Phenomena.* Princeton: Princeton University Press.

Gutenberg, B. & Richter, C.F. (1954). *Seismicity of the Earth and Associated Phenomena*, 2nd edn, Princeton: Princeton University Press.

Hallam, Anthony. (1973). *A Revolution in the Earth Sciences: From Continental Drift to Plate Tectonics*. Oxford: Clarendon Press.

Hallam, A. (1983). *Great Geological Controversies*. Oxford: Oxford University Press.

Harland, W.B. (1969). Contribution of Spitsbergen to understanding of tectonic evolution of North Atlantic region. In Kay (1969a), pp. 817–51.

Harrington, Horacio J. (1963). Deep focus earthquakes in South America and their possible relation to continental drift. In Munyan (1963a), pp. 55–73.

Harrison, Launcelot (1924). The migration route of the Australian marsupial fauna (Presidential address to the Royal Zoological Society of New South Wales). *The Australian Zoologist*, **3**, 247–63.

Harrison, L. (1926a). Crucial evidence for Antarctic radiation. *American Naturalist*, **60**, 374–83.

Harrison, L. (1926b). The Wegener hypothesis from a biological standpoint. *Sydney University Science Journal*, **10**, 10–19.

Harrison, L. (1928). The composition and origins of the Australian fauna, with special reference to the Wegener hypothesis. *Report of the Eighteenth Meeting of ANZAAS (Perth, 1926)*, **18**, 332–96.

Heezen, Bruce C. (1959a). Paleomagnetism, continental displacements, and the origin of submarine topography. In Sears (1959), pp. 26–8.

Heezen, B.C. (1959b). Géologie sous-marine et déplacements des continents. *Colloques Internationaux du CNRS*, **83**, 294–304.

Heezen, B.C. (1960). The rift in the ocean floor. *Scientific American*, **203**, 99–110.

Heezen, B.C. (1962). The deep-sea floor. In Runcorn (1962b), pp. 235–88.

Heirtzler, James R. (1968a). Evidence for ocean floor spreading across ocean basins. In Phinney (1968a), pp. 90–100.

Heirtzler, J.R. (1968b). Sea floor spreading. *Scientific American*, **219**, 66–9.

Heirtzler, J.R., Le Pichon, X. & Baron, J.G. (1966). Magnetic anomalies over the Reykjanes Ridge. *Deep-Sea Research*, **13**, 427–33.

Hess, Harry Hammond. (1938). Gravity anomalies and island arc structure with particular reference to the West Indies. *Proceedings of the American Philosophical Society*, **79**, 71–95.

Hess, H.H. (1954). Geological hypotheses and the earth's crust under the ocean. *Proceedings of the Royal Society*, A**222**, 341–7.

Hess, H.H. (1959). Nature of the great oceanic ridges. In Sears (1959), pp. 33–4.

Hess, H.H. (1962). History of ocean basins. In *Petrologic Studies*, ed. E.A.J. Engle, pp. 599–602. Geological Society of America.

Hess, H.H. (1968). Reply. *Journal of Geophysical Research*, **73**, 6569.

Hesse, Mary. (1974). *The Structure of Scientific Inference*. London: Macmillan.

Hills, George F.S. (1947). *The Formation of Continents by Convection*. London: Edward Arnold & Co.

Hinks, Arthur R., Simpson, G.C., Gregory, J.W. *et al.* (1931). Problems of the earth's crust: A discussion in section E (Geography) of the British Association. . . . *Geographical Journal*, **78**, 433–55.

Hobbs, William H. (1921). *Earth Evolution and its Facial Expression*. New York.

Holland, Thomas H. (1933). Presidential anniversary address: The geological age of the glacial horizon at the base of the Gondwana system. *Quarterly Journal of the Geological Society of London.* **89**, lxiv–lxxxvi.

Holland, T.H. (1937). The permanence of ocean basins and continental masses. *1937 Huxley Memorial Lecture at Imperial College of Science and Technology*, London: Macmillan.

Holland, T.H. (1941). The evolution of continents: A possible reconciliation of conflicting evidence. *Proceedings of the Royal Society of Edinburgh*, series B, **61**, 149–66.

Holland, T.H. (1943). The theory of continental drift. *Proceedings of the Linnean Society of London*, **155**, 112–25.

Holmes, Arthur. (1925a). Radioactivity and the earth's thermal history. Part IV: A criticism of Parts I, II and III. *Geological Magazine*, **62**, 504–15.

Holmes, A. (1925b). Radioactivity and the earth's thermal history, Part V: The control of geological history by radioactivity. *The Geological Magazine*, **62**, 529–44.

Holmes, A. (1928a). Radioactivity and continental drift. *The Geological Magazine*, **65**, 236–8.

Holmes, A. (1928b). Review of W.A.J.M. van Waterschoot van der Gracht, ed., *Theory of Continental Drift. Nature*, **122**, 431–3.

Holmes, A. (1929). A Review of the continental drift hypothesis. *The Mining Magazine*, **40**, 205–9; 286–8; 340–7.

Holmes, A. (1930). Radioactivity and earth movements. *Transactions of the Geological Society of Glasgow*, **18**, part iii (1928–29), 559–606.

Holmes, A. (1933). The thermal history of the earth. *Journal of the Washington Academy of Science*, **23**, 169–95.

Holmes, A. (1944). *Principles of Physical Geology.* London: Thomas Nelson & Sons Ltd.

Holmes, A. (1953). The south Atlantic: Land bridges or continental drift?. *Nature*, **171**, 669–71.

Holmes, A. (1965). *Principles of Physical Geology*, 2nd edn. New York: Ronald Press.

Hooykaas, Reijer. (1959). *Natural Law and Divine Miracle: A Historico-Critical Study of the Principle of Uniformity in Geology, Biology, and Theology.* Leiden: E.J. Brill.

Hooykaas, R. (1970). Catastrophism in geology, its scientific character in relation to actualism and uniformitarianism. *Medeeling Koninklijke Nederlandse Akademische Wetenschappelijke Afdeeling voor Letterkunde*, **33**.

Hopkins, Evan. (1844). *On the Connexion of Geology with Terrestrial Magnetism.* London: Richard and John Edward Taylor.

Hospers, J. (1955). Rock magnetism and polar wandering. *Journal of Geology*, **63**, 59–74.

Howell, Benjamin F. (1959). *Introduction to Geophysics.* New York: McGraw–Hill.

Hume, William F. (1948). *Terrestrial Theories: A Digest of Various Views as to the Origin and Development of the Earth and their Bearing on the Geology of Egypt.* Cairo: Government Press.

Hurley, P.M., Dealmeid, F.F., Melcher, G.C., Cordani, U.G., Rand, J.R., Kawashit, K., Vandoros, P., Pinson, W.H., Fairbair H.W. (1967). Test of continental drift by comparison of radiometric ages. *Science*, **157**, 495–500.

Hurley, P.M. & Rand, J.R. (1968). Review of age data in west Africa and South America relative to a test of continental drift. In Phinney (1968a), pp. 153–60.

Huxley, T.H. (1869). Presidential address: Geological reform. *Quarterly Journal of the Geological Society of London*, **25**, pp. xxxviii–liii.

Imbrie, John & Imbrie, K.P. (1979). *Ice Ages: Solving the Mystery.* Rep. 1986. Cambridge, Mass.: Harvard University Press.

Irving, Edward. (1956). Palaeomagnetic and palaeoclimatological aspects of polar wandering. *Geofisica Pura e Applicata*, **33**, 23–41.

Irving, E. (1957). The origin of the palaeomagnetism of the Torridonian sandstones of north-west Scotland. *Philosophical Transactions*, A**250**, 100–10.

Irving, E. (1958). Rock magnetism: A new approach to the problems of polar wandering and continental drift. In Carey (1958a), pp. 24–61.

Irving, E. & Green, R. (1958). Polar movement relative to Australia. *Geophysical Journal of the Royal Astronomical Society*, **1**, 64–72.

Irving, E. & Runcorn, S.K. (1957). Analysis of the palaeomagnetism of the Torridonian sandstone series of north-west Scotland. *Philosophical Transactions*, A**250**, 83–99.

Irwin, William P. (1972). Terranes of the western paleozoic and triassic belts in the southern Klamath Mountains, California. *US Geological Survey Professional Papers*, 800–C, 103–11.

Isacks, Bryan, Oliver, J. & Sykes, L.R. (1968). Seismology and the new global tectonics. *Journal of Geophysical Research*, **73**, 5855–99.

Jacobs, John A., Russell, R.D. & Wilson, J.T. (1959). *Physics and Geology*. New York: McGraw–Hill.

Jaworski, Erich. (1922). Die A. Wegenersche hypothese der kontinentalverschiebungen: Eine kritische zusammenstellung. *Geologische Rundschau*, **13**, 273–96.

Jeffreys, Harold. (1924). *The Earth: Its Origin, History and Physical Constitution*. Cambridge: Cambridge University Press.

Jeffreys, H. (1935). *Earthquakes and Mountains*. London: Methuen.

Jeffreys, H. (1952). *The Earth*, 3rd edn. Cambridge: Cambridge University Press.

Johnson, E.A., Murphy, T. & Torreson, O.W. (1948). Prehistory of the earth's magnetic field. *Journal of Geophysical Research*, **43**, 349–72.

Joleaud, Louis. (1923). Essai sur l'evolution des milieux géophysiques et biogéographiques (A propos de la théorie de Wegener sur l'origine des continents). *Bulletin de la Société Géologique de France*, 4th series, **23**, 205–70.

Joly, John. (1909). *Radioactivity and Geology*. London: Archibald Constable.

Joly, J. (1925). *The Surface-History of the Earth*. Oxford: Clarendon Press.

Joly, J. (1928). Continental movement. In van der Gracht (1928a), pp. 88–9.

Jones, David L., Cox, A., Coney, P. & Beck, M. (1982). The growth of western North America. *Scientific American*, **247**, no. 5, 50–64.

Jones, D.L., Irwin, W.P. & Ovenshine, A.T. (1972). Southeastern Alaska – a displaced continental fragment. *US Geological Survey Professional Papers*, 800–B, 211–17.

Jordan, Pascual. (1966t). *The Expanding Earth: Some Consequences of Dirac's Gravitation Hypothesis*. revised edn. 1971, trans. A. Beer, Oxford: Pergamon Press (first ed. Braunschweig: Friedrich Vieweg).

Just, Theodore. (1952). Fossil floras of the Southern Hemisphere and their phytogeographical significance. In Mayr (1952a), pp. 189–203.

Kasbeer, Tina. (1972). *Bibliography of Continental Drift and Plate Tectonics*. Boulder; Colorado: The Geological Society of America, Special Paper 142.

Kay, Marshall, ed. (1969a). *North Atlantic Geology and Continental Drift. A Symposium*. Tulsa: American Association of Petroleum Geologists.

Kay, M. (1969b). Continental drift in north Atlantic Ocean. In Kay (1969a), pp. 965–73.

Kay, M. & Colbert, E.H. (1965). *Stratigraphy and Life History*. New York: John Wiley.

Keith, Arthur. (1923). Outlines of Appalachian structure. *Bulletin of the Geological Society of America*, **34**, 309–80.

Kent, Peter, Bott, M.H.P., McKenzie, D.P. & Williams, C.A., eds. (1982). [Symposium on] the evolution of sedimentary basins. *Philosophical Transactions*, A**305**, 3–338.

Kern, Ernest L. (1972). For 'Believers'. *Geotimes*, **17**, no. 1, 9–10.

King, B.C. & King, G.C.P. (1972). The world rift system and plate tectonics or 1971 and all that. *Geotimes*, **17**, no. 1, 28.

King, Lester C. (1953). Necessity for continental drift. *Bulletin of the AAPG*, **37**, 2163–77.

King, L.C. (1958a). The origin and significance of the great sub-oceanic ridges. In Carey (1958a), pp. 62–102.

King, L.C. (1958b). A new reconstruction of Laurasia. In Carey (1958a), pp. 13–23.

King, L.C. (1958c). Basic palaeogeography of Gondwanaland during the late palaeozoic and mesozoic eras. *Quarterly Journal of the Geological Society of London*, **114**, 47–70.

King, L.C. (1961). The palaeoclimatology of Gondwanaland during the palaeozoic and mesozoic eras. In Nairn (1961a), pp. 307–31.

King, L.C. (1962). *The Morphology of the Earth: A Study and Synthesis of World Scenery.* New York: Hafner; London: Oliver & Boyd.

Kirkaldy, John F. (1954). *General Principles of Geology.* London: Hutchinson.

Kirkaldy, J.F. (1962). *General Principles of Geology*, 3rd edn. London: Hutchinson.

Kirsch, G., Wegener, K., van Waterschoot van der Gracht, W.A.J.M. *et al.* (1939). Frankfurt symposium on Atlantis and geophysics. *Geologische Rundschau*, **30**, Heft 1–4.

Kitts, David B. (1974). Continental drift and scientific revolution. *Bulletin of the American Association of Petroleum Geologists*, **58**, 2490–96.

Kitts, D.B. (1977). *The Structure of Geology.* Dallas: Southern Methodist University Press.

Kitts, D.B. (1981). Retrodiction in geology. In *Proceedings of the 1978 Biennial Meeting of the Philosophy of Science Association*, ed. P.D. Asquith & I. Hacking, 2 vols., vol. ii, pp. 213–26, East Lansing, Michigan: Philosophy of Science Association.

Kohler, Robert E. (1982). *From Medical Chemistry to Biochemistry: The Making of a Biomedical Discipline.* Cambridge: Cambridge University Press.

Kossmat, Franz. (1921). Erörterungen zu A. Wegeners theorie der kontinentalverschiebungen. *Zeitschrift der Gesellschaft für Erdkunde zu Berlin*, nr. 3/4, 103–10.

Krige, Leo J. (1926). On mountain building and continental sliding. *South African Journal of Science*, **23**, 206–15.

Krige, L.J. (1930). Magmatic cycles, continental drift and ice ages. *Proceedings of the Geological Society of South Africa*, **32**, xxi–xl.

Kubart, Bruno. (1926). Bemerkungen zu Alfred Wegeners verschiebungstheorie. *Arbeiten des Phytopalaeontologischen Laboratoriums der Universität Graz*, II (28 pp.).

Kuenen, Philip H. (1950). *Marine Geology.* New York: John Wiley.

Kuhn, Thomas S. (1962). *The Structure of Scientific Revolutions.* Chicago: University of Chicago Press.

Kuhn, T.S. (1970). *The Structure of Scientific Revolutions.* revised edn., Chicago: University of Chicago Press.

Kuhn, T.S. (1977). *The Essential Tension: Selected Studies in Scientific Tradition and Change.* Chicago: University of Chicago Press.

Kuhn T.S. (1983). Commensurability, comparability, communicability. In *Proceedings of* the 1982 Biennial Meeting of the Philosophy of Science Association, ed. P.D. Asquith & T. Nickels, 2 vols., vol. ii, pp. 669–88. East Lansing, Michigan: Philosophy of Science Association.

Kummel, Bernard. (1961). *History of the Earth: An Introduction to Historical Geology.* San Francisco: W.H. Freeman.

Lakatos, Imre (1978). *The Methodology of Scientific Research Programmes.* Cambridge: Cambridge University Press.

Lakatos, I. & Musgrave, A., eds. (1974). *Criticism and the Growth of Knowledge.* Cambridge: Cambridge University Press.

Lake, Philip. (1922). Wegener's displacement theory. *Geological Magazine*, **59**, 338–46.

Lake, P. (1923a). Wegener's hypothesis of continental drift. *Nature*, **111**, 226–8.

Lake, P. (1923b). Wegener's hypothesis of continental drift. *The Geographical Journal*, **61**, 179–94.

Lake, P. (1933). Gutenberg's fleisstheorie; A theory of continental spreading. *Geological Magazine*, **70**, 116–21.

Lake, P. & Rastall, R.H. (1910). *A Text-Book of Geology.* London: Edward Arnold.

Lake, P. & Rastall, R.H. (1920). *A Text-Book of Geology*, 3rd edn. London: Edward Arnold.

Lake, P. & Rastall, R.H. (1927). *A Text-Book of Geology*, 4th edn. London: Edward Arnold & Co.

Laming, D.J.C. (1958). Fossil winds. In Raasch (1958), pp. 35–9.

Landes, Kenneth K. (1952). Our shrinking globe. *Bulletin of the Geological Society of America*, **63**, 225–40.

Langseth, Marcus G., Le Pichon, X. & Ewing, M. (1966). Crustal structure of the mid-ocean ridges 5. Heat flow through the Atlantic Ocean floor and convection currents. *Journal of Geophysical Research*, **71**, 5321–54.

Laporte, Léo F. (1985). Wrong for the right reasons: G.G. Simpson and continental drift. *Geological Society of America: Centennial Special Volume*, I, 273–85.

Latour, Bruno (1985). Les 'vues' de l'esprit. *Culture technique*, no. 14, 4–29.

Latour, B. (1987). *Science in Action.* Cambridge, Mass.: Harvard Univerity Press.

Latour, B. & Woolgar, S. (1979). *Laboratory Life.* Beverly Hills: Sage.

Laudan, Larry (1977). *Progress and its Problems: Towards a Theory of Scientific Growth.* Berkeley: University of California Press.

Laudan, L. (1981a). The pseudo-science of science. *Philosophy of the Social Sciences*, **11**, 173–98.

Laudan, L. (1981b). *Science and Hypothesis.* Dordrecht: D. Reidel.

Laudan, L. (1984). *Science and Values: The Aims of Science and their Role in Scientific Debate.* Berkeley: University of California Press.

Laudan, L., Donovan, A. & Laudan, R. (1986). Scientific change: Philosophical models and historical research. *Synthese*, **69**, 141–223.

Laudan, Rachel. (1977). Ideas and organization in British geology: A case study in institutional history. *Isis*, **68**, 527–37.

Laudan, R. (1980a). The recent revolution in geology and Kuhn's theory of scientific change. In *Paradigms and Revolutions*, ed. Gary Gutting, pp. 284–96. South Bend: Notre Dame University Press.

Laudan, R. (1980b). The method of multiple working hypotheses and the discovery of plate tectonic theory in geology. In *Scientific Discovery: Case Studies*, ed. Thomas Nickels, pp. 331–43. Dordrecht: D. Reidel.

Laudan, R. (1980c). Oceanography and geophysical theory in the first half of the twentieth century: The Dutch school. In *Oceanography: The Past*, ed. Mary Sears & D. Merriman, pp. 656–66. New York: Springer-Verlag.

Laudan, R. (1982a). The role of methodology in Lyell's science. *Studies in the History and Philosophy of Science*, **13**, 215–49.

Laudan, R. (1982b). Tensions in the concept of geology: Natural history or natural philosophy? *Earth Sciences History*, 1, 7–13.

Laudan, R. (1983). Redefinitions of a discipline: Histories of geology and geological history. In *Functions and Uses of Disciplinary Histories*, ed. Loren Graham *et al.*, pp. 79–104. Dordrecht: D. Reidel.

Laudan, R. (1985). Frank Bursley Taylor's theory of continental drift. *Earth Sciences History*, 4, 118–21.

Laudan, R. (1987). The rationality of entertainment and pursuit. *Boston Studies in the Philosophy of Science* (forthcoming).

Laudan, R. (1987a). *From Mineralogy to Geology: The Foundations of Science. 1650–1830*. Chicago: University of Chicago Press.

Laudan, R. (1987b). Drifting interests and colliding continents. *Social Studies of Science*, **17** (forthcoming).

Le Grand, Homer E. (1986a). Steady as a rock: Methodology and moving continents. In Schuster & Yeo, pp. 97–138.

Le Grand, H.E. (1986b). Specialties, problems and localism: The reception of continental drift in Australia 1920–1940. *Earth Sciences History*, **5**, 84–95.

Le Pichon, Xavier. (1968). Sea-floor spreading and continental drift. *Journal of Geophysical Research*, **73**, 3661–97.

Le Pichon, X., Francheteau, J. & Bonin, J. (1973). *Plate Tectonics*. Amsterdam: Elsevier (Developments in Geotectonics, 6).

Lees, George M. (1953). The evolution of a shrinking earth. *The Quarterly Journal of the Geological Society of London*, **109**, 217–57.

Lees, G.M. (1954). The geological evidence on the nature of ocean floors. *Proceedings of the Royal Society*, A**222**, 400–2.

Leuba, John. (1925). *Introduction à la géologie*. Paris: Librairie Armand Colin.

Lister, G.S., Ethridge, M.A. & Symonds, P.A. (1986). Detachment faulting and the evolution of passive continental margins. *Geology*, **14**, 246–50.

Longwell, Chester R. (1928). Some physical tests of the displacement hypothesis. In van der Gracht (1928a), pp. 145–57.

Longwell, C.R. (1944a). Some thoughts on the evidence for continental drift. *American Journal of Science*, **242**, 218–31.

Longwell, C.R. (1944b). Further discussion of continental drift. *American Journal of Science*, **242**, 514–15.

Longwell, C.R. (1944c). The mobility of Greenland. *American Journal of Science*, **242**, 624.

Longwell, C.R. (1958a). My estimate of the continental drift concept. In Carey (1958a), pp. 1–12.

Longwell, C.R. (1958b). Epilogue. In Carey (1958a), pp. 356–58.

Longwell, C.R. & Flint, R. (1955). *Introduction to Physical Geology*, New York: John Wiley & Sons; London: Chapman & Hall.

Longwell, C.R. & Flint, R. (1962). *Introduction to Physical Geology*. New York: John Wiley.

Longwell, C.R., Flint, R. & Saunders, J.E. (1969). *Physical Geology*. New York: John Wiley.

Longwell, C.R., Knopf, A. & Flint, R. (1941). *Outlines of Physical Geology*. 2nd edn., New York: John Wiley & Sons.

Longwell, C.R., Knopf, A. & Flint, R. (1948). *A Textbook of Geology: Part 1 – Physical Geology*. 3rd edn., New York: Wiley; London: Chapman and Hall.

Losee, John. (1980). *A Historical Introduction to the Philosophy of Science*, 2nd edn. Oxford: Oxford University Press.

Lugeon, Maurice. (1940). Sur la formation des Alpes franco-suisses. *Compte Rendu sommaire des Séances de la Société géologique de France*, séance 22.1.40 (fasc. 2), 7–11.

Lugeon, M. (1941). Une hypothèse sur l'origine du Jura. *Bulletin des laboratoires de géologie, géophysique et du Musée géologique de l'Université de Lausanne*, 73.

Lugg, Andrew. (1978). Disagreement in science. *Zeitschrift für Allgemeine Wissenschaftstheorie*, 9, 276–92.

Lyell, Charles. (1830). *Principles of Geology*. I, London.

Lyttleton, R.A. (1983). *The Earth and its Mountains*. London: Wiley Interscience.

McArthur, Robert P. & Pestana, H.R. (1974). Is continental drift/plate tectonics a paradigm–theory? *Proceedings of the XIVth International Congress of the History of Science* (Tokyo & Kyoto), 3, 105–8.

McClelland, C.E. (1980). *State, Society and University in Germany 1700–1914*. Cambridge: Cambridge University Press.

McDougall, Ian & Tarling, D.H. (1963). Dating of polarity zones in the Hawaiian Islands. *Nature*, 200, 54–6.

McKenzie, Dan P. (1978). Some remarks on the development of sedimentary basins. *Earth and Planetary Science Letters*, 40, 25–32.

McKenzie, D.P. & Parker, R.L. (1967). The north Pacific: An example of tectonics on a sphere. *Nature*, 216, 1276–80.

McKenzie, Donald A. (1981). *Statistics in Britain 1865–1930: The Social Construction of Scientific Knowledge*. Edinburgh: Edinburgh University Press.

McPhee, John. (1983) *In Suspect Terrain*. New York: Farrar, Straus, Giroux.

MacDonald, Gordon J.F. (1964). The deep structure of continents. *Science*, 143, 921–9.

Marvin, Ursula B. (1973). *Continental Drift: Evolution of a Concept*. Washington: Smithsonian Institution Press.

Marvin, U.B. (1985). The British reception of Alfred Wegener's continental drift hypothesis. *Earth Sciences History*, 4, 138–59.

Mather, Kirtley F. (1967). *Source Book in Geology 1900–1950*. Cambridge, Mass.: Harvard University Press.

Matthew, William D. (1915). Climate and evolution. *Annals of the New York Academy of Science*, 24, 171–318.

Maxwell, Arthur E., Vonherzen. R.P., Hsu, K.J. *et al.* (1970). Deep sea drilling in the South Atlantic. *Science*, 168, 1047–59.

Mayr, Ernst, ed. (1952a). The Problem of Land Connections Across the South Atlantic, with Special Reference to the Mesozoic. Proceedings of the Symposium on The Role of the South Atlantic Basin in Biogeography and Evolution . . . published as *Bulletin of the American Museum of Natural History*, 99, 79–258.

Mayr, E. (1952b). Conclusion. In Mayr (1952a), pp. 255–8.

Mayr, E. & Provine, W.B., eds. (1980). *The Evolutionary Synthesis: Perspectives on the Unification of Biology*. Cambridge, Mass.: Harvard University Press.

Mears, Brainerd, Jr. (1970). *The Changing Earth: An Introductory Geology*. New York: Van Nostrand Reinhold.

Menard, H.W. (1968). Some remaining problems in sea floor spreading. In Phinney (1968a), pp. 109–18.

Menard, H.W. (1971). *Science: Growth and Change*. Cambridge, Mass.: Harvard University Press.

Menard, H.W. (1986). *The Ocean of Truth: A Personal History of Global Tectonics*. Princeton: Princeton University Press.

Mercanton, Paul. (1926a). Inversion de l'inclinaison magnétique terrestre aux âges géologiques. *Terrestrial Magnetism and Atmospheric Electricity*, **31**, 187–90.

Mercanton, P. (1926b). Magnétisme terrestre: Aimantation de basaltes groenlandais. *Comptes Rendus de l'Académie des Sciences de Paris*, **182**, 859–60.

Meyerhoff, Arthur A. (1968). Arthur Holmes: Originator of spreading ocean floor hypothesis. *Journal of Geophysical Research*, **73**, 6563–5.

Meyerhoff, A.A. (1972). Review (of Tarling & Tarling, 1971). *Geotimes*, **17**, no. 4, 34–6.

Meyerhoff, A.A. (1978). Problems of plate tectonics. *Canadian Society of Petroleum Geologists Annual Seminar*, 9 (68 pp).

Meyerhoff, A.A. (1980). Trilobites and other embarrassments from the deep Atlantic Ocean. *International Stop Continental Drift Society Newsletter*, **3**, no. 2, 3–4.

Meyerhoff, A.A. & Meyerhoff, H.A. (1972a). The new global tectonics: Major inconsistencies. *Bulletin of the American Association of Petroleum Geologists*, **56**, 269–336.

Meyerhoff, A.A. & Meyerhoff, H.A. (1972b). The new global tectonics: Age of linear magnetic anomalies of ocean basins. *Bulletin of the American Association of Petroleum Geologists*, **56**, 337–59.

Milanovskiy, Ye.Ye. (1983). Evolution and present state of the problems of the earth's expansion and pulsations. *International Geology Review*, **25**, 621–40.

Molengraaff, Gustaaf A.F. (1916). The coral reef problem and isostasy. *Nederlandse Akademie van Wetenschappen*, **19**, 610–27.

Molengraaff, G.A.F. (1928). Wegener's continental drift. In van der Gracht (1928a), pp. 90–2.

Monger, J.W.H. & Ross, C.A. (1971). Distribution of fusilinaceans in the western Canadian Cordillera. *Canadian Journal of Earth Sciences*, **8**, 259–78.

Moore, James R. (1986). Geologists and interpreters of Genesis in the nineteenth century. In *God & Nature*, ed. D.C. Lindberg & R.L. Numbers, pp. 332–50. Berkeley: University of California Press.

Moret, Léon. (1955). *Précis de géologie à l'usage des candidates à la licence ès sciences, au S.P.C.N. et aux grandes écoles*, 2nd. edn. Paris: Masson & Cie.

Morgan, W. Jason. (1968). Rises, trenches, great faults, and crustal blocks. *Journal of Geophysical Research*, **73**, 1959–82.

Mulkay, Michael. (1979). *Science and the Sociology of Knowledge*. London: George Allen & Unwin.

Mulkay, M. & Gilbert, G.N. (1982a). Accounting for error: How scientists construct their social world when they account for correct and incorrect belief. *Sociology*, **16**, 165–83.

Mulkay, M. & Gilbert, G.N. (1982b). Warranting scientific belief. *Social Studies of Science*, **12**, 383–408.

Mulkay, M., Gilbert, G.N. & Woolgar, S. (1975). Problem areas and research networks in science. *Sociology*, **9**, 187–203.

Munk, Walter H. (1956). Polar wandering: A marathon of errors. *Nature*, **177**, 551–4.

Munyan, Arthur C., ed. (1963a). *Polar Wandering and Continental Drift*. Tulsa: Society of

Economic Paleontologists and Mineralogists, Special Publication No. 10.

Munyan, A.C. (1963b). Introduction to polar wandering and continental drift. In Munyan (1963a), pp. 1–3.

Murphy, John. (1986). The voice of memory: History, autobiography and oral memory. *Historical Studies*, 22, 157–75.

Nairn, Alan E.M. (1956). Relevance of paleomagnetic studies of jurassic rocks to continental drift. *Nature*, 178, 935–6.

Nairn, A.E.M. (1957). Palaeomagnetic collections from Britain and South Africa illustrating two problems of weathering. *Advances in Physics*, 6, 162–8.

Nairn, A.E.M. (1960). Palaeomagnetic results from Europe. *Journal of Geology*, 68, 285–306.

Nairn, A.E.M., ed. (1961). *Descriptive Palaeoclimatology*. New York: Interscience Publishers.

Nairn, A.E.M. & Thorley, N. (1961). The application of geophysics to palaeoclimatology. In Nairn (1961a), pp. 156–82.

Néel, Louis E.F. (1951). L'inversion de l'aimantation permanente des roches. *Annales de Geophysique*, 7, 90–102.

Nevin, Charles M. (1936). *Principles of Structural Geology*, 2nd edn. New York: John Wiley.

Nevin, C.M. (1949). *Principles of Structural Geology*, 4th edn. New York: John Wiley.

Nicholls, George E. (1933). Presidential address (section D. Zoology): The composition and biogeographical relations of the fauna of Western Australia. *Report of the Twenty-first Meeting of ANZAAS* (Sydney, 1932), pp. 93–138.

Nitecki, Matthew H., Lemke, J.L. Pullman, H.W. and Johnson, M.E. (1978). Acceptance of plate tectonic theory by geologists. *Geology*, 6, 661–4.

Noble, G.K. (1925). The evolution and dispersal of the frogs. *The American Naturalist*, 59, 265–71.

Nölke, Friedrich. (1922). Physikalische bedenken gegen A. Wegeners hypothese der entstehung der kontinente und ozeane. *Petermanns geogr. Mitt.*, 114–15.

Nordeng, Stephen C. (1963). Precambrian stromatolites as indicators of polar shift. In Munyan (1963a), pp. 131–9.

Nörland, N.E. (1936). Astronomical longitude and azimuth determinations. *Monthly Notices of the Royal Astronomical Society*, 97, 489–506.

Nörland, N.E. (1936). Astronomical longitude and azimuth determinations. *Royal Astronomical Society Monthly Notices*, 97, 489–506.

Numbers, Ronald L. (1986). The creationists. In *God & Nature*, ed. D.C. Lindberg & R.L. Numbers, pp. 391–423. Berkeley: University of California Press.

Nunan, Richard (1984). Novel facts, bayesian rationality, and the history of continental drift. *Studies in History and Philosophy of Science*, 15, 267–307.

Nunan, R. (1986). Global expansion versus drifting continents: Theory appraisal in the geological underworld. MS. of talk given by author Blacksburg, Virginia, October 1986.

Nur, Amos (1983). Accreted terranes. *Review of Geophysics and Space Physics*, 21, 1779–85.

Oldham, Richard D. (1906). The constitution of the interior of the earth as revealed by earthquakes. *Quarterly Journal of the Geological Society*, 62, 456–75.

Oldroyd, David R. (1987). Punctuated equilibrium theory and time: A case study in problems of coherence in the measurement of geological time 'The 'KBS' Tuff controversy and the dating of rocks in the Turkana Basin, East Kenya. In *Measurement, Realism and Objectivity*, ed. J. Forge, pp. 89–152. Dordrecht: D. Reidel.

Opdyke, N.D. (1968). The paleomagnetism of oceanic cores. In Phinney (1968a), pp. 61–72.

Opdyke, N.D. & Runcorn, S.K. (1959). Palaeomagnetism and ancient wind directions. *Endeavour*, **18**, 26–34.

Ospovat, Alexander (1976). The distortion of Werner in Lyell's *Principles of Geology*. *British Journal of the History of Science*, **9**, 190–8.

Owen, H.G. (1976). Continental displacement and expansion of the earth during the mesozoic and cenozoic. *Philosophical Transactions of the Royal Society*, A**281**, 233–91.

Owen, H.G. (1981). Constant dimensions or an expanding earth?. In *The Evolving Earth*, ed. L.R.M. Cocks, pp. 179–92. Cambridge: British Museum and Cambridge University Press.

Owen, Richard. (1857). *Key to the Geology of the Globe*. Nashville, Tennesee: Stevenson & Owen; New York: A.S. Barnes & Co.

Pantin, Carl F.A. (1968). *The Relations between the Sciences*. Cambridge: Cambridge University Press.

Patterson, J.R. (1958). Address by Professor S.K. Runcorn on palaeomagnetism. In Raasch (1958a), pp. 4–8.

Penck, Albrecht. (1921). Wegeners hypothese der kontinentalen verschiebungen. *Zeitschrift der Gesellschaft für Erdkunde*, no. 3/4, 110–20.

Pepper, John H. (1861). *The Playbook of Metals*. London: Routledge, Warne, and Routledge.

Pestana, Harold R. (1979). Comment and reply on 'Acceptance of Plate Tectonic Theory by Geologists'. *Geology*, **7**, 163–4.

Phinney, Robert A., ed. (1968a). *The History of the Earth's Crust: A Symposium*. Princeton: Princeton University Press.

Phinney, R.A. (1968b). Introduction. In Phinney (1968a), pp. 3–13.

Pickering, Andrew (1984). *Constructing Quarks: A Sociological History of Particle Physics*. Edinburgh: Edinburgh University Press.

Pickering, William H. (1907). The place of origin of the moon: The volcanic problem. *Journal of Geology* (Chicago), **15**, 23–38.

Pitman, W.C. & Heirtzler, J.P. (1966). Magnetic anomalies over the Pacific–Antarctic Ridge. *Science*, **154**, 1164–71.

Poincaré, Henri. (1905). *Science and Hypothesis*. London.

Pyne, Stephen J. (1978). Methodologies for geology: G.K. Gilbert and T.C. Chamberlin. *Isis*, **69**, 413–24.

Raasch, Gilbert O., ed. (1958a). Polar wandering and continental drift, a symposium. *Journal of the Alberta Society of Petroleum Geologists*, **6**, 139–78. Issued as separate publication 1960; references are to re-issue.

Raasch, G.O. (1958b). Editorial forward. In Raasch (1958a), p. 3.

Raff, Arthur D. & Mason, R.G. (1961). Magnetic survey off the west coast of North America, 40° N. Latitude to 52° N. Latitude. *Bulletin of the Geological Society of America*, **72**, 1267–70.

Ramsay, William L. & Burckley, R.A. (1965). *Modern Earth Science*, 2nd edn. New York: Holt, Rinehart, Winston.

Rastall, Robert H. (1929). On continental drift and cognate subjects. *Geological Magazine*, **66**, 447–56.

Rastall, R.H. (1941). *Lake and Rastall's Textbook of Geology*, 5th edn. London: Edward Arnold.

Raup, David M. (1986). *The Nemesis Affair: A Story of the Death of Dinosaurs and the Ways of Science*. New York: W.W. Norton.

Read, Herbert H. (1949). *Geology: An Introduction to Earth History*. London: Oxford University Press.

Read, H.H. & Watson, J. (1962). *Introduction to Geology Vol. I: Principles*. London: MacMillan.

Read, H.H. & Watson, J. (1966). *Beginning Geology*. New York: St Martins Press.

Reid, Harry F. (1922). Drift of the earth's crust and displacement of the pole. *The Geographical Review*, **22**, 672–4.

Rhodes, Frank H.T. (1972). *Geology*. A Golden Science Guide. New York: Golden Press.

Ringer, Fritz K. (1969). *The Decline of the German Mandarins: The German Academic Community 1890–1933*. Cambridge, Mass.: Harvard University Press.

Roubault, Marcel. (1949). *La Genèse des Montagnes*. Paris: Presses Universitaires de France.

Rudwick, Martin J.S. (1963). The foundation of the Geological Society of London: Its scheme for cooperative research and its struggle for independence. *British Journal for the History of Science*, **1**, 325–55.

Rudwick, M.J.S. (1970). The strategy of Lyell's *Principles of Geology*. *Isis*, **61**, 5–33.

Rudwick, M.J.S. (1972). *The Meaning of Fossils: Episodes in the History of Palaeontology*. London: Macdonald.

Rudwick, M.J.S. (1976) The emergence of a visual language for geological science 1760–1840. *History of Science*, **14**, 149–95.

Rudwick, M.J.S. (1982a). Cognitive styles in geology. In *Essays in the Sociology of Perception*, ed. Mary Douglas, pp. 219–41. London: Routledge & Kegan Paul.

Rudwick, M.J.S. (1982b). Charles Darwin in London: The integration of public and private science. *Isis*, **73**, 186–206.

Rudwick, M.J.S. (1985). *The Great Devonian Controversy: The Shaping of Scientific Knowledge among Gentlemanly Specialists*. Chicago: University of Chicago Press.

Runcorn, Stanley K. (1954). The earth's core. *Transactions of the American Geophysical Union*, **35**, 49–63.

Runcorn, S.K. (1955a). The permanent magnetization of rocks. *Endeavour*, **14**, 152–9.

Runcorn, S.K. (1955b). Rock magnetism – geophysical aspects. *Advances in Physics*, **4**, 244–91.

Runcorn, S.K. (1955c). Palaeomagnetism of sediments from the Colorado plateau. *Nature*, **176**, 505–6.

Runcorn, S.K. (1956a). Paleomagnetic survey in Arizona and Utah: Preliminary results. *Bulletin of the Geological Society of America*, **67**, 301–16.

Runcorn, S.K. (1956b). Paleomagnetic comparisons between Europe and North America. *Proceedings of the Geological Association of Canada*, **8**, 77–85.

Runcorn, S.K. (1956c). Palaeomagnetism, polar wandering and continental drift. *Geologie en Mijnbouw*, n.s., **18**, 253–6.

Runcorn, S.K. (1959). Rock magnetism. *Science*, **129**, 1002–11.

Runcorn, S.K. (1962a). Towards a theory of continental drift. *Nature*, **193**, 311–14.

Runcorn, S.K., ed. (1962b). *Continental Drift*. New York: Academic Press.

Runcorn, S.K. (1963). Palaeomagnetic methods of investigating polar wandering and continental drift. In Munyan (1963a), pp. 47–54.

Runcorn, S.K., Benson, A.C., Moore, A.F. & Griffiths, D.H. (1951). Measurements of the variation with depth of the main geomagnetic field. *Philosophical Transactions*, A**244**, 113–51.

Runcorn, S.K. (1975). (Contribution to) Memorial meeting for Lord Blackett. *Notes and Records of the Royal Society of London*, **29**, 156–8.

Russo, Philibert (1930). Recherches sur les déplacements tectoniques des aires continentales. *Mémoires de la Société des sciences naturelles du Maroc*, **25**.

Sapp, Jan. (1983). The struggle for authority in the field of heredity, 1900–1932: New perspectives on the rise of genetics. *Journal of the History of Biology*, **16**, 311–42.

Sapp, J. (1987). *Beyond the Gene: Cytoplasmic Inheritance and the Struggle for Authority in Genetics*. New York: Oxford University Press.

Saxena, M.N., Gupta, V.J., Meyerhoff, A.A. & Archbold, N.W. (1985). Tectonic and spatial relations between India and Asia since proterozoic time. In *Contributions to Himalayan Geology*, vol. 3, ed. V.J. Gupta *et al.*, pp. 187–207. Hindustan Publishing Corporation.

Sayles, Robert W. (1914). The Squantum tillite. *Bulletin of the Museum of Comparative Zoology* (Harvard), 56, no. 2, (Geological Series, 10), 141–75.

Scheidegger, Adrian E. (1958a). *Principles of Geodynamics*. Berlin: Springer–Verlag.

Scheidegger, A.E. (1958b). Possible causes of continental drift. In Raasch (1958a), pp. 26–30.

Scheidegger, A.E. (1963). *Principles of Geodynamics*, 2nd edn. Berlin: Springer–Verlag.

Scherman, Elizabeth, Howell, D.G. & D.L. Jones. (1984). The origin of allochthonous terranes: Perspectives on the growth and shaping of continents. *Annual Review of Earth and Planetary Science*, **12**, 107–31.

Schlee, Susan (1973). *The Edge of an Unfamiliar World: A History of Oceanography*. New York: E.P. Dutton & Co., Inc.

Schroeder-Gudehus, Brigitte. (1978). *Les Scientifiques et la paix: La communauté scientifique internationale au cours des années 20*. Montréal: Les Presses de l'Université de Montréal.

Schuchert, Charles (1928). The hypothesis of continental displacement. In van der Gracht (1928a), pp. 104–44.

Schuchert, C. (1932). Gondwana land bridges. *Bulletin of the Geological Society of America*, **43**, 875–916.

Schulz, Bruno (1921). Die Alfred Wegenersche theorie der entstehung der kontinente und ozeane. *Die Naturwissenschaften*, **15**, 241–50.

Schuster, John & Yeo, R., eds. (1986). *The Politics and Rhetoric of Scientific Method: Historical Studies* (Australasian Studies in History and Philosophy of Science, 4). Dordrecht: D. Reidel.

Schwarzbach, Martin (1985t). *Wegener 1880–1930: Le Père de la dérive des continents*. trans. E. Buffetaut of (Stuttgart: 1980), Paris: Librairie classique Eugène Belin.

Schweydar, Wilhelm (1921). Bemerkungen zu Wegeners hypothese der verschiebung der kontinente. *Zeitschrift der Gesellschaft für Erdkunde*, nr. 3/4, 120–25.

Sears, Mary, ed. (1959). *International Oceanographic Congress 1959 Preprints*. Washington: American Association for the Advancement of Science.

Sears, M., ed. (1961). *Oceanography*. Invited papers presented at the 1959 International Oceanographic Congress. Washington: American Association for the Advancement of Science.

Sears, M. & Merriman, D., eds. (1980). *Oceanography: The past*. Proceedings of the 3rd International Congress on the History of Oceanography. New York: Springer–Verlag.

Secord, James A. (1986). *Controversy in Victorian Geology: The Cambrian–Silurian Dispute.* Princeton: Princeton University Press.

Semper, Max. (1917). Was ist eine arbeitshypothese?. *Zentralblatt für Mineralogie, Geologie, und Paläontologie,* 146–63.

Seward, Albert C. (1933). *Plant Life Through the Ages,* 2nd edn. Cambridge: Cambridge University Press.

Seward, A.C. (1943). Geology for Everyman. Cambridge: Cambridge University Press.

Sewell, R.B. Seymour & Wiseman, J.D.H. (1938). The relief of the ocean floor in the Southern Hemisphere. *Compte rendu du Congrès International de Géographie (Amsterdam),* 2, 135–40.

Shapin, Steven (1981). Of gods and kings: Natural philosophy and politics in the Leibniz–Clarke disputes. *Isis,* 72, 187–251.

Shapin, S. (1982). History of science and its sociological reconstructions. *History of Science,* 20, 157–211.

Shepard, Francis P. (1948). *Submarine Geology.* New York: Harper & Brothers.

Simpson, George G. (1943). Mammals and the nature of continents. *American Journal of Science,* 241, 1–31.

Singewald, Joseph (1928). Discussion of Wegener theory. In van der Gracht (1928a), pp. 189–93.

Smit Sibinga, Gerard L. (1927). Wegener's theorie en het ontstaan van den Oostelijken O. I. Archipel. *Tijdschrift van het Koninklijk Nederlandsch Aardijkskundig Genootschap,* (ser. 2) 44, 581–98.

Snider-Pellegrini, Antonio (1858). *La Création et ses Mystères dévoilés.* Paris: Franck et Dentu.

Soergel, Wolfgang (1916). Die atlantische 'Spalte'. Kritische bemerkungen zu A. Wegeners theorie von der kontinentalverschiebung. *Zeitschrift der Deutschen Geologischen Gesellschaft,* 68, 200–39.

Speis, Fred N. (1980). Some origins and perspectives in deep-ocean instrumentation development. In Sears & Merriman (1980), pp. 226–39.

Staub, Rudolf (1924). Der bau den Alpen. *Beitraege zur Geologischen Karte der Schweiz,* Neue Folge, no. 52.

Steers, James A. (1932). *The Unstable Earth.* London: Methuen.

Stehli, Francis G. (1970). A test of the earth's magnetic field during permian time. *Journal of Geophysical Research,* 75, 3325–42.

Steiner, J. (1977). An expanding earth on the basis of sea-floor spreading and subduction rates. *Geology,* 5, 313–18.

Stewart, John A. (1986). Drifting continents and colliding interests: A quantitative application of the interests perspective. *Social Studies of Science,* 16, 261–79.

Stille, Hans (1939). Kordillerisch-antlantische wechselbeziehungen. *Geologische Rundschau,* 30, 315–42.

Strahler, Arthur N. (1981). *Physical Geology.* New York: Harper & Row.

Sullivan, Walter (1974). *Continents in Motion.* New York: McGraw–Hill.

Sykes, Lynn R. (1968). Seismological evidence for transform faults, sea floor spreading, and continental drift. In Phinney (1968a), pp. 120–50.

Takeuchi, Hitoshi, Uyeda, S. & Kanamori, H. (1967). *Debate About the Earth.* revised edition, San Francisco: Freeman, Cooper & Co.

Tarling, D.H. & Tarling, P.M. (1971). *Continental Drift.* London: Bell; New York: Doubleday

(1971) and Harmondsworth: Penguin (1972).

Taylor, Frank B. (1910). Bearing of the tertiary mountain-belt on the origin of the earth's plan. *Bulletin of the Geological society of America*, 21, 179–226.

Taylor, F.B. (1928). Sliding continents and tidal and rotational forces. In van der Gracht (1928a), pp. 158–77.

Termier, H. & Termier, G. (1952). *Histoire géologique de la biosphère de la vie et les sédiments dans les géographies successives*. Paris.

Termier, H. & Termier, G. (1958). *The Geological Drama*. London: Hutchinson.

Theobald, Nicolas & Gama, A. (1956). *Géologie générale et pétrographie*. Paris: G. Doin & Cie.

Theobald, N. & Gama, A. (1961). *Géologie générale et pétrographie*, 2nd edn. Paris: G. Doin & Cie.

Thompson, Paul (1978). *The Voice of the Past: Oral History*. Oxford: Oxford University Press.

Tiercy, G. (1945). Réflexions sur la theorie des translations continentales. *Verhandlungen der Naturforschenden Gesellschaft in Basel*, 56, 531–43.

Torreson, O.W., Murphy, T. & Graham, J.W. (1949). Magnetic polarization of sedimentary rocks and the earth's magnetic history. *Journal of Geophysical Research*, 54, 111–29.

Umbgrove, Johannes H.F. (1942). *The Pulse of the Earth*. The Hague: Martinus Nijhoff.

Umbgrove, J.H.F. (1947). *The Pulse of the Earth*. 2nd ed., The Hague: Martinus Nijhoff.

Umbgrove, J.H.F., Good, R.D'O., Hinton, H.E. *et al.* (1951). The theory of continental drift (Symposium at joint meeting of sections of Geology, Zoology, Geography and Botany at the 1950 Birmingham meeting of BAAS). *The Advancement of Science*, 8, 67–88.

Vallance, T.G. (1981). The fuss about coal: Troubled relations between palaeobotany and geology. In *Plants and Man in Australia*, ed. D.J. & S.G.M. Carr, pp. 136–76. Sydney: Academic Press.

van Bemmelen, Rein W. (1939). Das permanenzproblem nach der undationstheorie. *Geologische Rundschau*, 30, 10–20.

van Bemmelen, R.W. (1954). *Mountain Building: A Study Primarily Based on Indonesia, Region of the World's Most Active Crustal Deformations*. The Hague: Martinus Nijhoff.

van der Gracht, W.A.J.M. van Waterschoot, ed. (1928a). *Theory of Continental Drift: A Symposium*. Tulsa: American Association of Petroleum Geologists.

van der Gracht, W.A.J.M. van W. (1928b). The problem of continental drift. In van der Gracht (1928a), pp. 1–75.

van der Gracht, W.A.J.M. van W. (1928c). Remarks regarding the papers offered by the other contributors to the symposium. In van der Gracht (1928a), pp. 197–226.

Vening-Meinesz, Felix A. (1934a). *Gravity Expeditions at Sea 1923–1932*. vol. 2, Delft: Waltman.

Vening-Meinesz, F.A. (1934b). Gravity and the hypothesis of convection currents in the earth. *Proceedings of Nederlandse akademie van wetenschaffen*, 37, 37–45.

Vening-Meinesz, F.A. (1962). Thermal convection in the earth's mantle. In Runcorn (1962b), pp. 145–76.

Verhoogen, John. (1985). North American paleomagnetism and geology. In Drake & Jordan (1985), 401–07.

Verhoogen, J., Turner, F.J., Weiss, L.E. *et al.* (1970). *The Earth: An Introduction to Physical Geology*. New York: Holt, Rinehart, Winston.

Vine, F.J. (1966). Spreading of the ocean floor: New evidence. *Science*, 154, 1405–15.

Vine, F.J. (1968). Magnetic anomalies associated with mid-ocean ridges. In Phinney (1968a), pp. 73–89.

Vine, F.J. (1977). The continental drift debate. *Nature*, **226**, 19–22.

Vine, F.J., Cox, A., Wilson, J.T. *et al.* (1977). Plate tectonics 1. 10 years after. *EOS*, **58**, 366–7.

Vine, F.J. & Matthews, D.H. (1963). Magnetic anomalies over oceanic ridges. *Nature*, **199**, 947–9.

Vine, F.J. & Wilson, J.T. (1965). Magnetic anomalies over a young ocean ridge off Vancouver Island. *Science*, **150**, 485–9.

Voisey, Alan H. (1958). Some comments on the hypothesis of continental drift. In Carey (1958a), pp. 162–72.

von Engeln, Oscar D. & Caster, K.E. (1952). *Geology*. New York: McGraw–Hill.

von Ubisch, Leopold. (1921). Wegeners kontinentalverschiebungstheorie und die tiergeographie. *Verhandlungen der Physikalischmedizinischen Gesellschaft zu Würzburg*, **46**, 57–69.

Wade, Arthur. (1927). The search for oil in New Guinea. *Bulletin of the American Association of Petroleum Geologists*, **11**, 157–76.

Wade, A. (1934). The distribution of oilfields from the view-point of the theory of continental spreading. *Proceedings of the World Petroleum Congress (London, 1933)*, 73–7.

Wade, A. (1935). New theory of continental spreading. *Bulletin of the American Association of Petroleum Geologists*, **19**, 1806–18.

Wade, A. (1941). The geology of the Antarctic continent and its relationship to neighbouring land areas. *Proceedings of the Royal Society of Queensland*, **52**, no. 1, 24–35.

Wallace, A.R. (1876). *The Geographical Distribution of Animals*. 2 vols. London.

Ward, L. Keith. (1927). The plan of the earth and its origin. *Proceedings of the Royal Geographical Society, South Australian Branch* (1926–1927). 1–29.

Wegener, Alfred L. (1912). Die entstehung der kontinente. *Geologische Rundschau*, **3**, 276–92.

Wegener, A.L. (1915). *Die Entstehung der Kontinente und Ozeane*. Brunswick: Friederich Vieweg & Sohn.

Wegener, A.L. (1921). Die theorie der kontinentalverschiebungen. *Zeitschrift der Gesellschaft für Erdkunde*, no. 3/4, 89–103; 125–30.

Wegener, A.L. (1922). The origin of continents and oceans. *Discovery*, **3**, no. 29, 114–18.

Wegener, A.L. (1924t). *The Origin of Continents and Oceans*. Trans. J.G.A. Skerl of 3rd German ed. London: Methuen & Co. Ltd.

Wegener, A.L. (1927). Die geophysikalischen Grundlagen der theorie der kontinentenverschiebung. *Scientia*, **61**, 103–16.

Wegener, A.L. (1928). Two notes concerning my theory of continental drift. In van der Gracht (1928a), pp. 97–103.

Wegener, A.L. (1929t). *The Origin of Continents and Oceans*. trans. J. Biram of 4th German edn., rept. New York, 1966: Dover.

Weiss, Frederick E. (1922). The displacement of continents: a new theory. *Manchester Guardian*, 16 March, p. 6.

Wernicke, Brian. (1981). Low-angle normal faults in the Basin and Range Province: Nappe tectonics in an extending orogen. *Nature*, **291**, 645–8.

Wertenbaker, William. (1974). *The Floor of the Sea: Maurice Ewing and the Search to Understand the Earth*. Boston: Little, Brown.

White, David. (1928). Discussion of floating continents. In van der Gracht (1928a), pp. 187–8.

White, Mary E. (1986). *The Greening of Gondwana*. Frenchs Forest, New South Wales: Reed Books.

Whitley, Richard. (1976). Umbrella and polytheistic scientific disciplines and their elites. *Social Studies of Science*, **6**, 471–97.

Whitmore, T.C. (1984). *Tropical rain forests of the Far East*, 2nd edn. Oxford: Oxford University Press.

Williams, L. Pearce. (1968). *Relativity Theory: Its Origin and Impact on Modern Thought*. New York: John Wiley.

Willis, Bailey. (1928). Continental drift. In van der Gracht (1928a), pp. 76–82.

Willis, B. (1932). Isthmian links. *Bulletin of the Geological Society of America*, **43**, 917–52.

Willis, B. (1944). Continental drift: Ein märchen. *American Journal of Science*, **242**, 509–13.

Wilson, D.W.R. (1958). Orocline concept and continental drift. In Raasch (1958), pp. 30–4.

Wilson, J. Tuzo. (1950). An analysis of the pattern and possible cause of young mountain ranges and island arcs. *Proceedings of the Geological Society of Canada*, **3**, 141–66.

Wilson, J.T. (1951). On the growth of continents. *Proceedings of the Royal Society of Tasmania for 1950*, 85–111.

Wilson, J.T. (1957). The crust. In *The Planet Earth*, ed. D.R. Bates, pp. 48–73. New York: Pergamon Press.

Wilson, J.T. (1960). Some consequences of expansion of the earth. *Nature*, **185**, 880–82.

Wilson, J.T. (1961). Letter to editor. *Nature*, **192**, 125–8.

Wilson, J.T. (1963a). Evidence from islands on the spreading of ocean floors. *Nature*, **197**, 536–8.

Wilson, J.T. (1963b). Hypothesis of earth's behaviour. *Nature*, **198**, 925–9.

Wilson, J.T. (1963c). Continental drift. *Scientific American*, **208**, no. 4, 83–100.

Wilson, J.T. (1964). Movement of continents. Address presented at Symposium on the Upper Mantle Project, XIII General Assembly of IUGG, Berkeley. *ICSU Review World Science*, **6**, 84–91.

Wilson, J.T. (1965a). A new class of faults and their bearing on continental drift. *Nature*, **207**, 343–7.

Wilson, J.T. (1965b). Transform faults, oceanic ridges, and magnetic anomalies southwest of Vancouver Island. *Science*, **150**, 482–5.

Wilson, J.T. (1966a). Some rules for continental drift. In Garland (1966), pp. 3–17.

Wilson, J.T. (1966b). Did the Atlantic close and then re-open?. *Nature*, **211**, 676–81.

Wilson, J.T. (1967). Some implications of new ideas on ocean-floor spreading upon the geology of the Appalachians. In *Appalachian Tectonics*, ed. T.H. Clark, pp. 94–7, Royal Society of Canada Special Publications No. 10. Toronto: University of Toronto Press.

Wilson, J.T. (1968a). Revolution in earth science. *Geotimes*, Dec., 10–16.

Wilson, J.T. (1968b). Reply to V. Beloussov. *Geotimes*, Dec., 21–2.

Wilson, J.T. (1968c). Static or mobile earth: The current scientific revolution. *Proceedings of the American Philosophical Society*, **112**, 309–20.

Wilson, J.T., ed. (1972). *Continents Adrift: Readings from Scientific American*. San Francisco: W.H. Freeman.

Wilson, J.T. (1976). *Continents Adrift and Continents Aground: Readings from Scientific American*. San Francisco: W.H. Freeman.

Wilson, J.T. (1985). Development of ideas about the Canadian Shield: A personal account. In Drake & Jordan (1985), 143–50.

Wing Easton, Nicolaas. (1921a). On some extensions of Wegener's hypothesis and their bearing upon the meaning of the terms geosynclines and isostasy. *Verhandelingen Geologisch Mijnbouwkundig Genootschap voor Nederland an Kolonien*, **5**, (Geol. ser.), 113–33.

Wing Easton, N. (1921b). Het ontstaan van den maleischen Archipel, bezien in het licht van Wegener's hypothesen. *Tidjschrift van het Koninklijk Nederlandsch Aardrijkskundig Genootschap*, **38**, 484–512.

Wood, Robert M. (1980). Geology versus dogma: The Russian rift. *New Scientist*, 12 June, 234–7.

Wood, R.M. (1985). *The Dark Side of the Earth*. London: George Allen & Unwin.

Wooldridge, S.W. & Morgan, R.S. (1937). *The Physical Basis of Geography: An Outline of Geomorphology*. London: Longman, Green & Co.

Wooldridge, S.W. & Morgan, R.S. (1959). *An Outline of Geomorphology: The Physical Basis of Geography*, 2nd edn (of 1937), London: Longman, Green & Co.

Wright, C.S. & Priestley, R.E. (1922). *British (Terra Nova) Antarctic Expedition 1910–1913: Glaciology*. London: Harrison & Sons.

Wright, William B. *et al.* (1923). The Wegener hypothesis: Discussion at the British Association, Hull. *Nature*, **111**, 30–31.

Wright, W.B., Dyson, F., Woodward Smith, A. *et al.* (1935). Discussion upon the hypothesis of continental drift: Joint meeting of the Geological Society and the Royal Astronomical Society. *Quarterly Journal of the Geological Society*, **91**, v–xi (Proceedings of the Meeting of 23 January 1935).

Wyllie, P.J. (1976). *The Way The Earth Works*. New York: John Wiley & Sons.

Yearley, Steven. (1981). Textual persuasion: The role of social accounting in the construction of scientific arguments. *Philosophy of the Social Sciences*, **11**, 409–35.

Zeuner, Frederick E. (1946). *Dating the Past: An Introduction to Geochronology*. London: Methuen.

Zeuner, F.E. (1950). *Dating the Past: An Introduction to Geochronology*, 2nd edn. London: Methuen.

Zeuner, F.E. (1952). *Dating the Past: An Introduction to Geochronology*, 3rd edn. London: Methuen.

Zeuner, F.E. (1958). *Dating the Past: An Introduction to Geochronoloy*, 4th edn. London: Methuen.

Zumberge, James H. (1963). *Elements of Geology*, 2nd edn. New York: John Wiley.

Index

---

actualism in geology, 19
aesthenosphere, 235–6
Ager, D.V., 242–3
Ahmad, F., 227, n. 18
Airy, G. B., 22
Alps, 1, 22–3, 26, 31, 42, 57, 61, 75, 111, 189, 237
  importance to European geologists, 24–6
American Association of Petroleum Geologists
  1926 symposium on Drift, 64–9, 71, 118, 123
American Geophysical Union, 162, 173, 215, 232, 234
Andes, 26, 42, 237
Andrews, E.C., 84–5, 92
Andrée, K., 57–8
anomalies, magnetic, see magnetic anomalies
anomaly, 47, 93, 132, 160, 223
  Kuhn & Laudan on, 6
Antarctica, 63, 206
  and Drift, 86–90, 123
Appalachians, 21–2, 42, 56, 75, 91, 111, 189, 230, 236–7, 258
Argand, E., 61, 77, n. 14
Arldt, T., 58
Australian fauna and Drift, 62, 81, 86–9, 103
autonomy of science, 8, 182–5

Bailey, E.B., 90–2, 236
Baker, H.B., 28, 30
bandwagons, 256–9
Barnes, B., 9
Barrell, J., 134, n. 6
Beloussov, V.V., 246, 262–3
Berry, E., 65–7, 71
Bertrand, M., 26

Betim, A., 83
biogeography, 22, 25–6, 86–9, 156, 247–8
  evidence for Drift inconclusive, 102–3, 151, 158–9
Bird, J., 237, 249
black boxes, 140, 143, 157, 165–6
Blackett, P.M.S., 140–68 passim, 183, 204
block-faulting, 181
Bloor, D., 9
Bourdieu, P., on technical and social interests in science, 10–11, 16, n. 21
Brannigan, A., 51, 227, n. 20
Brooks, C.E.P., 105, 134, n. 8
Bucher, W., 123, 135, n. 24
Bull, A.J., 112
Bullard, E.C., 142, 157, 161, 173, 188–9, 204–6, 226, n. 8, 230–1, 233, 256
  on Drift, 202–4
  on geomagnetism, 140, 164–5, 169, n. 23
  on geophysics versus geology, 184
  on resistance of scientists to change, 10
Bullard Fit, 203–5, 230, 260
Bullen, K., 243
Burmeister, F., 59–60

Caledonians, 56, 90–1, 230, 236
Carey, S.W., 92, 133, 143, 151–3, 168, n. 14, 192–5, 200, 204, 224, 253–4, 260, 262, 263, n. 5
  and Drift, 85, see Expansionism
Carlsberg Ridge, 206, 217
Carnegie group, see Department of Terrestrial Magnetism
Carozzi, A.V., 77, n. 12
Caster, K., 83, 92, 122
catastrophism, 17–19, 23–5, 29–30, 32
  and Drift, 66–7, 74

Challenger expedition, 171–2
Chamberlin, R.T., 1, 64–5, 67, 71
Chamberlin, T.C., 20–1, 30, 35, n. 7, 64,
    135, n. 20, 238
  planetesimal theory, 21
  on climatic cycles, 101
  see method of multiple hypotheses
Chaney, R.W., 153
Chelikowsky, J., 120
Chorley, R.J., 32
Cole, G., 20
Coleman, A.P., 101, 134, n. 2
Collet, L., 77, n. 13
Collinson, D.W., 157
competition among scientists, *see* struggle,
    interests
Comte, A., 130
continental drift *see* Drift
continental spreading, *see* Gutenberg, B.
continents
  fit of, 28–9, 39, 41–3, 56, 63, 65, 83,
    121–2, 151, 203, 206, 248, *see*
    Bullard Fit
  growth of, 21, 189–92, 201, 248
  stability of, *see* Permanentism
Contractionism, 1, 5, 19, 39, 40–1, 44,
    47, 51, 67, 72, 74, 82, 94, 110, 171,
    183, 197, 252
  and ocean basins, 189–92
  and radioactive heat, 104–5, 111–15
  summary of, 23–8
  improved versions of, 104–7
convection currents, 40, 60, 108, 111–17,
    120–2, 150, 158, 180–1, 192–204,
    213, 226, n. 10, 244, 248
  and geomagnetism, 140, 143–4, 158
  objections to, 116
coriolis force, 144
correspondences of geological features, *see*
    pattern matching, continents, fit of
Cotton, L.A., 85, 92
Cox, A., 155–7, 161, 163, 169, n. 19,
    213, 215, 217, 229, 231
Creer, K., 168, n. 11
crisis state, 6, 33–4, 47, 136, n. 28
Croneis, C., 121–2
crustal shortening, 26, 35, n. 8, 111
Cuvier, G., 23, 35, n. 2

Dacqué, E., 57
Dalrymple, G.B., 214, 229, 231
Daly, R.A., 109–10, 238

Dana, J.D., 20
Danish Geodetic Institute, 127
Darwinian evolution, 22, 74, 101–2
dating, radiometric, 111, 164, 213–4,
    227, n. 13, n. 14, 230
  reversal time scale, 155, 164, 211–14,
    230
David, T.W.E., 85
De la Beche, H., 23
Department of Terrestrial Magnetism
    (Carnegie Institution), 139–43, 161,
    167, n. 3
Deutsch, E.R., 152–3
Dewey, J., 230, 236, 243, 245, 249
diastrophism, 85, 120
Dicke, R.H., 195, 223–4, 226, n. 3, n. 4
Diener, C., 57–9, 77, n. 4
Dietz, R., 199–201, 215, 217, 223, 225
Dirac, P.A.M., 195
disagreement in science, resolution of,
    69–70, 76, 94–7, 127, 221
discovery, 51–2, 54, n. 11, *see* priority
Doell, R.R., 155–7, 161, 163, 169, n. 19,
    213, 215, 217, 229, 231
Dolphin Rise, 171–2
Drift, 1, 14, n. 1, 171, 183, 213, 225, 236
  and economic concerns, 86, 152
  and geophysics, 107
  and palaeomagnetism, 138–69
    a symbiotic relationship, 159–60
  as a holistic theory, 57, 63, 107
  as a research program, 69, 125–6, 133,
    255–6
  as a working hypothesis, 40, 58–60, 71,
    86, 88, 110, 119–20, 122–3, 202, *see*
    hypothesis, Wegener, methods
  becomes orthodoxy, 241–7, 253
  causes of, 28–31, 40, 108–9
  objections, 46, 60, 85, 109–12,
    121–2, 127, *see* convection currents
  Daly's version, 109–10
  du Toit's version, 82–3
  Gutenberg's version, 97, n.6, 187, n.9
  Holme's version, 111–16, 196, 225,
    198–9, 225, *see* Holmes, A.
  in textbooks, 63–4, 121–3, 240–7,
    250–1
    Australia, 83–9
    Brazil, 83
    Britain, 61–4
    France, 61, 244
    Germany and Austria, 57–60

Holland, 60–1
North America, 64–9
Switzerland, 61
plate tectonics version, *see* plate tectonics
pre-Wegener, 28–32, 46
relation to plate tectonics, 254–6
seafloor-spreading versions, (Hess, Dietz, Vine–Matthews, Vine–Wilson) *see* seafloor spreading
symposia on, 59, 61, 63–9, 77, n. 11, 122–3, 151–3, 204–6, 229–31, 237
Wegener's version, 40–6, 178, 182, 246
    forces, 129
    genesis, 39–40, 47–53
    limited in time, 78, n.20
    objections to, summarized, 55–9
    see Wegener, A.L., Taylor, F.B., Greenland, drift of
    social roots of, 49–50
du Toit, A.L., 1, 57, 82–3, 78, n. 20, 90, 92, 109–11, 119, 123, 150, 274
Durham, J.W., 153
Dutton, C., 22

Earth Science, *see* geology and Earth Science
earthquakes, 175
    deep focus, 178, 180, 189
    mechanisms, 233
    shallow focus, 178
East African Rift Valley, 178, 209, 226, n. 10
East Pacific Rise, 176, 197, 208, 209–10, 212, 215, 217
Edge, D., 169, n. 23
Egyed, L., 194–5, 223
Elie de Beaumont, L., 23
Elsasser, W.M., 139, 165
Eltanin-19 profile, 215–17, 220, 223, 229, 232, 260
empiricism, 18–19, 67, 80, 84, 118, 121, 162, 171–5, 184, 193, 202–3, 245
Eötvös, R., 53, n. 2
error, accounting for, 58–9, 62, 73, 256
Euler's theorem, 204, 233
evaporites, 97, n. 4
Ewing, M., 123, 164, 169, n. 18, 172–4, 176, 178, 184, 193, 224, 230, 234, 236
Expansionism, 5, 29, 171, 183, 193–5, 200–2, 235, 247, 250, 252–3, 263

and Drift, 151, 195, 223–4, 227, n. 15, n. 16, 252
and gravitation, 195, *see* Dicke, R.H., Jordan, P.
and seafloor spreading, 201, 227, n. 16, *see* Heezen, B.C.
Carey's version, 151, 168, n. 14, *see* Carey, S.W.
objections, 194, 198, 223, 252–3

fact–theory distinction, 35, n. 6, 53, 136, n. 26, 220–1, 270
facts, novel, 5, 99, n. 15, 128, 270
    Kuhn on, 162–3
faulting, extensional, 249
Feyerabend, P., 3, 15, n. 10, 96, 99, n. 16
Field, R., 164, 172–3, 186, n. 2
Fisher, O., 28–30, 66
Fisher, R.A., 104, 161, 168, n. 6
Frankel, H., 135, n. 16, 226, n. 6, 278, n. 2

Geikie, A., 17, 32
Geological Society
    of America, 101, 119, 162, 214, 229, 231–2
    of Berlin, 59
    of France, 61
    of London, 18, 112, 162, 188, 202–3, 243
Geological Survey
    Great Britain, 23, 90
    India, 77, n. 4, 90
    US, 66, 155
geology
    and Earth Science, 238–40, 244
    and hierarchy of sciences, 130–1
    and mathematics, 104–7, 134, n. 7, 204, 233–5
    and physics, 104–7, 109–10, 112, 115, 130–1, 162, 190, 195, 203, 238–9
    and structure of universities, 35, n. 7, 48–9
    as a descriptive science, 20, 23, 35, n. 5, 76, 93, 130–1
    as an explanatory science, 24, 35, n. 5, 48, 76, 93, 238–41, 250–1, 262, 271
    as an umbrella discipline, 92–3
    as collection of specialties, 72, 238
    as unified science, 72, 238, 240–1, 245, 247, 250–1
    changing images of, 2, 239–41

geology (*cont.*)
    prediction and retrodiction in, 46, 77,
        n. 6, 130
    versus geophysics, 184, 188, 203, 264,
        n. 15
geology and geophysics, marine, *see* marine
    geology and geophysics
geomagnetism
    axial dipole field, 147, 157, 168, n. 6
    fundamental theory of, 140, 142–3,
        145, 147
    internal theory of, 139–40, 142–5,
        164–5, 206–7, 213
geosyncline, 21–2, 180, 189–90, 192, 248
    248
Gicouate, M., 83
Gill, W.D., 203
Girdler, R., 226, n. 10, 227, n. 11
glaciation, great southern, 56, 63, 67, 82,
    90, 102, 119, 122, 134, n. 8, 135,
    n. 16, 143, 147, 160
Glen, W., 167, n. 4, 227, n. 12, 278, n. 7
Glossopteris, 44, 56, 63, 78, n. 18, 82, 89
Goddard Space Institute, 229
Goguel, J., 244
Gold, T., 168, n. 9
Gondwana, 26, 82–3, 87, 90, 114, 116,
    192, 206
    Symposium, 232
Gorshkov, G., 246
Grabau, A., 136, n. 27
Graham, J.W., 139–41, 147, 159, 161,
    167, n. 3
Graham's tests, 140–1, 165
Graham, W.T., 150
gravity anomalies, 116
Green, W.L., 36, n. 11
Greene, M.T., 28, 35, n. 3, 54, n. 5, n. 8,
    276
Greenland, drift of, 38, 44–6, 56–9, 64,
    67–8, 78, n. 24, 107, 120, 131
Gregory, J.W., 68, 72, 174
Griggs, D., 116–17
Gutenberg, B., 86, 110, 117, 123, 180,
    187, n. 9, 233, 262

Haldane, J.B.S., 104
Hales, A.L., 150
Hallam, A., 33
Harland, W.B., 263, n. 7
Harris, A., 249
Harrison, L., 87–9, 92

Heezen, B.C., 193–5, 200–1, 224
Heim, A., 77, n. 13
Heirtzler, J., 214–15, 217, 230
Hess, H.H., 116, 164, 171, 178, 186, n. 1,
    n. 2, 188–9, 195–201, 206, 208,
    210, 213, 215, 217, 223–5, 226,
    n. 3, 227, n. 21, 255, 267, *see*
    seafloor spreading
Hills, G.F.S., 136, n. 27
Himalayas, 21, 31, 42, 111, 147
Hinks, A., 72, 106–7
history
    participant, 185–6
    practitioners, 32–3
    views about science, 3–4, 13–14
    whiggish, 32, 36, n. 12
history of geology
    revisionist, 19–28, 33–4
    traditional, 17–18, 32–3
Hobbs, W.H., 136, n. 27
holarctic dispersion, 87, 101–3
Holland, T.H., 90, 92
Hollingsworth, S.E., 124
Holmes, A., 107, 109, 111–16, 143, 178,
    213, 225, 227, n. 21, 255, 262, *see*
    Drift, Holmes's version
Hopkins, E., 28–9
Hopkins, W., 134, n. 7
Hospers, J., 164, 168, n. 18
host-parasite technique, 87–8
Hurley, P., 230–1
Hutton, J., 17–18
Huxley's mathematical mill, 104, 106
hypothesis, method of, 20, 24, 53, 55, 62,
    70–1, 109, 162, *see* method

illustration, rôles in geology, 160, 236,
    260–2
incommensurability, 271
    Barnes & McKenzie on, 15n
    Feyerabend on, 75–6
    Kuhn on, 75
    Laudan on, 76
India, drift of, 147, 168, n. 8
inductivism, 18, 20, 23, 61–2, 65, 78,
    n. 16, n. 20, 106, 162
    and opposition to Drift, 71
instruments
    novel, *see* black boxes
    theory-laden, 165–6
interests
    and specialization, 92–7

broader social and geology, 9–10,
49–50, 273–4
internal, 10
social, 132–3, 222
social and cognitive, 8–12, 159–65,
184–5, 219, 254–5, 224–6, 259,
267, 269–70, 272–3, 275–6, *see*
struggle
International Geophysical year (IGY), 174,
186, n. 5, 192, 239, 246, 274
International Research Council, 77, n. 14
Irving, E., 146, 151, 168, n. 8, n. 10,
n. 11, 230
isostasy, 21–2, 27, 34, 41, 52, 55, 85, 87,
115
and landbridges, 40, 44, 58
isthmian links, 87, 101–3
IUGG, 139, 192, 246

Jaeger, J.C., 146, 168, n. 13
Jaramillo event, 214
Jaworski, E., 60
Jeffreys, H., 63, 109–15, 190, 234, 243,
251, 262–3
Johnson, E.A., 139–40, 159, 167, n. 3
JOIDES, 235–6
Joly, J., 27, 30, 64, 68, 79, n. 25, 107,
111
Jordan, P., 195, 223, 227, n. 17
Juan de Fuca Ridge, 179, 209–12,
214–15, 217, 229
Just, T., 123

Karroo, 82, 97, n. 4, 150
Kay, M., 230, 236–7
Keith, A., 66
Kelly, A.O., 256–7
Kelvin, 48
and age of earth, 104, 109
kimberlite, 42
King, L.C., 83, 123, 192
Kitts, D.B., 77n, 131, 136–7, n. 32, 265,
n. 26
Kossmat, F., 59
Köppen, W., 38, 40–1, 77, n. 2
Kubart, B., 60
Kuenen, P., 134, n. 12
Kuhn, T.S., 2, 8–9, 12, 54, n. 12, 69,
125–7, 160, 162–4, 267, 269–70,
272
model of science, 4
and specialization, 98, n. 13

and social analysis of science, 8
endorsed by some geologists, 4, 32–3,
117, 238, 241
gestalts, 262
history in textbooks, 32, 256
invention of research programs,
46–7
rationalist response to, 5–8
revolutionaries, 47
textbooks as shared beliefs, 120–1,
238
see crisis state, anomaly, progress,
paradigm, normal science
Kuhn–loss, 126, 271

Lakatos, I., 3, 7–8, 77, n. 5, 125–6, 128,
132, 136, n. 26, n. 28, 160, 165,
267, 269–70, 278, n. 6
see facts, novel, Research Programme
inventions of research programs, 47
model of science, 5
Lake, P., 62–3, 71, 73, 92
Laming, D.J.C., 152–3, 169, n. 16
Lamont–Doherty Geological Observatory,
123, 174, 184–5, 193, 208, 214–15,
217, 224, 229, 233–5, *see* Ewing, M.
landbridges, 26–7, 39, 43–4, 56, 61,
65, 81, 86, 187
and isostasy, 27, 100–4, 116, 134, n. 6
Landes, K., 189–90
Latour, B., 10–12, 97, n. 1, 165, 266,
n. 35
Laudan, L., 3, 8, 98, n. 14, 125–9, 136,
n. 28, 136–7, n. 32, 160, 267,
269–72, 275, 277–8
and inventions of research programs, 47
model of science, 5–7, *see* problems,
conceptual, and Research Tradition
my attitude toward, 13–14
on methodology, 73
Laudan, R., 14, n. 3, 35, n. 3, 136–7,
n. 32, 186, n. 5
Laurasia, 83, 114, 116
Le Pichon, X., 230, 234–6, 245
Lees, G., 188
Leuba, J., 10, 54, n. 9
lithosphere, 235–6
localism, 75, 80–9, 97, n. 1, 131, 174,
213, 219, 221, 250, 253, 257–9,
263, 270, 272, 276
Longwell, C.A., 68, 119–21, 151, 153
Lugeon, M., 26, 77, n. 13

Lugg, A., 98, n. 14
lunar craters, 38
Lyell, C., 17–20, 32, 35, n. 5, 66, 104, 170, 243
  and inductivism, 78, n. 16
Lyttleton, R.A., 252

MacDonald, G.J.F., 208, 231
McDougall, I., 214
McKenzie, Dan, 231, 233, 243
McKenzie, Donald A., 9
magnetic anomalies (seafloor), 178–9, 196–7, 200–2, 206–9, 214, 227, n. 10, 229–30
  bilateral symmetry, 210–12, 215–18, 220, see *Eltanin*-19
magnetic surveys, 178–9
magnetism, 28
  see geomagnetism and palaeomagnetism
magnetometers
  astatic, 142–3, 149, 159, 165–6
  MAD, 175
  spinner, 139, 165–6
  towed, 175, see black boxes
marine geology and geophysics, 64, 118, 123, 221, 239
  difficulties of, 171–2, 174, 192
  dominance of geophysics, 174, 183–4
  empirical character of, 171–5
  techniques and instruments, 72–3, 178–9
Marine Physical Laboratory, 174
Marvin, U.B., 33, 78, n. 15, 256
mass extinction, 250
Mather, K.F., 241
Matthew Effect, 227, n. 21, 257
Matthew, W.D., 87–8, 100–2
Matthews, D., 206–8, 210, 213, 215, 223, 226
Mayr, E., 123, 135, n. 23
Menard, H.W., 176, 197, 200, 221, 230–1, 258
Mercanton, P.L., 138–9
Mesosaurus, 43–4
Meteor expedition, 172
method
  as rhetoric, 72–3
  formal literature on, 79, n. 27
  of multiple hypotheses, 20, 79, n. 27, 119, 127, 268
  parasitic on practice, 73
Meyerhoff, A.A., 225, 251–2, 265, n. 23

Meyerhoff, H., 251–2
Mid Atlantic Ridge, 31, 83, 172, 176–7, 192, 196, 198
Milankovich cycles, 77, n. 2
Mohorovicic Discontinuity (Moho), 192, 226, n. 2
Molengraaff, G.A.F., 69
Morgan, J., 233–4
Morley, L.W., 226, 226, n. 9
mountains, 26, 30–1, see orogeny, Alps, Himalayas, Appalachians
Munk, W.H., 147
Munyan, A.C., 153

Nairn, A.E.M., 158
National Research Council, 67, 119
National Science Foundation (NSF), 170, 173
Naval Research Laboratory, 173
networks, research, 93, 97, n. 1, 164–5, 259
Nevin, C., 108, 122
Nicholls, G.E., 89, 98, n. 9
Nitecki, M., 135, n. 25, 264, n. 13, 266, n. 32
Nörland, N.E., 127
normal science, 11, 33–4, 247
Nölke, F., 60, 70
Néel, L., 145, 155, 164

ocean basins
  and land-based theories, 182, 188, 191–2, 197–8, 246
  different from continents, 176–7, 188–9, 192, 208, see sial, sima
  sediments, 176–8, 180, 196–7, 215, 226, n. 7, n. 9, 230–1
  topography, 176, see ridges, trenches
oceanography, 118, see marine geology and geophysics
Office of Naval Research (ONR), 170, 173, 199, 274–5
Opdyke, N., 201, 214–15, 230
orocline, 151–2, 168, n. 14
orogeny, 22, 24, 26, 40, 55, 57, 59, 105, 110, 180, 189, 191–2, 196, 199, 244
  causes of, 42, 78, n. 20, 84, 112–14, 116–17, 236–7, 249, see convection currents, crustal shortening, geosyncline
Owen, R., 28–9, 66

palaeobiogeography, 22, 25–6, 39, 42–4,
  81, 90, 100–4, 122–3
  inconclusive, 56
  of Australia, 86–9
palaeoclimatology, 29, 38, 41, 44, 56,
  105, 146–9, 152, 155, 158, 201, *see*
  evaporite, glaciation, great southern,
  Glossopteris, tillite
palaeomagnetism, 133, 138–69, 201, *see*
  polar wandering, geomagnetism
  and Drift, 182–3, 193, 203–6
  and palaeoclimatology, 146–9, 152,
    155, 158
  and prediction in geology, 160–1
  and reversals, 138–9, 145, 149, 155,
    164, 167, n. 5, 206–7, 211, 214,
    217, *see* dating
  as a black box, *see* black boxes
  causes of, 138, 145, 149, 154, 164
  problems of, 139, 145, 154, 157, 166
  social and technical interests in, 159–65
  stability of, 139–41, 145, 149, 154,
    167, n. 5
Pangaea, 40, 42, 44, 63, 82, 91, 116, 204
Pantin, C.F.A., 131
paradigm, 4, 8–9, 32–4, 47, 54, n. 12, 70,
  76, 121, 125, 163, 254, 258, 269,
  272, 275
Parker, R., 233
pattern-matching, 41–2, 56, 59, 60,
  121–2, 204
  across Atlantic, 42, 82, 91, 119, 230,
    236–7, 274
  difficulties of, 50
Penck, A., 59
Permanentism, 1, 5, 28, 30, 41, 44, 47,
  51, 57, 61, 64, 67, 74, 82, 85, 87,
  93, 94, 110, 118, 171, 183, 189,
  197, 202, 214, 229–30, 248, 251–2,
  257
  general description of, 19–23
  improved versions, 100–4
Pickering, W.H., 28, 30, 66
Pitman, W., 215–17
plate tectonics, 2, 208, 218, 229, 238,
  241–2, 232–6, 247, 249–51, 254–5
  and continent geology, 236–7, 248–50
  and plate motion, 233–4
  as a version of Drift, *see* Drift, *see* seafloor
    spreading, Vine–Wilson version
  opposition to, 245–6, 251–3, 262–3
Playfair, J., 18

Poincaré, H., 72
polar wandering, 44, 85, 143–8, 150–3,
  193–4, 214
Polflucht, 40, 98, n. 6
priority, concern for, 225–6, 227,
  n. 19–21, 228, n. 22
problem-field, 85, 91, 93, 96, 97, n. 1,
  158, 163–4, 185, 222, 246, 259
problem-solving model of science, 6–7, 95,
  223, 225, 269, 271–3, 275, 278,
  n. 5
problems
  conceptual, 15, n. 10, 93, 104, 124–5,
    269–70
    and Drift, 51, 60
    examples of, 129–32
    Laudan on, 6, 54, n. 10, 129
  empirical, 6, 131, 269
progress, 95–6, 271
  as cumulation, 126
  Laudan, 7, 16, n. 21
  problems of defining, 16, n. 21
  struggle model, 11–12
progressionism, 19
Project Mohole, 174, 186, n. 4, 239

radioactive heating, 27, 34, 41, 72, 87,
  178
  and Drift, 111–16
  Jeffreys on, 104–6
rafting, 101, 103
Rastall, R.H., 109, 203
Red Sea rift, 217, 226, n. 10
Reid, H.F., 62, 71
research program
  and theories, 125, 128–9
  choice within, 79, n. 26
Research Programme, Lakatos, 5, 125,
  269
Research Tradition, Laudan, 5–6, 76, 125,
  269, 271
Revelle, R., 134, n. 12, 182, 186, n. 2
revolution
  Kuhn, 2, 6
  Laudan, 6–7
Reykjanes Ridge, 208, 214, 217, 229
rhetoric in science, 54, n. 12
Richter, C.F., 180
ridges, 182, 189, 191–3, 195, 198–202,
  208–9
  and fracture zones, 196–7, 208–9
  median rifts in, 193, 198

ridges (*cont.*)
  origin of, 177–8, 193, 196–8, *see* East
    Pacific Rise, Mid Atlantic Ridge, Juan
    de Fuca
Rockies, 26, 111, 237
Roubault, M., 244
Royal Astronomical Society, 162, 243
Royal Geographical Society, 62
Royal Society, 162, 188, 204, 208, 243
Rudwick, M.J.S., 10, 15, n. 16, 19, 35,
    n. 3, 54, n. 12, 266, n. 34, n. 37,
    276
Runcorn, S.K., 142–69 *passim*, 183, 204,
    210, 214
Rutherford, E., 190
Rutten, M.G., 204–5

San Andreas Fault, 150, 202, 209–10
Sapp, J., 10
Sayles, R., 78, n. 21
Scablands debate, 78, n. 19
Scheidegger, A., 152
Schuchert, C., 65–7, 71–3, 92, 102
Schulz, B., 60
Schweydar, W., 59
Scopes trial, 74
Scripps Institution of Oceanography, 118,
    172–6, 184–5, 197, 199, 232–3
seafloor spreading, 202, 206, 212, 223–4,
    227, n. 21, 231, 241
  see Hess, H.H., Dietz, R., Matthews, D.,
    Vine, F., Wilson, J.T.
  Hess's and Dietz's version, 197–200,
    225
  Vine–Matthews version, 206–7, 214,
    217, 227, n. 10
    initial response, 207–8, 226, n. 9
  Vine–Wilson version, 208–13, 217, 230,
    232, *see* plate tectonics
Secord, J.A., 15, n. 16, 19, 35, n. 3, 276
seismology, 41, 175, 230, 233, *see*
    earthquakes
Semper, M., 58–60, 70
Seward, A.C., 90
Sewell, R.B.S., 92, 98, n. 10, 172
Shapin, S., 9
Shepard, F., 118
shield, 189–90, 248
  Canadian, 189–91
sial, 22, 27, 40–1, 57, 97, n. 6, 187,
    n. 9

sima, 22, 27, 40–1, 57, 187, n. 9, 235
Simpson, G.G., 100, 102–4, 118–19
Singewald, J.T., 71
Skeats, E.W., 84
Snider–Pellegrini, A., 28–9, 66
social analysis of science, 8–12, 47–8,
    182–5, *see* interests, struggle
  funding of science, 182–4
Soergel, W., 58–9, 70, 77, n. 5
specialization, 47–8
  and Drift, 70–2, 75, 80–9, 203, 257–9
  and theory change, 183–5
  and theory choice, 34–5, 92–7, 131,
    164, 219, 221, 270, 272, 276
  Kuhn on, 98, n. 13
  Lugg on, 94, 98, n. 14
  see localism
species distribution, *see* biogeography,
    palaeobiogeography
Stafford, R., 36, n. 10
Staub, R., 61
Steers, J.A., 108–9
Stehli, F.G., 251
Stewart, J.A., 15n
Stille, H., 136n
stripes, magnetic, *see* magnetic anomalies
struggle for authority (or credibility),
    10–12, 94, 128, 130, 162, 219, 224,
    267, 270, 272–3, 275–8
  and inventions of research programs,
    47–8
  and specialization, 12
  internal, 10, 34
  my view of, 14, 275–6,
Strutt, R.J., 111
subduction, 197–8, 201
Suess, E., 24, 30–1, 36, n. 9, 50, 57, 61,
    90, 192
Suessian synthesis, 24–6, 27–8, 31
Sykes, L., 230–1, 233

Tarling, D., 214
Taylor, F.B., 28, 30–2, 46, 52, 64, 66, 68,
    225
technical interests, *see* interests
techniques, novel, *see* black boxes
tectonics, vertical, 246, 250, 263
Temple, F., 74
terranes (accreted, suspect or
    allochthonous), 248, 250
theoretical pluralism, 4, 19, 33–4, 95–6,

126–9, 131, 163, 247, 251, 263, 269

Theories

as parts of complex structures, 5, 69–70, 269

epistemic stances, 6, 91–2, 272

theory and practice, geologists' views on, 19, 76, 83–4, 126–7, 250, 257, 272

theory choice as comparative, 4–8, 69–70, 121, 124, 126–9, 219–22, 269–71

Tiercy, G., 77, n. 13

tillite, 44

Dwyka, 82

Squantum, 67, 78, n. 21, 146

Torreson, O.W., 139–40

transform faults, 208–9, 210, 231, 233

trenches, oceanic, 116, 177–8, 189, 208

origin of, 180–1, 193, 196–9

Umbgrove, J.H.F., 136, n. 27

underdetermination of theories, 116, 221–2

uniformitarianism, 17–18, 19, 29, 35, n. 4, 66–7, 74

as rhetoric, 20

Vacquier, V., 175, 197, 201, 208

van Bemmelen, R.W., 136, n. 27

van der Gracht, W., 64, 68, 109

Vening-Meinesz, F., 60–1, 116, 171, 178, 180, 186, n. 1, 195–6, 201

Verhoogen, J., 168, n. 11, 213

Vine, F., 199, 206–8, 210–13, 214–15, 217, 223, 226, 227, n. 11, 229–31

Vine–Matthews Hypothesis, *see* seafloor spreading

Voice-Over, 3–14, 32–5, 46–53, 69–76, 124–33, 159–67, 218–26, 254–63

function of, 2–3

von Ubisch, L., 60

Wade, A., 85–6, 92

Wallace Line, 81, 97, n. 3

war, effect on geology, 50, 36, n. 9, 61, 74, 77, n. 14, 117–18, 170–5

Ward, L.K., 84, 92

Wegener, A.L., 10, 15, n. 18, 17, 33, 37–54, 55–6, 225, 241, 255–6, 260

and *The Origin of Continents and Oceans*, 38, 40–1, 46, 52–3, 58–9, 62, 77, n. 14, 82, 108

as a discoverer, 33, 46, 51–2

as a specialist, 54, n. 9, 94

defense of Drift, 59, 67–8, 107–9

methods, 41, 72, 108

criticized, 52, 60, 62–3, 65–6, 71, 108–9

on mathematics, 54, n. 7

on the causes of Drift, 40, 108

Weiss, F., 61

Werner, A., 17–18, 23

White, D., 67

Whitley, R., 98, n. 12

Willis, B., 65, 71, 100, 102, 118–21, 136, n. 29

Wilson, J.T., 136, n. 30, 164, 186 n2, 186, n. 7, 208–10, 215, 223–4, 232–3, 237–8, 240, 247–8, 260

and Contractionism, 189–92

and Drift, 200–1

and Expansionism, 195, 201

Wilson, D.W.R., 152

Wiseman, J., 172

Wood, R.M., 14, n. 3, 35, n. 3, 135, n. 13, 168, n. 9

Woods Hole Oceanographic Institution, 172–4, 184–5

Woolgar, S., 10–11, 97, n. 1

Wright, C.S., 63, 89–90, 92

Wright, Sewall, 104

Wright, W.B., 61

WWSSN, 175, 181, 233

Wyllie, P.J., 256, 262

Yakushova, A., 246

Yearley, S., 54, n. 12